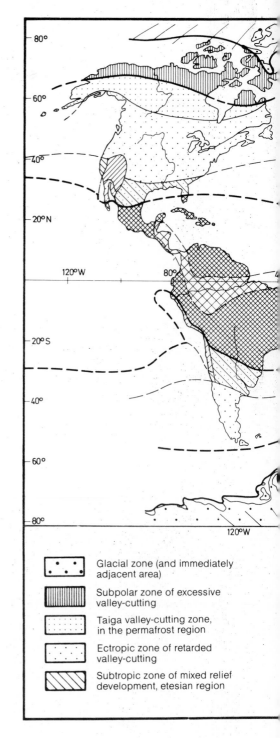

Fig. 13. The morphoclimatic zones of the present (excluding the high mountains).

Glacial zone (and immediately adjacent area)

Subpolar zone of excessive valley-cutting

Taiga valley-cutting zone, in the permafrost region

Ectropic zone of retarded valley-cutting

Subtropic zone of mixed relief development, etesian region

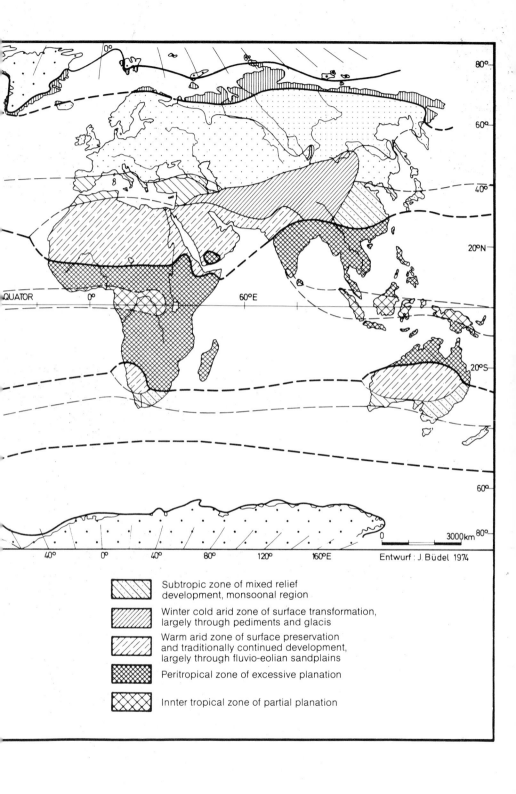

	Subtropic zone of mixed relief development, monsoonal region
	Winter cold arid zone of surface transformation, largely through pediments and glacis
	Warm arid zone of surface preservation and traditionally continued development, largely through fluvio-eolian sandplains
	Peritropical zone of excessive planation
	Innter tropical zone of partial planation

Entwurf: J. Büdel 1974

80°
60°
40°
20°N
EQUATOR   0°   60°E
20°S
60°
3000 km   80°

40°   0°   40°   80°   120°   160°E

Climatic Geomorphology

Climatic Geomorphology

by Julius Büdel

Translated by
Lenore Fischer and
Detlef Busche

Princeton University Press
Princeton, New Jersey

Translated from the German: Julius Büdel, *Klima-Geomorphologie*,
©1977 by Gebrüder Borntraeger, 1 Berlin–7 Stuttgart
Library of Congress Cataloging in Publication Data will be
found on the last printed page of this book

This book has been composed in Linotron Times Roman and Gill Sans

Clothbound editions of Princeton University Press books
are printed on acid-free paper, and binding materials are
chosen for strength and durability

Printed in the United States of America by Princeton
University Press, Princeton, New Jersey

In memory of my teachers
Eduard Brückner (1862-1927) and
Albrecht Penck (1858-1945)

Eduard Brückner (left) and Albrecht Penck (right) in the summer of 1893 near Flims (Vorder Rhein Valley, Graubünden) during joint field-work for the *Alpen im Eiszeitalter*. This fundamental work on glacial effects on a major mountain system and its forelands provided the first insights into the events of the Ice Age Pleistocene relief generation, the most striking relief generation in Central Europe. This was geomorphology's first step toward acquiring the rank of a natural science independent both in goal and method. (The picture, named "Albrecht the Squashed" [*Der erdrückte Albrecht*] was dedicated by Brückner to the wife of his friend Penck. It was very kindly placed at my disposal by the latter's daughter-in-law, Mrs. Anny Penck.)

Contents

Author's Foreword

This book discusses new insights concerning the influence of the *climate* upon the earth's relief, in particular upon the continents. It has been found that this climatic, this exogenic influence is much stronger than any endogenous influence. Endogenous forces, as any geologic map will show, are distributed randomly over the entire earth, while the exogenic influence of climate follows strict rules in its distribution from the pole to the equator, both today as in the past. Their distribution and intensity make climatic influences the best basis upon which to found a *system of geomorphology*, such as the one presented here.

The *relief sphere* of the continents, in all its many variations, is the boundary between the solid crust (the *lithosphere*) and the *atmosphere*. The internal structures of the crust are determined largely by endogenous processes, but the shape of its outer surface, the relief sphere, is actively molded down to the last minutiae by exogenous processes. These all derive in one way or another from the effects of *solar energy*.

Endogenous influences on the relief sphere are of secondary importance, and can be divided into three types. Local active effects are fairly rare, consisting of fresh faults, grabens, and volcanoes, all of which are speedily subjected to exogenous transformation. This is even more true of regional active endogenous effects, namely, regional uplift and downwarping of the crust (*epeirovariance*). Epeirogenic movements occur extremely slowly, and delineate only the general plan of lowlands, basins, and mountains. The most important features of these geographic products, namely, their vertical height and finely chiseled shape, are completely due to exogenous processes. Thus epeirovariance is not of great importance. The morphologic hardness of the substrate (*petrovariance*) works passively against exogenous erosion, and a comparison of the two shows that it is the exogenous erosional processes which are decisive in determining the appearance of the relief.

Since the exogenous processes are all set in motion by solar energy and by the interacting mechanisms subsumed under the word "climate," it follows that the appearance of the relief must vary decisively with climate. Until recently, many authors divided the climatically affected portions of the relief into three areas: the *arid region*, in which wind was considered the primary factor, the *glacial region* of glacial effects, and the fluviatile or *fluvial region*, in which the landforms were ascribed to the rivers. The relief itself was ignored in favor of the more striking hydrologic component of the relief-building process. Instead of a categorization of the relief, the result was a categorization of the terrestrial water cycle, which was assumed to be equivalent.

Careful investigation has produced a picture differing in two ways. For one thing, a precise study of the relief itself, comparing reliefs of different climatic zones, shows a far greater variety than the above tripartite categorization allows for, even when comparing only features of unquestionably modern, i.e., Holocene origin. Secondly, it has been found that the processes at work on the present relief create very characteristic, though highly diverse assemblages which vary according to climatic region. Rivers are only one component in a complex mechanism, and behave differently with regard to their companion mechanisms in each climatic region.

According to this point of view, only the glacial zones remain as an independent region. Even in the driest of deserts, the effects of the wind are not as important for the major relief forms as the occasional effects of running water. Thus in the morphological sense, the arid regions can also be reckoned as part of the "fluvial" zone.

Careful investigation of the present relief, and of its currently active relief-forming mechanisms, has shown that the fluvial region (including the arid zone) may be broken down into *nine major morphoclimatic zones*. These, along with the glacial zone, are shown in Fig. 13 (see endpaper). The relief assemblages of the nine zones are hardly less distinctive than those of the glacial zone itself. Given these zones (though it may be necessary in the future to subdivide the tropics), the main task facing geomorphology is twofold.

I. Climatic geomorphology is concerned with the exact explanation of today's relief-forming mechanisms and of their relief product, the subject of the first section in this book. It is extremely important to realize that it is the interplay of all the component processes of any given relief-forming mechanism which produce the relief itself. Naturally one should understand the components in all their subtlety, but it is even more important to understand the way in which they interact in nature. A large number of today's geomorphologists devote themselves to the isolation and increasingly specialized analysis of individual component elements, something which can be done purely analytically with physical and chemical apparatus. The study of the complex interaction of all the component processes tends to be neglected, though it is this interplay alone which leads to the creation of the relief. This book is primarily dedicated to the study of this interaction and the resulting relief.

Only the natural relief can be used for this, and it cannot be replaced by even the cleverest of simulations. It is the natural relief alone which can show what the entire chain of processes has produced over long periods of time. Short-term observations of isolated component processes (in the most artificial environment possible) can scarcely be applied to the time spans necessary to create the relief.

Figure 5 approaches this same theme from a different angle. Investigations on a basic integration level, be they never so exact, can never produce conclusions on the order of the more highly complex levels. Like behavioral studies, geomorphology is a discipline which, in contrast to many others, must proceed from analysis to synthesis in order to obtain results. Investigations on a basal level, pursued not for their own sake but directed toward the study of more complex chains of processes, are being used increasingly in climatic geomorphology today. These include, e.g., clay and heavy mineral analyses, palynology, paleontology, radiometric dating, dendrochronology, and paleomagnetic studies. Many of the conclusions in this book are based on such studies.

The historical, synthetic viewpoint and the quantitative, analytic viewpoint do not contradict each other in geomorphology. As a whole, the science must aim at synthesis, which in turn must be based on analysis. Analysis alone can never hope to explain the relief, but must be incorporated into the synthesis, the overall picture of the present development and the history of the relief.

II. The explanation of the relief-forming mechanisms at work today and of their product, the relief currently being produced in the major morphoclimatic zones, is but the precursor to the second major part of modern geomorphology. This is concerned with explaining *all parts* of the total relief in any given zone. The total relief is not simply a product of the current relief-forming mechanisms. The Holocene relief in the mid-latitudes does not constitute even 5% of the total relief, most of which consists of the remains of older relief generations preserved in the hard lithosphere. Relief generations from the past are the product of ancient processes governed by past, often completely different, climatic conditions. Now that these are long gone, they can no longer be observed and measured, but they have left much evidence of their former activity, including such things as paleosols, duricrusts, and correlated sediments. Above all, however, it is the *shape of the paleo-relief itself* whose remnants attest to the long-term, uniform activity of past relief-forming mechanisms. This fact has often been overlooked.

Chrono-geographic comparison may be used to reconstruct past relief-forming mechanisms. Our knowledge of the relief types being created in the various morphoclimatic zones of today has advanced to the point that analogs can be found among them for most paleo-reliefs. By studying the relief-forming mechanisms at work now, we may draw conclusions regarding those of the analogous paleo-relief. Qualitative analyses of paleosols and sedimentary remains may serve to refine and support such conclusions. In this manner, it is often possible to reconstruct the former relief-forming mechanisms in detail.

The second part of the book divides up the total relief into individual relief generations, corresponding to the various climatic phases through which the total relief has lived during the latter part of earth history. Some of the relief generations reach back as far as the start of the Late Cretaceous, or around 80 MY; we therefore call this time period the geomorphic era. Recognizing the relief generations, separating their remains, and unraveling their developmental environment is the task of climato-genetic geomorphology. Both of the main objectives of geomorphology are summed up under the name *climatic geomorphology*, the title of this volume.

The conclusions which were obtained in this fashion have led to a deepened understanding of geomorphology as a whole. A description of this has been included as Chapter 1 of the text.

As this book came out in 1977 (German text), the bibliography does not go beyond that year. I am extremely grateful to all those who, over the forty years during which these ideas were maturing, have helped me through parallel research and through oral exchange of ideas. These include teachers, friends, colleagues, and students; in particular: G. Angenheister (Munich), H. Bremer (Cologne), A. Cailleux (Paris), T. Czudek (Brno), R. Dehm (Munich), R. W. Fairbridge (New York), R. W. Feyling-Hansen (Arhus), J. Fink (Vienna), G. Furrer (Zurich), R. Galon (Torun), T. Gjelsvik (Oslo), U. Glaser (Würzburg), H. Hagedorn (Würzburg), J. Hövermann (Göttingen), G. Hoppe (Stockholm), F. Isachsen (Oslo), A. Journaux (Caen), K. Kayser (Cologne), W. Kubiena (†Hamburg), H. Louis (Munich), H. Mensching (Hamburg), H. Mortensen (†Göttingen), M. Pecsi (Budapest), M. J. Proudfoot (†Chicago), A. Semmel (Frankfurt), H. Späth (Cologne), A. Székely (Budapest), G. Stäblein (Berlin), M. F. Thomas (St. Andrews), C. Troll (†Bonn), A. L. Washburn (Seattle), H. Wilhelmy (Tübingen), and A. Wirthmann (Karlsruhe). For the majority of my travels I am grateful for the support of the German Research Council, the *Deutsche Forschungsgemeinschaft*.

I am particularly indebted to D. Busche (Würzburg) and L. Fischer (Princeton) for their translation of the text into English, and to D. Busche for his addition of a glossary of new terms introduced by climatic geomorphology into the English language. I also wish to express my sincere thanks to the German publisher, Borntraeger-Schweizerbart in Stuttgart, and to Princeton University Press for the time and care they have lavished on the production of this work.

Translators' Foreword

Translating is not simply a matter of stringing parallel rows of one-to-one equivalents. If this were so, international communication would long since have been turned over to machines. Rather, each language is unique, and shapes its users, just as its users in turn shape it. This truism is rarely so well exemplified as in the present work.

Any science, in investigating known or unknown phenomena, builds with a structure of words, tagging those characteristics which it chooses to observe, ignoring those which it considers unimportant. When therefore, scientists of different language communities separately investigate the same phenomenon, it is not surprising that they should see different things, describe them in different ways, and come to different conclusions. It is, perhaps, not fair in our present example to speak of the English-speaking and German-speaking geomorphologists as completely isolated from one another, for a few Americans, and some Britons, do read German, while many Germans are well-versed in English. Yet German geomorphologists, though far from being a homogeneous group, have taken their own path in research, and as a result, look at even such apparently simple things as valleys in a way which will be unfamiliar to their Anglo-American colleagues. It is inevitable, then, that German geomorphology will have developed a considerable vocabulary for which standard English equivalents are either inappropriate or lacking altogether. This is especially the case in Julius Büdel's work, for in over forty years of research he has developed a new system of climatic and climato-genetic geomorphology, necessitating the coining of terms to describe new insights. It would thus be but a slight exaggeration to speak of a distinctive Büdel language.

We were therefore faced with the choice of either importing words wholesale from German, or translating each term as it arose. The first of these alternatives would have produced a book, each page of which bristled porcupine-like with foreign terms, sufficing to daunt the most determined of readers. The second alternative is also not without its drawbacks. These are superbly exemplified by the history of the word "pediment." Originally used in architecture to describe the triangular structure crowning the portico of a Grecian or Renaissance building (most likely a corruption of the word "pyramid"), it was used by McGee (1897) to describe the silhouette typically formed in the SW United States by gentle slopes splaying out from the feet of sharply defined desert ranges. Soon the term became attached to the surfaces as seen from above, and was introduced in this sense into German. On the assumption that "pedi-" was derived from Latin and referred to the position of the feature at the mountain's foot, "pediment" was translated as "*Fussfläche.*" The latter term is used by most workers as a general descriptive term, whereas "*Pediment*," though also used in German, has become freighted with so many diverse genetic implications as to render it almost useless. Thus neither "*Fussfläche*" nor "*Pediment*" as used by Büdel refers to the same thing as the English "pediment" (even charitably assuming the existence of a standard English definition for this). In this search for a neutral term we have committed perhaps the crowning folly of retranslating "*Fussfläche*" into English as "piedmont surface"; "pediment" is employed here only where specifically used by Büdel.

Thus the danger of translating terms into another language is that they will come to be misapplied. Yet this is a danger which all terms run, even within the framework of one language, and is an unavoidable outgrowth of the fact that our concepts and usages do evolve.

In other cases a term in German may correspond to several terms in English, all meaning slightly different things, belonging to so different a conceptual framework as to render them unsuitable. An example of this is the German word "*Rumpffläche.*" To use a literal translation such as "terminal torso" or "torso plain" seems futile, as these have never really been adopted by English speakers. To translate it as "peneplain" or "peneplane" would be to confuse inextricably the theories of Büdel with those of Davis. To translate it as "erosional surface," following the French generic term "*surface d'érosion*" would be incorrect, as the surface, according to Büdel, is as much the result of chemical deep weathering as it is of erosive wash processes. We hope that our solution, "etchplain," is sufficiently correct in definition and neutral in connotation to be satisfactory.

Our solution, on the whole, then, has been to translate the terms involved as meaningfully as possible, so that the English-speaking reader will be better able to comprehend their intended significance. For easier identification of the original term, the German is placed in parentheses after its English equivalent upon first appearance in the text. We have further provided a glossary, listing the English terms which we have used, stating their German equivalents, and briefly defining them. In this way we hope to link the English and German terms as closely as possible, and prevent the lamentable splintering of definitions described in the examples above.

Even were terms well-established, one-to-one equivalents, however, translators could still not throw up their jobs and turn to computer programming. For terms are but building blocks, which may be haphazardly rammed together into a slipshod hovel, or lovingly cemented into an imposing edifice. Now it is the sad fate of scientific literature in the English-speaking world, or so many German scientists complain, that while laudably aspiring toward clarity and brevity, it is also flat, featureless, and dull. Some German scientific writing, on the other hand (and Büdel is perhaps a good example of this), has developed a style which is equally unappealing to English speakers, but for the opposite reason. In attempting to resolve these seemingly incompatible styles, to steer a steady course between prolixity and brevity, between floweriness and flatness, we have chosen clarity and harmony as our guides.

For further clarity we have added three maps of Central Germany to the discussions in Chapter 3, and inserted additional place names on some of the maps already present.

A few remarks should be added as to why the translation was undertaken at all. As pointed out at the start, the differences in terminology reflect the great disparity of the German and English approaches to the field. While the ideas of the English-speaking workers could easily enter German geomorphology, at times to great effect, flow in the reverse direction has been weaker, and has frequently occurred only via French. The present translation was therefore undertaken to increase direct exchange of information, and to further the understanding of at least one German approach to the subject. Certainly a number of German workers would object to regarding this text as the definitive summary of German geomorphology, though they would probably agree that it is at least representative.

We are pleased that Princeton University Press has undertaken the publication of this translation, and that they decided to hand the work over to two translators: an American professional translator with a background in the earth sciences, and a German geomorphologist whose close contact with Professor Büdel has helped to ensure that the translation preserves Büdel's intended meaning as closely as possible.

We should like to thank all those who have helped us in unearthing suitable terms, and who have warned us against pitfalls. In the case of conflicting suggestions, choices were made only after thorough debate between the two translators. Special thanks are due to Michael F. Thomas for his copious advice, and to the British Geomorphological Research Group, who participated in a joint conference with the German Arbeitskreis für Geomorphologie shortly before the completion of the translation.

Finally we should like to thank Professor Büdel himself, who benevolently encouraged us at all times, and who kindly clarified many a difficult passage.

August 1980

Lenore Fischer   Detlef Busche
Princeton, N.J.   Geographisches Institut
                                                  Universität Würzburg

Climatic Geomorphology

1

The Subject of Geomorphology: The Terrestrial Relief Sphere and Its Development

1.1 The Importance of the Terrestrial Relief Sphere and Its Main Phases of Development

Geomorphology studies the configurations of the land surface, that is, of the relief sphere or outer boundary of the earth's crust. This is an independent structure formed of mountains and valleys, hills and slopes, and expansive plains sprinkled with a host of minor features. All these forms are generated largely by exogenic forces, that is, by solar radiation and by the atmosphere.

The most recent features of any relief are formed by processes governed by the present climate. In the mid-latitudes these features are usually in the minority. The older elements of the relief, however, have been formed by successive sets of processes according to their climatic history, and can therefore be assigned to various former relief generations. These two observations are fundamental to this book.

Endogenic factors produced by the earth's glowing interior interact with exogenic processes and affect relief formation. The extent to which this occurs will be discussed below (primarily in Section 1.4)

About 71% of the earth's relief sphere is covered by oceans. This oceanic hydrosphere includes the Caspian Sea, which, although isolated today, resembles the smaller seas in size, depth, and basin configuration. The submarine relief concerns us only marginally. The topography of the deep-sea bottoms, recently elucidated by modern sounding methods, is, unlike the continents, largely of endogenous origin. The coastal forms are particularly interesting to a study of the oceans, as are the shelf seas which flood the continental platforms. The shelf seas are bounded on their marine side at more or less the 200 m isobath, where the continental platforms plunge down steep continental slopes to the deep-sea bottom. The deep-sea floor consists mostly of huge plains at depths of 4000-5000 m, interrupted only by rare isolated features and by systems of narrow, often surprisingly linear and closely set ridges and trenches.

Barely 3% of the relief sphere is covered by the continental ice sheets of the polar regions (Antarctica and Greenland), as well as by the smaller mountain glaciers in the high and mid-latitudes and at very high elevation in the tropics. This terrestrial cryosphere can be over 3000 m thick in the continental ice caps. Among the mountain glaciers, the Aletsch glacier (the largest of the Alpine glaciers), 800 m thick, is exceptional. But the *subglacial relief sphere* is of far greater interest to the geomorphologist. In the recent geologic past, during the Pleistocene glacial stages, around 8.8% of the continents (over one-quarter of the present land surface) was covered by ice caps whose traces are still very evident, especially along the margins of the former and present glaciers. Thus the relief of the glacial fringes is an important concern of geomorphology. Yet it has become increasingly evident in the last few years that the pre-glacial features of the former landscape are much more visible through their glacial overlay than had been previously assumed.

But the basic and most important topic of geomorphology concerns the region exposed directly to the atmosphere: the *subaerial relief sphere*. This covers over a quarter (26%) of the earth's surface today, and includes the ice-free 90% of the continents on which the higher plants and animals have evolved and which form the human habitat. The distribution of peoples on the earth, their prehistoric and historic development, their entire range of activities in trade, settlement, social life, and politics, are necessarily and inevitably determined by the changing conditions of the continental relief sphere, by the distribution of the mountains, rivers, plains, and coasts. Still more important than the subaerial relief sphere are its four companion spheres, which we will describe below. These expand the relief sphere three-dimensionally, and at the same time generate the highly complex processes causing the trinity of denudation, transportation, and sedimentation.

These closely interwoven features of the companion spheres and of the just-mentioned trinity are governed almost entirely by climate, that is, by the effects of solar energy. Naturally these exogenic influences vary from one climatic zone to another: this change is the subject of climatic geomorphology.

The relief sphere therefore differs very substantially from the lithosphere (the crust), which is affected largely by endogenic influences, by conditions and movements within the earth's glowing interior.

We therefore distinguish between the lithosphere and its problems, dealt with by geology, and the relief sphere and its problems, dealt with by geomorphology. These two spheres are separate, a point which is important both despite and because of their undeniably interrelated nature.

Crust and relief have one characteristic in common, however, though manifested in diverse ways: both are due only partly (often in fact, only to a very minor extent) to current processes. They both preserve the lasting marks of ancient processes through successive generations.

Soon after the earth formed as an incandescent body five to six billion years ago, it began to cool and form a crust. This crust undoubtedly went through many stages of agglutination, remelting, and displacement, stages whose appearance must remain conjectural. The oldest preserved crustal fragments are from at least two billion years ago, when the "anhydrous" period was long over. Solid land was formed (perhaps as one primordial continent), surrounded by seas and covered by an atmosphere. These subjected the crust to external attack and alteration. The oldest buried remains of ancient land surfaces in Central Europe (the pre-Permian etchplain, covered by laterite-like soil of the lower New Red Sandstone) date from the Late Paleozoic (around 200 MY ago). The sediments and scouring marks of the great Permo-Carboniferous Ice Age on the southern continents (which at that time were still connected) are of similar age. Today these buried land surfaces are exposed in the same way as bedding planes, and have no influence on the shaping of the landscape. They may be covered by newer sediments and volcanic extrusions, lifted or worn away, or downwarped and orogenetically kneaded and reincorporated into the crust.

The oldest relief assemblage of non-alpine Europe still forming the relief today, the pre-Cenomanian landsurface, dates to the transition between the Early and Late Cretaceous, and is best seen in the Kuppenalb[1] of the Franconian Alb in S Germany (see Sections 3.3.1.5 and Fig. 72 below). Otherwise such remains are only exposed in small scattered areas, where they display the characteristic shape and parent material or even soil of tropical etchplains. More evident are the remains of the oldest Tertiary relief generations. While Paleocene and Eocene remains are only fragmentarily preserved (see Fig. 1), Oligocene relict surfaces extend even into regions previously folded in the Alpidic orogeny. We therefore designate the time which has elapsed since then as the *geomorphic era*. Since the Cretaceous-Tertiary revolution in plant and animal life and the relief development up to Mid-Pliocene times (about three to four MY ago) both took place under conditions governed exclusively (though with local crustal movements and occasional dry intervals) by climates resembling those of today's seasonally tropical and subtropical areas, this major portion of the geomorphic era may be designated as the time of the *tropicoid paleo-earth*. This is followed by the relief generations of the Late Pliocene, and by the three main epochs of the Quaternary: the Earliest Pleistocene, the Ice Age Pleistocene, and the Holocene. Recognizing, ordering, and distinguishing these relief generations, so as to analyze today's highly complex relief, is the task of climato-genetic geomorphology.

[1] Kuppenalb: a hummocky Alb surface. Kuppen are rounded knolls 50-110 m in height.

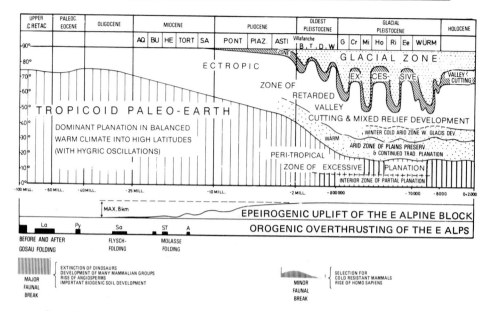

Fig. 1. The development of the morphoclimatic zones during the course of the "geomorphic era" from the Late Cretaceous to the present. Abbreviations: geological stages since the start of the Late Tertiary (Miocene): Aquitanian, Burdigalian, Helvetian, Tortonian, Sarmatian, Pontian, Piacenzian, Astian, Villefranche, Biber, Tiglian, Donau, Waalian, Günz, Mindel, Holstein, Riss, Eem, Würm. Below: Laramian, Pyrenean, Savian, Styrian, and Attican orogenies.

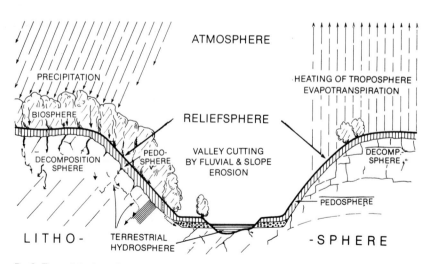

Fig. 2. The relief sphere (heavy line) as the major energy conversion surface of the earth.

By studying the development of the crust and of its fossil-bearing strata, geologists and paleontologists have reconstructed the impressive history of the earth and the evolution of life. But this story is incomplete and cannot properly describe our present world without being supplemented by a history of landforms.

1.2   The Relief Sphere and the Earth's Layered Structure

The earth's layers fall into three basic sets: the outer envelope, the crust, and the interior. The interior and the outer envelope, despite enormous differences in mass, pressure, and temperature, are similar in two important ways. Firstly, they lack any stable structure, consisting of amorphous masses of molecules (or ions in the outer atmosphere) in poorly defined concentric shells. The molecules of the interior, of course, are packed to an extraordinary density. Secondly, both these shells exist in a state of timeless flux, bearing no permanent record of their former states. Such a record is preserved only by the crust and its two subspheres, the lithosphere, determining its inner structure, and the relief sphere, determining its outer structure. These two subspheres, together having an average crustal thickness of 20 km (6-12 km below the ocean floor and 25-80 km beneath the continental plates), form a very thin hard shell round a glowing filling of stellar material. An earth model the size of a hen's egg would have a highly fragile shell of only 0.15 mm, one third the thickness of a normal eggshell.

This spherical structure is shown in Fig. 3. The area between the earth's center (around 6370 km below the surface) to a boundary currently set at 2800-2900 km deep is occupied, according to the geophysicists, by an iron-rich core. This controls the earth's magnetic field, a wide, wavering envelope that surrounds the earth and protects it against the solar wind. This magnetic field, visible in the form of the aurora borealis, is also responsible for aligning compass needles and all magnetizable particles in cooling rocks or compacting

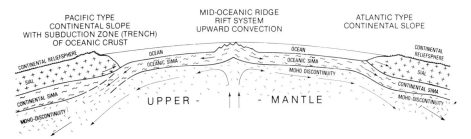

Fig. 3. Plate tectonics and continental drift.

sediments. Frequent magnetic reversals have occurred during the earth's history, and their record, preserved by such particles in the cooling or compacting rock, can be used for stratigraphic correlation (see Fig. 61). The high iron content of the earth's core has also been deduced from the composition of many meteorites (fragments of other planets) which have fallen upon the earth.

The earth's core is conjectured to have two layers, and subdivisions are certainly present in the next major shell, the *mantle*. The lower mantle reaches from about 2900 to 900 km in depth, and a middle layer (probably also bilayered) from about 900-200 km in depth. Nearest to us lies the upper mantle, extending from around 200 km depth to the crust.

This very important boundary between the mantle and the crust is called the Mohorovicic Discontinuity after its discoverer, shortened to Moho-discontinuity or Mo-zone. This is actually made up of a number of closely packed shells, but as a whole it forms a very clear boundary between the solid crust and the vastly different molten material or *magma* of the upper mantle.

We have not yet succeeded in drilling through to the upper mantle. Only indirect evidence attests to the conditions there, so foreign to us who live on the earth's exterior. At the Moho-discontinuity the speed of seismic waves climbs rather sharply from 6-6.5 km/sec to 7.8-8.8 km/sec. All matter would be fluid in the high temperatures of the upper mantle (around 1800°C; we may compare this with the melting point of iron at 1540°C), if the weight of the overlying crust did not raise the melting point. The magmatic substance contains certain forms which may be identified as heavy "ultrabasic" rocks (e.g., eclogite, peridotite, and dunite), though this is structurally inappropriate, as no stable and clearly defined rock units exist there (as do Rhön basalt, Julier granite, or Bolzano porphyry on the earth's surface), but only closely packed individual molecules. These behave solidly in reaction to sudden thrusts, yet are latently fluid. This means that when the pressure of the crust is suddenly removed, as along a crack, the magma becomes a highly mobile incandescent mass, and, accompanied by gas eruptions, shoots forcefully into the crust or along volcanic vents to the earth's surface. Tremendous gas explosions may accompany such outbreaks. Basaltic lava from great depth is particularly fluid, and can build incandescent lava lakes, as in Hawaii, or pour forth in flaming torrents, as in Iceland. The sources of such magmatic outbreaks may lie at various depths, and the composition of the fresh volcanic rock varies accordingly.

The magma beneath the Moho-discontinuity is not only latently fluid in this sense, but is also secularly plastic as a whole, being subject to slow, locally varying movements. These are partly generated passively by changing pressures in the overlying crust (e.g., by crustal burdening and unburdening through the build-up and melting of the great inland ice caps), and partly self-generated by forces within the upper mantle or in the layers below. We may assume speeds of between 0.1 and 3 cm/yr, reaching a maximum of probably 10 cm/yr (Angenheister, 1970; Seibold, 1973). Such movement may be horizontal or vertical, and is connected with the movement of the continental plates, first recognized by Wegener (1915).

The continental plates (including the continental shelves) are 20-30 km thick on the average; the plate under the Alps is up to 40 km thick, that under the Himalayas up to 80 km thick (see Fig. 4). The construction of such continental plates is tripartite. The *superstructure* (not present everywhere), consists of relatively undisturbed clastic sediments, consolidated if at all through diagenesis (primarily through induration under pressure). Thus calcareous mud turns to limestone, sand turns to sandstone or arkose, clay to shale, and peat to lignite. Such a superstructure is common in zones of recent subsidence, where it can be several km thick.

Beneath this and penetrating it in the form of uplifted blocks, lies the far thicker *geologic infrastructure* or *basement*, which may be up to 20 or 30 km thick. Areas of recent but intense tectonic disturbance, such as the Alps, whose nappe structure consists partly of series belonging to the superstructure, are exceptional (see Fig. 4). Otherwise the majority of the infrastructure consists of sediments which are transformed (by great pressure combined with tectonic stress, by high temperature, contact with plutons, or permeation with magmatic-plutonic gases) into *metamorphic* or *crystalline* rocks (e.g., schist or para-gneiss). Magmatite which has risen to the surface also undergoes such a transformation. Metamorphosis occurs largely because the majority of the infrastructure is at times locally subject to major internal movements or folding phases. These folding phases (*Faltungsphasen*, after Stille, 1924, 1929, 1950) take place almost without exception in the semi-plastic depths, not at the surface or in the mountains. The resulting continental blocks were at some time further rigidified by magmatites penetrating the entire structure.

Superstructure and infrastructure both consist of lighter material, the magmatic fraction being largely acidic granite. For this reason, these two layers are together designated as the *granitic crust* or *SIAL*. The lowest part of the crust consists of heavier basic rock, such as basalt, and is therefore called the *basaltic crust,* or *SIMA*. Even this material is still lighter than the superheated structureless magma beneath the Moho-discontinuity.

Such is the overall structure of the continental plates. The crust beneath the oceans is much thinner, being only 6-12 km thick. Here only the deeper basalt crust is present, covered by a thin sheet of loose, geologically recent sediments.

Modern international exploration of the oceanic deeps has thus supported Wegener's theory (1915), showing that the continental plates float upon the heavier magma like icebergs, and have gradually drifted apart. This is true especially of the margins of the S Atlantic. The majority of this work has been performed by Americans, the German contributors at the time of writing

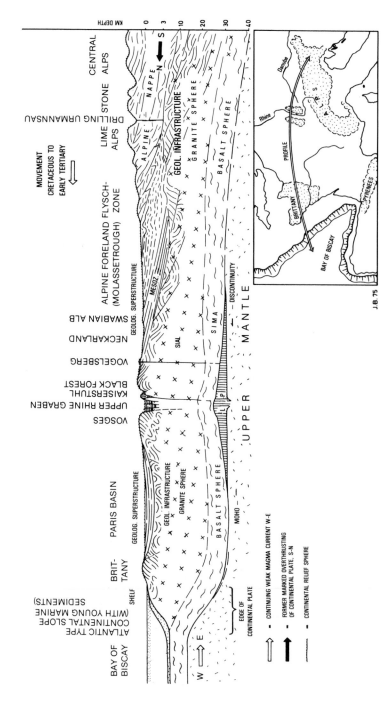

Fig. 4. Continental structure along a profile from the Bay of Biscay to the Alps. L P: laccolith pillow of recently intruded mantle material (mantle diapir). This introduces taphrogenesis, and is the source of synorogenic volcanism, according to Illies (1970).

being primarily Seibold (1973, 1975) and Tollmann (1974). Figure 3 is drawn after the latter.

The American E coast and the European and African W coasts still show a striking resemblance in outline and structure. Formerly they were joined, their former contact being marked today by the mid-Atlantic ridge (Fig. 3). This ridge, like the similar rift systems of the other oceans, represents a line of upwelling magmatic convection from which the Old and New Worlds have drifted away since the Cretaceous at a rate approximately that of the above-mentioned rates of magmatic convection. Areas of descending magmatic convection are presumed to lie along the deep-sea trenches and on the margins of many oceans. Here crustal material is subducted, that is, one plate is thrust under the other, and is gradually melted into the magma. After decades of scornful rejection of Wegener's theory, its offspring, the newly formulated theory of plate tectonics, plays a major role in geophysics today.

One may now conjecture that the major orogenes have also been formed by the collision and subduction of two continental plates. "Orogenes" here are understood as formerly deeply sunken oceanic troughs and/or continental foredeeps filled with sediments and then reduced in width by tens or even hundreds of km by lateral compression, folding, and sheet-like overthrusting. This interpretation seems the more probable as the scrambling of rock units now present in rigid form on the earth's surface could only have occurred in the hot semi-plastic deeps of the earth's crust.[2]

Orogenic compression leads to thickening of the light crustal material (up to 40 km thick under the Alps, and up to 80 km under the Himalayas); due to its greater buoyancy in the magma this material is then gradually raised. This large-scale vertical up- and downwarping of the crust is called *epeiro-genesis* (literally "continent building"). The alternation and varying degrees of such epeirogenesis on the continents is called *epeirovariance*. This determines the raw endogenous form of the mountain ranges (see Section 1.4 below), whose finer modeling is then wrought almost exclusively by exogenic processes. Epeirogenic uplift and subsidence do not always follow orogenic folding at depth, but this is usually the case in those mountain ranges which have undergone the greatest uplift.

Another form of continental crustal movement with very striking effects on the relief sphere is *taphrogenesis* (Illies, 1964, 1965a, 1965b, 1970, 1972).

---

[2] The term "orogenesis" (literally "mountain building") for such paroxysmal crustal movement is a misnomer. After being kneaded and deformed through extreme lateral pressure, units were, for reasons to be explained, raised by simple epeirogenic uplift. Since the folding was best revealed in high mountains, it was until recently thought to represent the cause of mountain development. This is erroneous. No mountain range on earth was ever upfolded. Traces of extreme deformation are also found at low altitude, as in the African basement. In the Alps it has long been proven that the present mountain ranges are much younger than the orogenic folding which took place at depth, and which has only been revealed through subsequent uplift and dissection.

This consists of down-faulted rift systems such as the Upper Rhine Graben with its N and S extensions, the Red Sea, and the Central and E African Rifts. These long, down-faulted crustal splinters, together with their diverging fracture rims, show the location of deep fracture lines in the crust (see Fig. 4). They mark either places where two continental blocks are beginning to pull apart or places where such movement has been aborted. Often, as with the deep-sea trenches, they are accompanied along their length by volcanic and seismic activity.

Not only such small grabens, but also wide basins and mountain foredeeps such as the European Alpine forelands can sink deeply and rapidly, and this can take place within mountain arcs, as in the Upper and Lower Hungarian Basin or in Wallachia (see Sections 3.3.1.6 and 3.3.2.4 below).

Stratigraphic disturbances in the more or less horizontal superstructure and disruptions of the infrastructure are often visible at the surface and are frequently truncated by the relief sphere, which may cut across km of strata, sometimes even exposing the roots of endogenic structures with complete disregard for tectonic patterns. In this manner all epeirogenic structures, whether uplifted or downwarped, are transformed by the exogenic processes of relief building. Even recent taphrogenic fracture zones experience erosion of their rims and deposition in their grabens, processes which may lead to their complete filling in. Volcanoes are similarly converted into volcanic ruins.

The relief sphere of the earth is covered by its outer envelope or *atmosphere* (including the magnetosphere), which surrounds the planet to a thickness of at most around 800 km. The oceans, which in accordance with gravity rest in the earth's great basins, cover 71% of its surface.

These two major members of the outer sphere clearly resemble the earth's interior in that all processes of radiation, wind movement, pressure waves, and other rapidly succeeding changes in the atmosphere, and all surface and deep-water currents, tidal movements, swells, convection, internal waves, etc. in the ocean, are purely ephemeral processes leaving no lasting traces behind them. One ocean swell is absorbed by another in a matter of seconds, leaving not the slightest trace of the first. In the liquid and atmospheric shells the molecules (or in the highest atmospheric layers, the ions) have complete freedom of motion. This also characterizes the interior filling, with the difference that under the tremendous pressures of the latter, the compact molecular masses move far more slowly. The oceans and the atmosphere are linked by the hydrologic cycle and the three states of water. Evaporation from the sea and from the terrestrial hydrosphere sends water vapor into the atmosphere; this vapor then cools and returns as precipitation to the sea and the relief sphere.

This cycle also includes the third and smallest of the outer spheres, the *cryosphere*, formed by the glaciers and permafrost. In the accumulation zone above the snow line, firn collects in consecutive layers and is mobilized by gravity and by its own pressure to stream down to the ablation zone, where it is consumed by melting (evaporation occurs in all stages). All glaciers are simply transitory stages of the hydrologic cycle, involving only a certain delay in the cycle. As a phenomenon they may be long-lived, but the ice itself as a substance changes continually through accumulation and ablation. The individual molecules are incapable of free motion, yet the fine grains of firn and ice move relative to each other, particularly along shear planes. Thus the cryosphere, like the outer spheres and the interior filling, consists of transitory forms which neither create nor leave behind any lasting traces.[3] Such traces are left only by the processes of the lithosphere within the crust, and by the exogenically governed relief sphere (along with its companion spheres) at the crustal surface. These two spheres alone can be studied to investigate the earth's history.

1.3 Geomorphology's Place in the Natural Sciences

Only in and on the crust do past events significantly affect crustal form, so that geology and geomorphology must rely on a methodology different from that of the other sciences. We will discuss this with reference to geomorphology. Such methodological reflections have in the past been rare, so that the nature of geomorphology has been much misunderstood. This naturally pertains, *mutatis mutandis*, to geological methods as well.

Geomorphology, the study of the genesis and history of landforms, differs from the other natural sciences in three main ways. The first difference lies in its genetic character. This means that products or traces of earlier relief-forming processes must be understood in relation to the processes dominating relief-formation today. The second difference is the resulting highly complex nature of the subject, requiring that the most diverse influences be disentangled in careful chronological fashion, in order that their mutual influences be properly evaluated. The third difference, resulting directly from the other two, is that the total relief visible today is in all ways a completely unique product; its lengthy and complicated developmental process cannot be recreated in a laboratory or described in terms of simple formulas or graphs.

We have already seen that conditions are quite different in the earth's interior filling and outer envelope. These spheres are largely the concern of physics, and particularly of geophysics (terrestrial geophysics, hydrology, glaciology, meteorology, and astrophysics). Geochemistry is also concerned with these spheres, as well as with many geological questions (including the processes of diagenesis and metamorphism, the formation of volcanic rocks, etc.), and covers a large part of mineralogy.

[3] The marks carved by a glacier into the underlying rock are not preserved within the glacier's interior. Like the work of the sea upon its coastline, this is a special case of exogenic attack upon the lithosphere, molding its surface into the relief sphere.

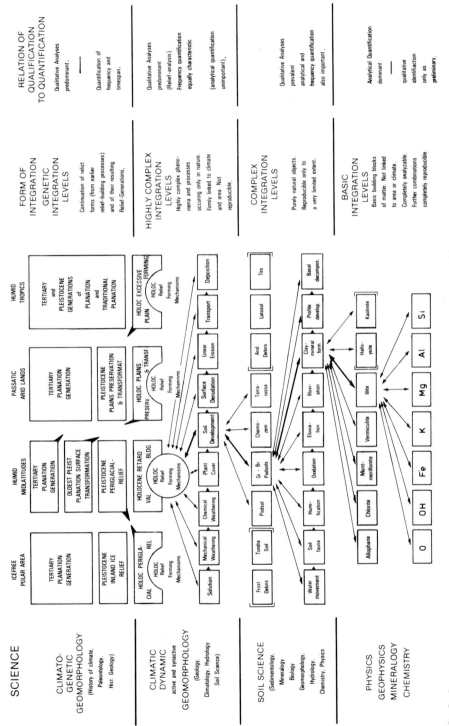

Fig. 5. Geomorphology in the framework of the natural sciences, working on the highly complex integrational level.

Geomorphology's independent position is due to its recent reevaluation of climato-genetics, permitting the entire subject to be studied from all sides. It thereby becomes clear that this very complex field cannot be subsumed under any other natural science. This is briefly illustrated in Fig. 5 (Büdel, 1975b).

This figure has four vertical columns. The two outer, narrow columns (that on the left indicating the various natural sciences involved, that on the right showing the approximate relationship of quantification to qualification within each science) will be readily understood once the two inner columns have been discussed. We will begin with the narrower of these, entitled "integrational levels."

This column portrays a major characteristic of sciences (shown to the extreme left) such as physics, chemistry, mineralogy, cell physiology, etc., which, with the help of increasingly refined analyses, endeavor to explain the basic components of matter and the nature of existence. Processes taking place at this basic integration level can be arbitrarily isolated under specific laboratory conditions, and not only can be completely and artificially reproduced, but also varied *ad infinitum*. These variations, produced *en masse* by our chemical and pharmaceutical industries, are, despite their complexities, fully comprehensible, and can be represented by simple formulas, graphs, or multi-dimensional models. Time is as little concerned here as are natural conditions such as climate. Most of the reactions are completed in a matter of seconds.

With natural objects the situation is completely different: these develop over very long periods of time in an open natural environment with specific climatic conditions. Complex conjunctions of extremely varied influences result in completely unique relief products. These can be partially reproduced, but as wholes they cannot be recreated in a laboratory experiment and cannot be described by simple formulas. On such complex integrational levels, successful explanations are yielded only by tracing the natural influences involved and by a comparative study of related cases.

For example, we may take a soil blanket, e.g., the gray-brown podzolic soil found widely in Central Europe (Fig. 5, center left, fourth line from below; also the profile in Fig. 6). Where present (as an example of a "natural test site") on Würmian sediments, its exact age is known: the circa 12,000 years of the Holocene. But it is also found under analogous climatic conditions over many other substrata, showing here the same two to three A and two to three B horizons. Here the country rock has been physically and materially fully converted into part of the pedosphere. Only underneath this do the B/C and C_1 horizons form the decomposition sphere, transitional to C_2, the unweathered lithosphere.

Far more numerous and varied are the influences which work over long time periods to produce the continental relief sphere. About a dozen elements

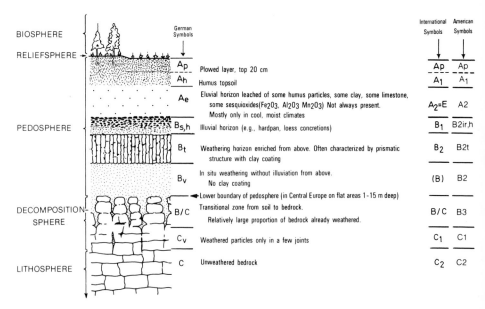

Fig. 6. Typical structure of an ectropical relief-covering soil (based on a 1-1.5 m section of gray-brown podzolic soil).

combine in relief-forming mechanisms (called by some authors the "morphodynamic system") to produce a landscape. Of the dozen elements, only nine are represented in Fig. 5 (center left, fifth line from below). One of these is soil development, of whose variants the gray-brown podzolic soil development is only one complex subordinate element. The other elements and subelements, however, are all equally complex. Thus, compared with the already complex soil science, geomorphology works on an extremely comprehensive integrational level. At the same time, the series of processes working together to form the relief are also dependent on special conditions of time and space. Investigations must be based on very careful observation, and on the strict mental reenactment of the interacting processes involved. This is done, on the one hand, by finding convenient natural test sites, where one can trace isolated elements of the system. Equally important, however, is the subsequent comparison with areas whose natural conditions differ in one way or another.[4]

While each climatic zone has different relief-forming mechanisms, only four particularly important ones have been shown in Fig. 5 (wide column,

[4] Details of the methodology to be used here will be discussed in the anticipated third (German) edition under the title "dynamic geomorphology." A further area of our subject, "active geomorphology," dealing with the creation of the relief products, that is, the landform development itself (not at all a self-evident subject), will also be treated there.

center left). Further subdivision of the elements and subelements has been performed, by way of example, for the humid mid-latitudes alone, but one could just as well have identified the mechanisms of any other climatic zone. Of the relief-forming mechanisms of the mid-latitudes, only one element, soil development, has been further analyzed. Following the thick arrows, the breakdown continues, always following one member of the chain, down to the last basic integration level of the chemical components. All arrows in this system point in two directions: downward toward analysis, and upward toward synthesis.[5]

Once the entire complex involved in a single element or subelement (e.g., the gray-brown podzolic soil) has been established on one of the integration levels (through use of an appropriate natural test site, through comparison of similar and deviating cases under changing climatic conditions, and through building up of a logical chain of evidence), then, should there remain any specific questions relating to the less complex, that is, the basic integration levels, *ad hoc* analyses of laboratory experiments may be used to good effect. This may be true for questions relating, for instance, to the analysis of heavy and clay minerals, transformation of clay minerals, osmotic binding of water in the soil, preservation and verification of presumed paleosol components, questions of fossil content, pollen analysis, all kinds of weathering crusts, reversals of the magnetic field, etc. Questions posed to neighboring branches of science, such as soils science, petrography, paleontology, and palynology, have, simply by being asked, often suggested new viewpoints or lines of research; thus other sciences may profit from work done in geomorphology, the central discipline. Such enrichment is nearly always mutual and has often proved its value, as in Quaternary studies, or in the identification of soil relicts from various relief generations. When individual analyses or mathematical models are applied without recognizing their place in the multi-leveled integrational system of geomorphology, no matter how rigorous or logical the attempt in itself, unnatural models and inappropriate simplifications may result. For this reason I refuse to consider such randomly performed analyses as building blocks for an exact geomorphology.

On most of the earth, the relief which we seek to explain was not produced by Holocene relief-forming mechanisms. This means that even the most precise study of the currently active mechanisms (through dynamic and active geomorphology) cannot explain the relief we see today. We cannot project the activity of the current relief-forming mechanisms indefinitely into the past, for it would require hundreds of thousands, even millions of years for them to create their mature forms. But in such a long period of time the terrestrial

---

[5] In the fifth and third rows from below, the boxes containing the individual elements or subelements of the level are connected by small arrows pointing left to right, indicating that these processes, though interrelated, can also be viewed as an evolutionary sequence running from left to right.

climate has changed markedly everywhere, and in the ectropics, with their pronounced ice age hiatus, even repeatedly. Often up to 95% of the currently visible relief is the product of earlier geologic periods which had quite different climates, and correspondingly owes its existence to quite different relief-forming mechanisms. We therefore have not only the complex current mechanisms to interpret, but also the aftereffects of the older genetic integration levels as well (Fig. 5, second column from the right, top). The diverse, frequently sparse traces are often difficult to establish as synchronous and related, but logical and historical clues may be used to identify the various relief generations, much as archeology and prehistory identify ancient cultures with unknown languages. This purely genetic analysis is a prerequisite to any explanation of today's landforms. This analysis is a part of climato-genetic geomorphology.

The relief sphere is so encompassing and so important for all life and for humans themselves, that the task of explaining it affords geomorphology an independent and indispensable position among the natural sciences, a position which cannot be filled by any other science.

1.4   Endogenic Raw Form and Exogenic Actual Form

The great discoveries of the early geologists concerning the construction of the upper lithosphere, crowned by Lyell (1830-1833) and leading to the foundation of modern geology, had the effect that for two generations it was considered (and in some ways still is today) that the earth's surface was also exclusively or at least primarily formed by these endogenic processes: the relief sphere was thought to be completely dependent on the lithosphere. The great change was introduced, slowly enough, by the work of Rütimeyer (1869). He was the first to demonstrate in detail that valleys are not cracks in the crust, but hollows carved out by rivers. Earlier suggestions had certainly pointed in this direction, but it required nearly a century for the realization to become widespread that the relief sphere is in all major aspects a completely separate and purely exogenic product.

Of the various endogenous crustal movements, recent epeirogenic vertical movements may still be considered to play a major role in creating the present-day relief sphere. Until raised above sea level, no continental relief can be formed. Thus epeirogenic movements roughly define the distribution and plan of the continents and islands, of mountainous and hilly country, of domes and basins. Were they alone at work, they would create an *endogenic raw form*, which we must picture as a smooth dome over all present-day uplifted areas (Fig. 7). But this is completely imaginary, and has never existed. Of course orogenic movements which occurred at great depths, and whose consolidated, jumbled folds were later (usually epeirogenically) uplifted, may be

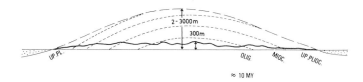

Fig. 7. Endogenous crude shape and exogenic actual shape of a gently domed Tertiary sequence. Taken more or less from the Ukraine. See legend in Fig. 8.

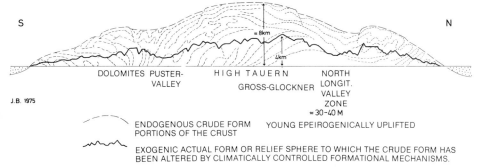

Fig. 8. Endogenous crude shape and exogenic actual shape along a S-N profile through the E Alps. Without exogenic influences the endogenic crude shape would consist of an internally highly folded, externally gently rounded dome about eight km high. During uplift exogenic processes reduced this crude form by about half (by more in the valleys), transforming it qualitatively into the finely chiseled actual shape of today. The *relief sphere* is *actively* formed by exogenic mechanisms.

used, as in the Alps, to reconstruct an endogenous raw form arching high over today's peaks. This is done by reconstructing the individual tectonic units by means of "aerial arches" and "aerial nappes," shown in Fig. 8 (upper broken line). This would suggest that about five km of sediments are "missing" over the peak of Mount Blanc (itself five km high), and four km over the Grossglockner (also four km high); that is, the crude form indicated by the aerial arch would about double the thickness of the present mountain. This was never actually the case, for no sooner does an area rise above sea level than erosion immediately sets in, accompanied by corresponding deposition in the surrounding seas and basins. Exogenic processes alone are responsible for actively changing the gently domed, endogenically uplifted areas into the finely chiseled *actual exogenic forms* we see today in mountain ranges such as the Alps.

This effect increases with height. The slopes and stream profiles become steeper, the erosion more energetic. This is especially true where mountains, despite simultaneous weathering, are lifted above forest and snow lines into regions where erosion is accelerated by frost action. This is a major limiting factor for mountain growth.

Exogenic modification has two effects. Firstly, it results in the very considerable *quantitative reduction* of the endogenically conceived raw block. This is far more than the mere reduction of the mountain peaks would indicate, for the valleys between are carved more deeply. The latter is part of the second, more striking, and important *qualitative alteration* of the crude arch into the finely chiseled and manifold shapes of the mountain ranges today.

Naturally this actual form corresponds in some places to the endogenous character of the substrate, reflecting the *petrovariance*, or differences between the various kinds of rock. But the substrate exercises a purely passive and coincidental influence on the active exogenic processes, which work independently of the local rock type. The resistance of the substrate does not depend on the hardness scale, but rather on the morphologic resistance of the stone, which in turn varies with climate, i.e., with exogenic conditions. While granite is very hard in the polar frost-debris zone, where weathering is largely mechanical, chemical decomposition in the humid tropics transforms this rock into disaggregated grus tens of m deep, which can be crumbled with the hand and which offers little resistance to the processes of mass wasting.

A morphologically hard layer may also assume different shapes even within one climate, according to the manner and environment in which it crops out. Along a jagged mountain ridge, especially where the beds are vertical, the lithologic differences are visible even from a distance. Moderate slopes cutting across almost horizontal beds often clearly display the lithologic structure, forming the much-debated escarpments.[6] But on flat areas, be these low-lying valley floors or high relict plains above petrographically stepped slopes (tread and riser slopes), the same lithologic differences will be morphologically hardly evident, if at all.

Similarly, the many up-ended beds and reefs of the Dolomites and the heavily folded strata of the Dachstein limestone (see Fig. 8) are completely truncated and have little effect on the appearance of their respective relict plains. For the field geologist it is a blessing when a hard layer is revealed in the relief; but for the geomorphologist it is important to determine how and to what extent this occurs, and to find where it is absent.

In poorly consolidated sediments with simple endogenic up-warping, as shown in Fig. 7 for a dome in Tertiary beds such as might be found in the Ukraine, exogenic effects are equally important, both for the quantitative reduction and the qualitative transformation of the broad, gently domed area into a hill country.

In morphoclimatic zones of dominant planation, such as the seasonal tropics, planation may fully counteract endogenic uplift, where this is not too strong. The exogenic effect resembles a carpenter's plane moving across a

---

[6] Level areas on slopes caused by the cropping out of harder beds are called "benches" or "treads." Exogenically formed remains of old river floors are called "terraces."

board slowly rising from the carpenter's bench. The board rises, but its surface does not. This is how the broad "peneplain"-like planation surfaces were formed. These cut across rocks of the most diverse hardnesses, including tightly folded and consolidated basement structures. "Aerial arches" reconstructed from them would lead to crude forms as much as several km high, which however, have never existed. Here we are dealing with steady planation, which over a very long period of time wore away the thick rock units, layer by layer, parallel to the original surface. The relief remained flat throughout the entire time. Earlier one assumed the pre-existence of high mountains, which only later were supposedly reduced to their "torso" (*Rumpf* in German, hence the German term *Rumpffläche*). Thus Davis (1899) and his German translator Rühl (Davis-Rühl, 1912) as well as his active followers today refer to peneplains. Use was also made of the terms "end-peneplain" and "terminal torso" in literal translation from the German *Endrumpf*. Today we know that such areas were never higher. This was first recognized by W. Penck (1924), who introduced the term *Primärrumpf*, i.e., a planation surface which was continually eroded away as fast as it was lifted up.

Only two endogenous processes, highly localized in effect, have any direct influence on the exogenic relief, creating recent volcanic forms (cones, craters, diatremes, and calderas), and fresh forms of taphrogenesis (fracture scarps of rift valleys). As we have already mentioned, however, rapid exogenic modification takes place even here. Volcanoes soon become volcanic necks, and the degree of their alteration can be used to distinguish relief generations (Büdel, 1954b, 1955a). Etchplains can erode away even hard basaltic dikes, beveling them off until no trace remains. Fracture rims are soon worn down, their flanks gullied, their grabens filled in with sediments. The faults along the Rhine graben, recently traced to a depth of seven km (Illies, 1970), have displaced the once continuous geologic beds along a 60°-65° shear surface by about 3000 m. Yet the present height of the graben walls is little over 300 m, and their slope is not nearly as steep. Only their rectilinearity still reflects their origin as fault scarps. Endogenous causes and exogenic countereffects have worked simultaneously at varying intensities for many millions of years.

But even these cases are exceptions in the much broader land masses, where the endogenous structures have been far more thoroughly altered and truncated by exogenic processes.

1.5 The Relief Sphere as the Main Energy Conversion Surface of the Earth

Where do the exogenic relief-forming mechanisms get the energy to transform the crustal surface so thoroughly and independently into the completely different relief sphere?

All these processes and their consequences (to be discussed in the following three sections) are set in motion by solar energy. True, about half of the solar constant (the amount of short-wave visible light hitting the atmosphere at its outer edge, around 800 km away) is absorbed without noticeably warming the atmosphere, and, more importantly, without particularly affecting the exogenic processes on the earth's surface.

But the other half of the short-wave radiation, the "global radiation," hits the earth's surface directly or indirectly, illuminating and heating it. Above the continents this light-heat conversion takes place in a thin layer, but it is far more intense here than on the sea, where the light radiation penetrates far deeper. Waves, convection, and currents act as mixing agents, warming far more water to a correspondingly lesser degree. As water has also the greatest heat capacity of all known compounds, the ocean serves excellently for thermal storage. It warms the air masses above it, which in turn exert an ameliorating effect on the ectropic land masses when moving over them in winter. On the continents a thinner layer is heated much more intensely during the day, especially in summer and in the lower latitudes; the area in which diurnal or annual temperature fluctuations occur there extends no deeper than 10-12 m. (At this depth the annual temperature fluctuations cease. The average temperature pertains here, known in Central Europe as the "cellar temperature" of about 11°C.)

This heating surface of the earth is what warms the lowest layer of our atmosphere, the troposphere, in which those phenomena classed as "weather" take place. For the long-wave thermal radiation sent back by the earth's surface (particularly by the continental surface) does not penetrate the atmosphere, but rather is to a great extent absorbed, largely by tropospheric water vapor, producing the well-known "greenhouse effect." The result is that the temperature of our atmosphere drops with distance from the heating surface at an average rate of 0.5°C per 100 m in humid air.[7]

The transformation of light into heat (a process which works as insolation weathering on bare stone), makes the relief sphere the *main energy conversion surface of the earth*, causing it to govern all processes of weather and climate in the overlying air masses. The great temperature differences between the equatorial belt and the polar caps cause a pressure differential, which, along with the coriolis effect, keeps today's atmospheric circulation in motion. Heating of the land, and especially of the sea, causes evaporation which maintains the water vapor content of the troposphere. Variously induced cooling of the air masses leads to precipitation of all kinds, which in turn causes a differentiation between humid and arid regions of the earth, and

[7] The upper boundary of the troposphere in the polar regions lies 8 to 9 km up. Above this lies the multilayered tropopause, after which follows the stratosphere, whose base has a temperature of about −55°C. Above the equator the tropopause lies at a height of 17-18 km, and the temperature is about −85°C.

creates the continental hydrosphere and the cryosphere of the polar regions and high mountains. Intermittent freezing on the surface, especially on bare rock, causes frost-shattering; permafrost below the surface has other, more marked effects.

These climatic effects set off the far-reaching energy conversions, which in the form of *relief-forming mechanisms*, work on and shape the relief sphere. The highly complex system of these relief-forming mechanisms can be divided into a dozen self-contained stable and mobile elements, whose effects are closely interrelated. The *stable elements* (to be discussed individually in Sections 1.6.1-1.6.6) create the four *companion spheres* (the hydro- and cryosphere, the biosphere, the pedosphere, and the decomposition sphere) of the relief sphere, which lend the relief sphere its three-dimensional character. The *mobile elements*, a trinity of denudation, transportation, and deposition, lead eventually to the overall form of the relief, and thereby to the relief sphere's most extreme energy conversions, whose primary impetus springs from the sun.

1.6 The Stable Elements of the Relief-Forming Mechanism

1.6.1 GROUND TEMPERATURE, GROUNDWATER, AND PERMAFROST

The most immediate contributions of the exogenic processes to the relief-forming mechanisms consist of insolation, which strikes the relief sphere and warms it (as well as terrestrial radiation which then cools it), and precipitation which falls on the continental masses. The earth's thermal conductivity (in the soil, regolith, and bare rock) is very low, especially in bog soils, so that over most of the earth the daily and annual temperature fluctuations even out at a depth of 12 m, where almost the exact mean annual temperature is present. This increases with the geothermal gradient toward the earth's interior by an average of around 3°C per 100 m.

Precipitation (rain, melt water, and condensation water) penetrates far deeper into the earth. Together with heat, it causes mechanical and chemical weathering, which along with the biosphere, are responsible for soil development (see below). The water's influence, however, ranges even deeper, for it is the main agent forming the decomposition sphere, i.e., the zone of partial bedrock weathering (formed primarily along joints, see Fig. 6). Water also creates the two maxima of this sphere (see Fig. 9) in areas of extremely differing climates: through chemical weathering in the humid tropics, and through permafrost in the upper middle and polar latitudes. Both maxima are relict or partly relict (see below). Today's area of continuous deep permafrost is delineated in Siberia and in arctic N America by the −4°C isotherm

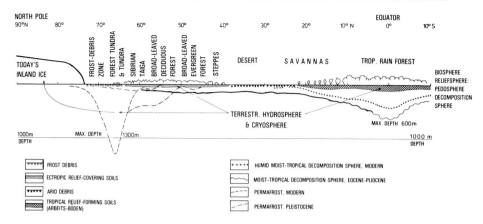

Fig. 9. The pedosphere and decomposition sphere in their present sequence from the equator to the pole. Also the maximum depth of chemical decomposition (formed in the Tertiary) and the maximum depth of permafrost (formed in the Pleistocene).

(see Fig. 37). Here the permafrost still maintains itself, but its origins reach back into the Ice Age Pleistocene. Its maximum depth occurs in the coldest part of E Siberia within the Cherski Mountain Arc (with mean January temperatures of under −50°C, and mean annual temperatures of less than −16°C), where it reaches a depth of 1300 m. The similarly relict tropical decomposition sphere reaches a maximum depth of 600 m.

Infiltrating water, a major agent in the creation of the pedosphere and decomposition sphere, seeps into the joints and bedding planes of the substrate. As the water becomes saturated with dissolved particles, its effectiveness as a solvent is reduced. Only in the cavernous mazes of deep-reaching karst is continued widening of underground chambers possible, for here the dissolved particles often drain to deeper outlets, flushed out by fresh water from above. In other rocks the water may be stored in the deepest open joints, and may, after much time and long journeying, return again to the surface. These deepest waters were once assumed to be ''juvenile,'' arising from the earth's interior. Today we know that all water, even the hot mineral-rich water from great depths, arrived there by infiltration from the earth's surface, even though this infiltration took place long ago. In the Simplon tunnel in Switzerland, whose highest point lies at 700 m, highly productive hot springs were tapped nearly two km below the average surface of the Simplon Massif (2400 m). These springs, producing 320 l/sec of water at 50°C, would undoubtedly also have been struck at greater depth (Pressel, 1928). Artesian water, which has infiltrated along permeable beds dipping between impermeable layers, and which may be under such pressure that when drilled, it may shoot to the surface along the bore-hole, has been encountered down to depths of 1100 m in S Algeria. On reaching the surface there, it had a temperature of around 35°C. In the Egyptian Sahara W of the Nile, Knetsch (1966) has dated the upper stories (down to 400 km below the surface) of artesian water in the Nubian sandstone sequence to 25,000 to 40,000 years B.P., i.e., to the pluvials of the last ice age. Deeper stories of this artesian water (reaching to perhaps triple the depth) may well be stored here from the older Pleistocene pluvials. Similar situations have been found in the Negev (S Israel), in Saudi Arabia, and in Australia. The deepest artesian water drilled to date in the Sahara was found at 2500 m near Marsa Matruh (Knetsch, oral communication). These sources of relict water are no longer being replenished by precipitation. Their exploitation for oasis irrigation is therefore short-sighted, as they can be depleted like any other resource.

So far we have dealt with groundwater in its broadest sense, and with permafrost. But atmospheric water also creates and feeds the surficial members of the terrestrial hydrosphere, i.e., lakes and all forms of surface drainage, from the smallest rills and tropical sheet runoff to streams and rivers. As we will see later, this water which runs over the earth's surface is the main cause of changes in the relief sphere, and is the prime contributor to relief development. Depending on climate, this occurs in varied ways.

1.6.2 SOLUTION

Movement in the terrestrial hydrosphere is usually accompanied by solution. Solution takes place in surface runoff, in the ground, and in the decomposition sphere. Solution is particularly important in the humid tropics, because of the high temperature and acidity of the soil water there. Due to the thickness of the soil blankets in the humid and seasonal tropics, the rivers carry dissolved matter derived mainly from solution near the surface. Since chemical decomposition is also important, most tropical rivers (apart from clear-water and black-water rivers) carry a large suspended load. In the seasonal tropics one must take care to compare rivers only during their flood stages in the rainy season, for in the dry season they usually evaporate down to small rivulets (Photo. 17), carrying proportionally more suspended and dissolved matter. This is why many desert rivers also show such high values for both these loads. In all climatic zones the maximum dissolved loads are carried by rivers draining areas rich in gypsum and limestone, as for instance, the Kocher River in S Germany (Wagner, 1950).

In limestones throughout the world, solution creates deep-reaching karst forms which affect the surface relief. These forms may be considered exceptional, lithologically controlled cases of deep-reaching decomposition. Karst topography is largely controlled by subterranean drainage through widely branching cave systems. Where karst areas border the sea, as in the Yugoslavian Adriatic and in other areas of the Mediterranean, the cavernous labyrinths may carry fresh water below sea level. This water may be under hydrostatic or hydrodynamic pressure, and may rise along tubes to the sea floor. H. Lehmann (1936, 1954, 1955, 1957) has repeatedly emphasized the basic differences between ectropic and tropic surface karst. Others such as Gerstenhauer (1967) and Blume (1970) have verified and elaborated on his results. As with glacial and coastal landforms (also clearly defined geographic and thematic forms in the overall relief sphere), an old and very wide literature exists on the subject of karst. We will come back to this in other contexts (see, e.g., Sections 2.4.4, 3.3.1.5, and 3.4.3).

In the ectropics, such intense solution and karstification is possible only in limestone. In the tropics such features appear in other rocks as well, though to a far lesser extent. Wirthmann (1970) found *sinkholes* 30 m deep in the peridotite of New Caledonia. Basalt and other basic rocks there are also subject to moderate solution in the form of *joint widening*. This overlaps with the closely related process of chemical *deep weathering along joints*, with subsequent washing-out of the detritus from above. Wilhelmy (1975 and earlier) has described *karren* in the tropics (granite and other crystalline karren) etched 50 cm deep even into the surface of acid rocks.

1.6.3 MECHANICAL WEATHERING

In pure form, mechanical weathering breaks down the *in situ* rock without altering it chemically. Mechanical weathering is most effective on bare rock surfaces, particularly where these are steep, and is therefore greatest in climatic zones where the rocks are exposed over large areas, or are covered with only a thin regolith veneer. These conditions are most common in the polar regions, in high mountains, and in arid deserts.

The most important form of mechanical weathering is *frost weathering*. Water trickles into the finest crevices of the rock, and upon freezing, expands in volume by 1/9. The extreme pressure thus produced can never fully be relieved at the open side of the crack, particularly when located deep underground. The pressure is therefore directed mainly against the sides of the joint, prying the rock apart. In ice wedges and in the *ice rind* (the upper part of the permafrost layer beneath the active layer; see Section 2.2.11) the cumulative effect of many winters (and particularly of winters with deep frosts) leads to the creation of deep, ice-filled joint systems. The rock becomes totally fragmented and is shot through with horizontal layers of relatively pure ice. The resulting *ice rind effect* has far-reaching consequences for the total morphological development of the permafrost region (Büdel, 1969b). Steep rock faces in all frost-prone climates have rock fall heaps produced by frost shattering.

In frost-free arid regions *salt wedging* occurs when salt crystals precipitated along joints absorb water and expand. The effect may resemble that produced by frost shattering, and like the latter, it can also work even under a blanket of debris. But it is neither vertically nor horizontally as widespread as frost shattering, and therefore has nothing like the same quantitative effect. Nor does salt wedging have the extensive consequences of frost shattering or permafrost.

In some respects the effects of insolation weathering have also been over-estimated. Due to the low thermal conductivity of rock, heat cannot penetrate very far, and cannot be responsible for the spalling or exfoliation of thicker plates of rock (Wilhelmy, 1958a, 1975). Nevertheless, insolation weathering is evident on the surface of granular rocks such as granite. Here the variously pigmented mineral grains expand at different rates in response to heating, causing small-scale crumbling of the rock surface. Insolation weathering is an important cause of *sanding,* which plays a major role in frost-free deserts and on the bare rock slopes of inselbergs in the humid tropics. In the Sahara (Photos. 24 and 43) and in the seasonal tropics of S India (Photo. 20) I know of many rock slopes, smooth or partly block veneered, where none of the loosely strewn, precariously balanced rocks tumbles down. At the foot of such a wall there is often a steep haldenhang (basal slope strewn with blocks), but never a gently concave foot slope (Photo. 22), and the steep rock wall or block talus slope ends at an abrupt piedmont angle. In the seasonal tropics the surrounding area is a latosol blanket, which in the desert has usually been transformed into a *fluvio-eolian sand plain* (*Sandschwemmebene*). This frequently occurring assemblage shows that for a long time the most important form of erosion occurring on the rock cliff was sanding.

It is probable that *salt wedging* and *hydration* also help loosen mineral grains or fractions of grains. In hydration, water molecules attach to the surfaces of various minerals, or penetrate and adhere to their crystal lattices. The resulting expansion may be very disruptive. In most cases this presupposes a chemical weathering of the stone or of its individual mineral components. According to Wilhelmy (1975), however, hydration can set in even before any marked chemical weathering has taken place. Hydration and salt wedging, as well as the resultant sanding or grus weathering, are thus typical examples of weathering processes in which physical and chemical influences work together. The relative importance of these two influences varies climato-genetically, as well as in time and space.

Insolation may also be involved in scaling thin 1-3 cm thick plates from a rock surface, but it cannot be as important in this process as is chemical weathering. According to Klaer (1956) and Wilhelmy (1958a, 1975), chemical processes play an increasing role over time in the creation of tafonis. These start off as larger or smaller hollows which work into a rock wall from without; they then form a hard external rind, and work from inside outwards. The same is true of large hollow blocks (*Hohlblöcke*), which, protected by an external crust, are hollowed out internally in the moisture-preserving shade.

The flaking of thin platelets of rock from within such hollows is also accomplished chemically.

Weathering rinds, necessary for the creation of tafonis, are also of chemical origin, as are other forms of rinds and desert varnishes. These have been ably reviewed by Hagedorn (1971, p. 24), based on his observations in the Tibesti Mountains of the Central Sahara. Both chemical and physical processes are involved in forming iron, laterite, and calcareous crusts. We will come back to these crusts later (see Section 2.3.2.2)

Exfoliation (previously called desquamation) of rock slabs up to several dm or even m thick, according to Kieslinger (1958) and Louis (1968a), is purely physical in origin. This was described fairly early on the slopes of steep inselbergs, especially in homogenous plutonic rocks lacking pre-existing planes of weakness. In their original environment, these rocks were subjected to extreme pressure. Where they are lopsidedly relieved (by planation exposing an inselberg, by tropical valley-cutting, ectropical erosion, or even by deep quarrying), *unloading joints* develop parallel to the slope. These may eventually lead to the separation and sheeting off of very thick slabs (Photos. 20, 22, and 43).

1.6.4 CHEMICAL DECOMPOSITION AND TRANSFORMATION

Unlike mechanical weathering, chemical decomposition can lead to thorough or even complete transformation of the parent rock. Part of the material is carried away in solution. Another part (especially the quartz grains) is preserved for a long time in the residual mass. The majority, however, is gradually changed into completely different mineral components. This occurs through increasingly effective chemical transformations, and is accompanied by various processes of mobilization and redeposition in the soil. Two conditions are necessary for this change: heat and water. The higher the temperature and the more acidic the water (the lower its pH), the more intense the chemical disintegration. Intensity of chemical decomposition also depends on whether both factors are present for longer or shorter periods of the year, and whether the hot and humid periods occur simultaneously or not. This itself shows how very dependent this type of weathering is on climate.

A major chemical process involved is hydration, discussed above. Under the influence of slightly acidic water especially, this can gradually lead to *hydrolysis*, i.e., to a peripheral exchange between the cations of the mineral surface and the H+ and OH− ions of the water. Hydrolysis can even attack otherwise highly insoluble silicates such as feldspars. This is especially true of the dark components of plutonites and metamorphics, and of the components of diagenetically altered sediments. Weathering of different kinds, then, occurs to various degrees, according to the direct or indirect effects of the climate. Oxygen, an ever-present component of water, leads to oxidation, and especially to the formation of Fe- and Al-oxides (earlier grouped under the name sesquioxides). The Fe-oxides, according to their composition, cause the brown to red staining of most soils. At the same time, many of the rock components are converted into clay minerals (particles $>2\mu m$), which form the majority of the so-called "soil." Their crystal lattices expand in water and adsorb water coatings which permit the clay particles to slide against each other. These two features lend soil its characteristic plasticity. When dry, the same soil will have some porosity.

This chain of disintegration and transformation is not entirely the result of inorganic chemical processes dependent on specific climatic conditions. Organo-chemical and biochemical influences of the biosphere are also important, and only in conjunction with these can a soil be formed. Soil development is a decisive element in the relief-forming mechanisms of any climatic zone. Before discussing these relief-forming mechanisms, therefore, we must first look at the effects of the biosphere, and especially of the plant cover.

1.6.5 The Plant Cover and the Animal Biosphere

The biosphere affects the soil in four stages. The first of these consists of the plant cover and its roots. The second is the total plant and animal life within the soil, called the *edaphon*. The faunal component consists of insects (primarily termites, ants, and mites), burrowing mammals, various species of worms, and a few snails. The floral component consists, along with fungi and algae, largely of bacteria. A gram of soil can contain hundreds of thousands, even millions of these smallest of all living creatures. Their effect on the soil is marked and diverse. The CO_2 content of the soil air, which is largely responsible for the acidity and chemical aggressiveness of the soil water, is about one-third due to root respiration and about two-thirds to the activity of the soil bacteria as decomposers. Where aeration is poor, as in tropical clays during the rainy season, the CO_2 content increases with depth. On the whole, as the soil thickness and biologic population increase, the CO_2 content of the soil air rises from the polar areas (0.02-0.05%) through the mid-latitudes (around 1%) to the humid tropics (up to 20%), or, in other words, by up to a thousand times.

The third stage consists of dead plant matter. In its original, chemically unaltered form in forest litter, this non-humic organic constituent forms a humous reserve and plays little role in the soil budget.

This changes in the fourth stage, in which the soil bacteria mineralize and break down the non-humic component into humic matter, making it once more available to the plants as nutritive matter. A series of organic acids produced at the same time (e.g., fulvic acid and humic acid) increase the chemical aggressiveness of the soil water. The humic matter becomes increasingly resistant to microbial activity, leading to enrichment of the dark organic matter in the soil. This is one of the factors responsible for the characteristic staining of the A_1 horizon in many ectropic soils (see Fig. 6).

It is thus the bacteria whose decompositional activities render the soil useful to the biosphere above ground. The soil flora increases the chemical aggressiveness of the soil water and enlarges the pore volume of the soil. Worms and termites riddle and mechanically rework the soil, bringing finer particles to the surface, a factor which is particularly important in the soils of the humid tropics. Thus soil development is intimately bound not only to the life within the soil, but also to the biosphere upon its surface: the one sphere cannot live without the other.

1.6.6 SOIL DEVELOPMENT

The interdependence of biosphere and soil is even clearer when one compares the spatial distribution of the soil zones with that of the denser vegetation zones. Both are bound to the major climatic zones where water and heat are present. Both therefore avoid the deserts and the polar areas, and are instead associated with the non-polar humid areas. Here further gradations reveal close links between the variety of plant cover and the soil type. This is especially true of the humid ectropics: we may roughly equate tundra, taiga, mixed deciduous forest, parkland, tall-grass steppe, and short-grass steppe with tundra soils, podzols and lessivés, brown forest soils, gray-brown podzolic soils and pseudogleys, chernozems, and chestnut soils respectively.

All these ectropic soils (like their lithologic and other local variations) are marked by clear horizons (see Fig. 6). Such horizons are strikingly absent from the soils of the humid tropics, despite their great thickness, or are only very weakly indicated. At the same time, tropical soils are far less varied, though the plant communities range according to hygric climate from desert savannas through various wetter kinds of savanna woodland and tropical scrub and woodland communities to the evergreen equatorial rainforest. Ectropic soils, about one to two m thick, eventually achieve a (usually very standardized) *mature profile*. Once this has formed, further encroachment of the soils deeper into the bedrock occurs only at an extremely reduced rate. In tropical lowland soils, which may be up to 30 m thick, weathering of the underlying rock continues over very long times with only slight periods of retardation. Beneath the actual soil blanket exists a much larger sphere of partial weathering, or decomposition sphere. These differences lead us to call the ectropic soils *relief-covering soils* (*Ortsböden*), and the tropic soils *relief-forming soils* (*Arbeitsböden*) (see Section 2.3.2.2). The latter play an exceedingly important role for denudation processes in the tropical lowlands.

Mechanical weathering is the first to attack fresh rock (exposed by glacial retreat or in fresh volcanic rock, in fault scarps, or undercut slopes), and is therefore often regarded as the forerunner of soil development. But when the deeper parts of an ectropic soil profile (the deep B horizon, B/C horizon, or the C horizon; see Fig. 6) contain fragments of unweathered bedrock, these are probably not fragments of incipient soils, but rather Pleistocene frost debris or relict tundra soil, in which remains of old frost structures can sometimes be recognized. Deeper crevices and pipes filled with pre-Pleistocene red loam may occur, deriving from an even older (Tertiary) soil generation. Such pipe fillings in the Franconian Alb have been paleontologically dated into the Eocene (Dehm, 1961a and b). Pockets of grus are also Tertiary relicts and are found down to depths of 70 m in the Frankenwald, according to Wurm (oral communication). In such cases, at least three superimposed soil generations are present. Naturally they are no longer clearly divided, having in many cases been mixed by later soil development and displacement processes. While the investigation of paleosols in stratified deposits has progressed well, especially in connection with loess studies, the classification of soil generations within one and the same profile has advanced but little. A wide field for future investigation lies before us in such polygenetic soil studies.

But it must be remembered that any unusual soil component (e.g., kaolin in an ectropic soil profile) may arise in many ways: (a) it may be due to recent soil development; (b) it may derive from a past (interglacial or pre-glacial) period of soil development, which was then incorporated into the present soil (remains of polygenetic soil development); (c) it may derive from the bedrock (in our example, from Triassic red clays); or (d) it may derive from sources such as (b) or (c), but is not *in situ*, being transported from higher areas by erosion (of what kind and age?). Only when all these questions have been answered can the correct climato-genetic evaluation of such a profile be attempted.

1.7   The Four Companion Spheres:
the Terrestrial Hydrosphere and Cryosphere,
the Biosphere, the Pedosphere, and
the Decomposition Sphere

The stable elements create the relief sphere's four companion spheres. These intensify the effects of the light-to-heat conversion upon the relief sphere (which we may provisionally imagine as flat; see Fig. 9) and expand the relief sphere from a two-dimensional surface into a thicker, three-dimensional layer in the earth's structure. This layer, whose processes are controlled entirely by exogenic climatic influences, is unique in character.

This sets the stage for the mobile elements, which create the relief itself. The companion spheres, therefore, while not part of the relief sphere, are of interest to geomorphology. In many ways, geomorphology takes a medial, even key position between the neighboring subjects of terrestrial hydrology, plant geography and ecology, soil science, geology, paleontology, and climatic history. Geomorphology's closest ties are with soil science. It is a welcome development that in many European universities today, the study of geomorphology is accompanied by a suitably oriented study of soil science.

The distribution of the companion spheres in and on the relief sphere (provisionally imagined as flat) from the equator to the pole is shown in Fig. 9 in a simplified profile with vertical exaggeration. This emphasizes how each of the individual spheres changes with the major climatic zones.

The biosphere is best developed in the humid tropics (from the equator to 15° latitude). Here it still retains the character of the ''Tertiary plant and animal paradise.'' The pedosphere, with its characteristic relief-forming soils (mainly reddish-brown tropical loams or latosols), is also thickest here: while averaging six m, it can be up to 30 m in depth, though such extreme cases may result from long development. While such soils remain active where the rainy season is four to four-and-a-half months long, many authors consider that they can only originate where the rainy season is at least seven to nine months long. The relief-forming processes connected with the existence or origin of these soils are still active N and S of the equator to regions with as little as four-and-a-half to four humid months (at the desert fringe of the tropical savannas), as is clearly shown by the presence of the appropriate assemblage of landforms, i.e., inselberg-studded etchplains. In the underlying decomposition sphere, depths of up to 200 m may be regarded as modern. Where this reaches 600 m, as in some itabirites (Tienhaus, 1964a and b), it is certainly relict, and may have taken the entire Tertiary to develop.

In the desert, latosol soils inherited from a humid-tropical past have only escaped being reworked by modern arid weathering debris sheets in a few areas. Usually the latosols have been replaced by fluvio-eolian sandplains in the warm trade-wind deserts, and by coarse debris sheets covering the pediments and glacis in the winter-cold inland deserts.

Remains of past humid-tropical decomposition spheres underlie many deserts far into the higher latitudes. Such remains also occur in the mid-latitudes, e.g., in the form of deep weathering along joints, or as relict (Tertiary) grus, traces of which may be found even in the polar areas. Here relicts of the humid-tropical decomposition sphere overlap with the current permafrost region, and even more so with the region of relict Pleistocene permafrost. Modern permafrost reaches depths of 1300 m in the coldest parts of Siberia (in the Cherski Mountain Arc), where ice wedges may run 30 m deep. Traces of a continuous (Würm) permafrost region are found in Central Europe as far S as the Alps, and outposts of the same are to be encountered in SE Europe and even in the mountains of the N Mediterranean area.

This entire region is covered today by one to two m of ectropic, relief-covering soils closely associated with the present vegetation belts. The un-

derlying relict decomposition spheres indicate that polygenetic remnants of
Pleistocene and Tertiary soils are often to be found in and under these Holocene
soils.

Finally, beyond the forest and tundra belt, these relief-covering soils are
replaced by subpolar blankets of frost debris. Where land is present (e.g., on
Greenland or the Arctic islands) we meet continental ice by latitudes of at
most 70° to 80°. The Antarctic cap, over 3700 m thick, covers the pole itself.

I.8   The Mobile Elements
of the Relief-Forming Mechanism

Any piece of dry land, even a newly risen skerry in the Gulf of Bothnia,
or a fresh, steep-sided volcanic island off the coast of Iceland or Java, is in
theory immediately attacked by both types of relief-forming mechanisms: by
the stable elements, which give rise to the companion spheres, and by the
mobile elements, which, with the help of gravity, instantly subject any piece
of land above sea level to the mobile trinity of erosion, transportation, and
deposition. This attack is rarely immediately effective (exceptions might consist of a fresh cinder tuff cone or a large sand dune). In over 99% of all cases,
the mobile forces become effective only after the companion spheres, particularly the pedosphere and the terrestrial hydrosphere, have been formed. The
mobile elements of the relief-forming mechanism represent the ultimate stage
in the energy conversions of the relief sphere, and are responsible for fully
transforming the endogenous raw form into the exogenic actual form.

1.8.1 AREAL AND SLOPE EROSION

Areal and slope erosion refer to denudation processes occurring on the
interfluves between the narrow erosional lines of the rivers. The active fluvial
work of attacking the substrate and carving the valley courses will be designated here as *fluvial erosion*.[8] On much of the earth, in all the ectropics and
in mountains of all zones, this *linear erosion* cuts down faster than *areal
erosion (Breitenabtragung)* on the interfluves. Wherever linear erosion and
valley-cutting are dominant, the landscape becomes characterized by slopes,
ranging from gentle inclines of less than 2° to vertical cliffs. Areal erosion
here becomes *slope erosion (Hangabtragung)*. In these areas denudation intensifies with increasing slope, while the soil cover concurrently becomes thin
and discontinuous. In the humid tropics a slope of 65° may still be covered

---

[8] Only in German is this narrower meaning of linear or fluvial erosion conveyed by the word
"erosion" alone. In French and English "erosion" has a much broader meaning, referring to
denudation in general.

by soil, but in the mid-latitudes this cover becomes discontinuous at around 55°. In the polar regions bedrock may become visible, according to lithic type, at little over 30°, and in fact, nearly horizontal rock surfaces (ice-scoured surfaces or *Rundhöckerfluren*) may still be exposed due to the retreat of continental ice. Flat bedrock may also be exposed in deserts. The quite different conditions pertaining on the inselberg-studded etchplains of the humid and especially of the seasonal tropics will be discussed in the next section.

Concave slopes can intersect in mountain ridges, but this is not common. Rather, a steep young slope usually terminates at a more or less sharp *working rim* or convex break in slope (*konvexe Arbeitskante*; abbreviated *Cx B/S* in the figures) where it intersects an old plain crowning the mountain summit (Fig. 10). This plain we call a paleo-relief or *relict surface*. The paleo-relief is controlled by its own system of shallow swales (S in Fig. 10). Young V-shaped ravines encroaching from below (V in Fig. 10) often do not meet the swales at this working rim, and do not constitute extensions of them. In such cases, the rim completely interrupts erosion (*Abreissen der Denudation*, Büdel, 1954b; Wirthmann, 1964). In the frost debris zone of Spitzbergen, Wirthmann found that the active layer (the layer of seasonal thaw, here the erosional agent) dries out above such a working rim. The rim does not ac-

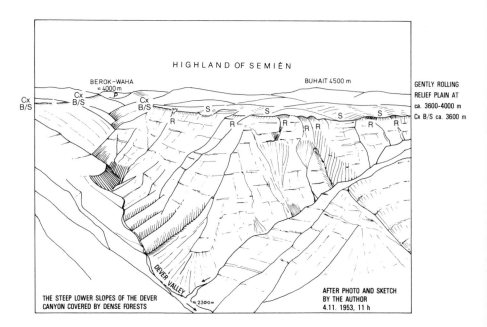

Fig. 10. Interruption of erosion along the sharp rims (convex breaks in slope: Cx B/S) of the Dever Canyon in the Highlands of Semién (N Ethiopia). Many of the young V-shaped ravines (R) encroaching from below upon the rims do not meet the hanging swales (S) of the relict plain: the valley generations are completely independent of each other. Drawn after Photo. 29.

celerate erosion of the relict surface (where denudation is slight in any case), but even retards it. In the humid tropics such a rim can, through soil desiccation, form a hard laterite crust: the paleo-relief thus protects itself from excessive denudation by forming an indurated edge.

The development of slopes in general, and especially of gentle footslopes, is covered in an extensive literature, replete with theoretical models and mathematical formulae, but only too often remote from natural conditions, i.e., saying little about the actual processes at work. Summaries by such authors as Carson and Kirkby (1972) ignore such important works as Macar (1970). This topic, whose study has centered too much on current processes, will be treated in the third edition. Here our aim is simply to point out the climatically determined differences in areal erosion or slope development as seen in the ectropics and in the tropical mountains.

Erosion works fastest in the active layer of the frost debris zone where the entire sheet of soil or regolith is subject to solifluction. The denudational efficiency of solifluction is aided by meltwater collecting above the basal permafrost, by fracturing of the stone in the upper permafrost layer (or ice rind), and by episodes of heavy sheet wash occurring once or twice a decade. These processes are replaced by even more effective agents on slopes of over 25°. The combined effect of all these processes and their interaction with linear erosion in this "zone of excessive valley-cutting" will be described in greater detail below.

In the tundra these processes are very much reduced. In the taiga and in the forested mid-latitudes (including the better vegetated of the ectropic steppes), a continuous sheet of relief-covering soils overlies the bedrock. This soil is not in motion as a whole, but has formed through Holocene *in situ* weathering (hence the German term *Ortsboden*). Areal erosion here is confined today to a rather inefficient surface wash; this was even less effective under the forest or steppe cover of the virgin landscape. The change to a cultural landscape, which occurred at about 3500 B.C. in Central Europe, has increased the effectiveness of surface wash, especially on places like the steep slopes of vineyards. Nevertheless, the relief-covering soils of the cool and rainy mid-latitudes have been almost entirely preserved. In the Mediterranean, earlier introduction of the plow (as early as 5000-6000 B.C.) led to greater and longer deforestation; coupled with torrential winter rains and major destruction of the soil flora in arid summers, this has so contributed to surface wash, that little of the old relief-covering soil is left on the steeper slopes. Further influences acting upon this "subtropic zone of mixed relief-development" will be discussed in Chapter 3.

The arid deserts are divided into the winter-cold interior deserts of the higher mid-latitudes and the essentially frost-free trade-wind deserts. In the latter especially, areal erosion and relief development today are far less effective than has hitherto been assumed, and tend to conserve the assemblage of forms inherited from a past humid-tropical climate.

Continuing into today's seasonal tropics, we find that relief-forming processes in the mountains resemble those of the subtropics. The lowlands, however, present a radically different picture, to which we will now turn our attention.

1.8.2. PLANATION

The humid tropics consist mainly of the seasonal tropics, defined as all areas with four to nine humid months and characterized by savanna and savanna woodland belts, including the Miombo forests of Africa. All gently uplifted areas of this zone, and especially the formerly united shields of the southern continents (e.g., Australia, S India, much of Africa, and tropical S America), are covered by wide, inselberg-studded *etchplains* (*Rumpfflächen*). These lie today at varying heights, separated by *etchplain escarpments* (*Rumpfstufen*). Where actively forming, they are invariably covered by thick relief-forming soils, mostly tropical red loams or latosols.

Along the basal surface of weathering, chemical decomposition works ceaselessly downwards, while in the rainy season finely worked material is correspondingly removed from above by highly effective *sheet wash*. This *mechanism of double planation surfaces* (*Mechanismus der doppelten Einebnungsflächen*, Büdel, 1957a, 1965) is alone responsible for creating these etchplains over long geologic periods. During this time, even the largest rivers have remained fully integrated into the process of *planation* (*flächenbildende Denudation*), so that the etchplains have remained undissected. The rivers here are entirely passive, contributing to the downwearing of the land merely by transporting the worked material. The accompanying erosion differs from that of all other regions (especially from that of the ectropics) in one important respect: whereas in the ectropics erosion becomes more effective with increasing slope, in the tropics chemical decomposition (the major erosional agent) is most active where the slope is gentlest and where soil and ground water can collect and stagnate, enabling prolonged attack on the underlying surface. This was expressed by Bremer (1971, 1972, 1973b, 1975), following up observations by Jessen (1936), as *divergent weathering and erosion (divergierende Verwitterung und Abtragung)*.

We will discuss this phenomenon, along with the slightly different conditions found in the inner, perhumid tropics, in greater detail in Sections 2.3 and 2.4

1.8.3 LINEAR EROSION AND VALLEY-CUTTING. MATURE FORMS OF RELIEF. THE ROLE OF CLIMATIC CHANGE

In all other regions, i.e., in all of the ectropics and in the mountains of the deserts and of the humid tropics, linear fluvial erosion, because of the active scouring of its vertical component, cuts down faster than does denudation on the interfluves. This produces valley-cutting. Fluvial erosion thus determines the angle of the slopes, which in turn determines the particular variant of slope erosion, according to the climatic zone concerned.

In the resulting valley the river not only cuts actively downwards, but also passively transports both the material which it has torn loose from the river bed, and the far greater quantity of material brought laterally through slope erosion—the latter being, as described above, a process controlled by the river's own down-cutting (see Fig. 11).

The concept of a valley is defined in this context as follows: "a valley is a large, elongated, open-ended concavity, created by the active linear erosion of a river and by the slope erosion controlled thereby, both processes working faster than the areal erosion of the interfluves. This may be happening today, or may be the still visible result of an older relief-generation" (Büdel, 1970a and b).

Valleys defined in this manner differ greatly among today's various morphoclimatic zones, even when other conditions are equal. Such differences involve valley cross-profile (steepness of the slopes, presence, width, and debris-content of a valley floor), longitudinal profile (climbing steeply or curving gently upward; smooth or broken by rapids or waterfalls), drainage pattern (greater or lesser joint control), and finally, rate of down-cutting, including capacity for headward erosion.

By far the most vigorous valley-cutting occurs in today's periglacial region. Here preparatory loosening by the ice rind so increases the active work of the rivers that this region may be called the "polar zone of excessive valley-cutting." Large portions of the temperate mid-latitudes still display valley features inherited from the periglacial regime of the Pleistocene cold periods; their further Holocene development, however, has been greatly retarded.

When a given set of climatic conditions remains stable for a long time, then the valleys created during that time will develop a certain *mature form*,

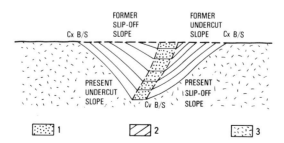

Fig. 11. Creation of a simple valley by fluvial erosion. (1) Matter removed by fluvial linear erosion. (2) Matter synchronously removed by slope lowering. (3) Bedrock. Cx B/S: convex break in slope (slope rim). Cv B/S: concave break in slope (piedmont angle).

fully adapted to the given climate. Once this mature form is achieved, valley-cutting slows considerably. We may compare this state to the mature profile of an ectropic soil. Should the climate change, drastic changes will quickly occur in the linear erosion, areal denudation, and above all, in the relationship between the two, until they have adapted to the new climatic conditions and a new balance has been achieved. This concept was developed by Troll (1924) and Bremer (1959) through quite different field observations. Thus after the end of the Eem interglacial, the onset of the early Würm Glacial was accompanied by especially marked changes. This is similarly true for the rapid change to the postglacial climatic optimum after the end of the Würm Glacial.

1.8.4 Transportation

All relief development is involved with creating concavities. These include everything from the shallowest of swales and the steepest of ravines, through broad, open valleys, to the extreme case of wide etchplains which are constantly worn down parallel to themselves. To create these concavities, material must be removed, which means that bedrock must be prepared, i.e., converted into a condition in which it can be eroded. This is done by the various weathering processes, which loosen and break down the rock.

In the polar zone, the frost debris sheet is moved in its entirety by solifluction, though this occurs less rapidly toward the base of the blanket. With the help of the ice rind effect, this passive transport simultaneously attacks the ground beneath, and is thus itself an important agent in breaking down the bedrock. Thus both the passive removal of material and much of the active attack on the substrate occur in one single process.

The other extreme is presented by the mechanism of double planation surfaces in the seasonal tropics. Here only the processes of preparing and rendering the bedrock movable take place at the weathering front. No transportation occurs there, other than through ground water solution. The chemical preparation produces a very fine-grained material (see Fig. 12), which lies fairly uniformly in a blanket many meters thick of stable relief-forming soil. On the surface of this blanket, material is removed by sheet wash, which of course is particularly effective here, since the fine material can be taken up and washed away by the smallest of the rainy season rivulets. Thus active preparation and passive removal of the material take place here on two widely separated levels.

It is particularly striking that roughly the same amount of material is removed at the surface as is created at the base of the relief-forming soil. We may therefore define relief-forming soil as a transitional stage between preparation and removal. The unweathered bedrock remains at a fairly uniform distance below the surface, and its exposure to external influence (heat and

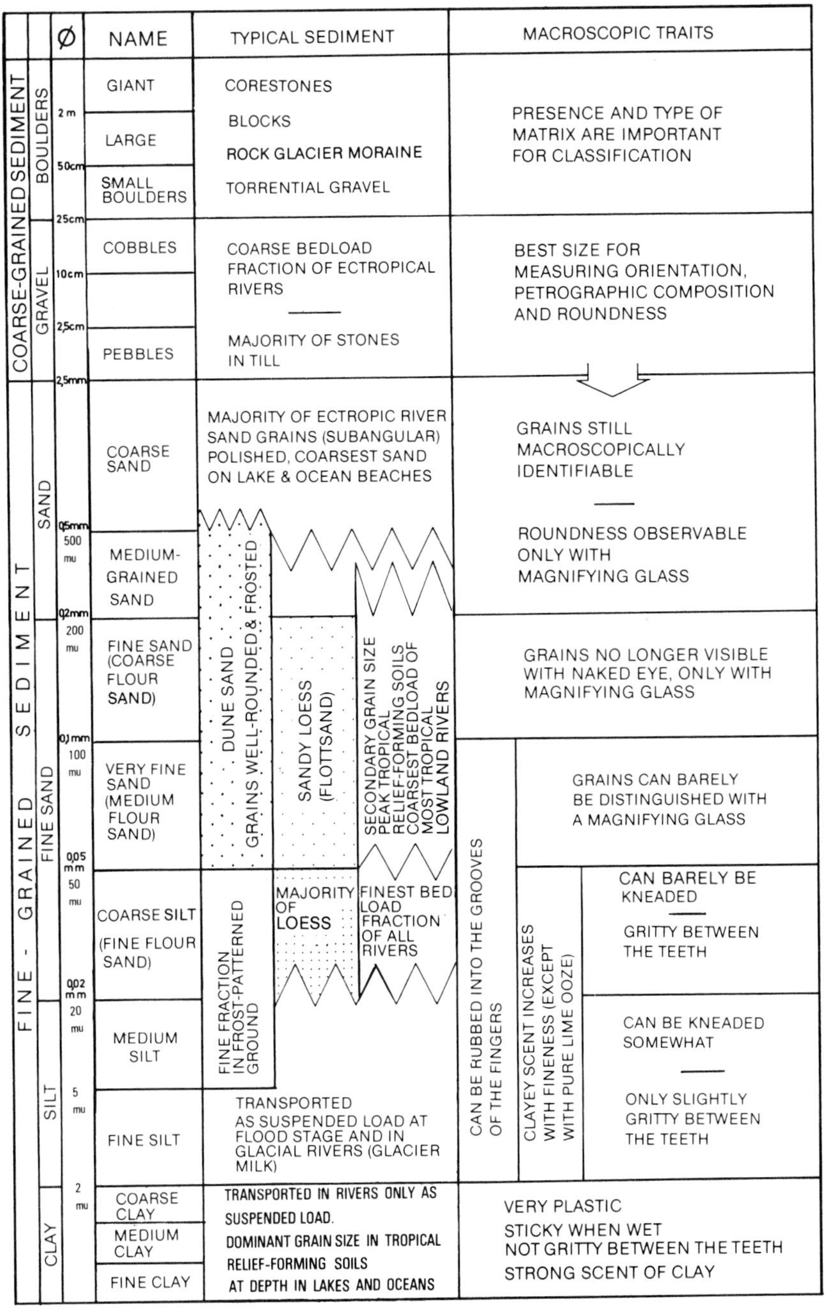

	Ø	NAME	TYPICAL SEDIMENT	MACROSCOPIC TRAITS
COARSE-GRAINED SEDIMENT — BOULDERS	2 m	GIANT	CORESTONES	PRESENCE AND TYPE OF MATRIX ARE IMPORTANT FOR CLASSIFICATION
	50cm	LARGE	BLOCKS / ROCK GLACIER MORAINE	
	25cm	SMALL BOULDERS	TORRENTIAL GRAVEL	
COARSE-GRAINED SEDIMENT — GRAVEL	10cm	COBBLES	COARSE BEDLOAD FRACTION OF ECTROPICAL RIVERS	BEST SIZE FOR MEASURING ORIENTATION, PETROGRAPHIC COMPOSITION AND ROUNDNESS
	2,5cm	PEBBLES	MAJORITY OF STONES IN TILL	
FINE-GRAINED SEDIMENT — SAND	2,5mm	COARSE SAND	MAJORITY OF ECTROPIC RIVER SAND GRAINS (SUBANGULAR) POLISHED, COARSEST SAND ON LAKE & OCEAN BEACHES	GRAINS STILL MACROSCOPICALLY IDENTIFIABLE
	0,5mm / 500 mu	MEDIUM-GRAINED SAND		ROUNDNESS OBSERVABLE ONLY WITH MAGNIFYING GLASS
FINE-GRAINED SEDIMENT — FINE SAND	0,2mm / 200 mu	FINE SAND (COARSE FLOUR SAND)		GRAINS NO LONGER VISIBLE WITH NAKED EYE, ONLY WITH MAGNIFYING GLASS
	0,1 mm / 100 mu	VERY FINE SAND (MEDIUM FLOUR SAND)		GRAINS CAN BARELY BE DISTINGUISHED WITH A MAGNIFYING GLASS
FINE-GRAINED SEDIMENT — SILT	0,05 mm / 50 mu	COARSE SILT (FINE FLOUR SAND)		CAN BARELY BE KNEADED / GRITTY BETWEEN THE TEETH
	0,02 mm / 20 mu	MEDIUM SILT		CAN BE KNEADED SOMEWHAT
	5 mu	FINE SILT	TRANSPORTED AS SUSPENDED LOAD AT FLOOD STAGE AND IN GLACIAL RIVERS (GLACIER MILK)	ONLY SLIGHTLY GRITTY BETWEEN THE TEETH
FINE-GRAINED SEDIMENT — CLAY	2 mu	COARSE CLAY	TRANSPORTED IN RIVERS ONLY AS SUSPENDED LOAD.	VERY PLASTIC STICKY WHEN WET NOT GRITTY BETWEEN THE TEETH STRONG SCENT OF CLAY
		MEDIUM CLAY	DOMINANT GRAIN SIZE IN TROPICAL RELIEF-FORMING SOILS	
		FINE CLAY	AT DEPTH IN LAKES AND OCEANS	

Overlaid labels across the SAND/SILT columns: GRAINS WELL-ROUNDED & FROSTED · DUNE SAND · SANDY LOESS (FLOTTSAND) · SECONDARY GRAIN SIZE PEAK TROPICAL RELIEF-FORMING SOILS COARSEST BEDLOAD OF MOST TROPICAL LOWLAND RIVERS · MAJORITY OF LOESS · FINEST BED LOAD FRACTION OF ALL RIVERS · FINE FRACTION IN FROST-PATTERNED GROUND · CAN BE RUBBED INTO THE GROOVES OF THE FINGERS · CLAYEY SCENT INCREASES WITH FINENESS (EXCEPT WITH PURE LIME OOZE)

Fig. 12. Table of grain sizes.

water supply) is kept more or less constant. This is similar to what W. Penck (1924) had in mind with his concept of "renewal of exposure" (*Erneuerung der Exposition*).

Fluvial work is similarly twofold. The flowing process itself, whether laminar to streaming or turbulent to shooting, is inherently a passive form of transportation. The additional energy expended in carrying *solutes* is nil, in carrying the *suspended load* (suspended particles up to 20 μm in diameter, see Fig. 12), is minor, and in carrying the coarse *bedload* is considerable. The river transports not only that which it wrests from its own bed: this would produce the stippled central band in the valley cross-section of Fig. 11. Rather the river must at the same time passively remove the far greater amount of material (fine-grained in many climatic zones) brought to it by slope erosion and by its tributaries.

In etchplain development, the rivers are fully integrated with surface wash, and behave almost entirely as passive conveyor belts. Any active down-cutting here occurs no faster than planation on the interfluves. The fact that the rivers are incapable of greater down-cutting is due less to the energy consumed in carrying the load of fine sand and suspended particles than to the fact that the far more energy-consuming bedload of coarser gravel is absent. Lacking this, the river is left without effective cutting tools. Moreover, the rivers flow for long stretches at gentle slopes in the soil blanket, and are not in contact with the bedrock. Yet where they cross a bed of unweathered rock protruding through the soil blanket, they fail to cut through this even over long periods of time, but cross it in the form of rapids or waterfalls. An example of this is discussed in 2.3.2.3 [6].

The very existence of these striking etchplains as a relief type is the main proof that these rivers cannot cut downwards faster than planation, and cannot carve valleys even over very long periods of time. Otherwise such etchplains could never have been created; they would have been dissected long ago. Setting aside all theoretical models, which can be based only on a select few of the actual influences and their climatically controlled variants, it is, in the framework of geomorphology, always the type of relief itself, created by all the various influences involved, which proves the total effectiveness of these influences. In ectropic rivers, the energy consumed on the bedload certainly influences their capacity for active erosion. The rivers in the lowlands of the seasonal tropics are completely different in this and in all other characteristics.

Even in the ectropics, the occurrence of down-cutting is not ruled solely by the amount of energy consumed on the load, for the resistance of the substrate is equally important. Outside of the permafrost region, this is controlled by the geomorphic hardness of the rock, which again is dependent upon climate. All rock types are least resistant in the permafrost zone of excessive valley-cutting, where the ice rind prepares the way for down-cutting.

1.8.5. TRANSITIONAL DEPOSITION AND MINOR CATASTROPHES

The dissolved load of a river is invariably carried to the sea or to a terminal lake, as is most of the suspended load (see the grain-size table, Fig. 12). Part of the suspended matter is dropped upon contact with the salt water, but a large portion (including most of the clay components) is carried on into the ocean by currents. The light-colored "plumes" of such water, rich in suspended elements, are visible from the air, and can be traced far into the open sea. Coral reefs in the tropics are interrupted by these plumes, as coral can only tolerate very clear sea-water. The finest clay particles sink only in the ocean deeps.

The bedload carried by a river varies greatly with the different morpho-climatic zones. In humid tropical lowland rivers it consists mostly of fine sand around 20-50 μm to 200 μm in diamater. In the dry season, the unsorted fine sand beds of smaller streams lie quite dry, and those of medium-sized rivers likewise, though here one last rivulet may still be active in the middle or along the margin of the bed (Photo. 17). Even large rivers dwindle to a fraction of their capacity.

Even outside the humid tropics, rivers with broad, active gravel beds are restricted to very specific areas. They are found above all in the polar zone of excessive valley-cutting; also in the ectropic mountain ranges and their forelands; in the torrential winter rivers of the etesian subtropics; and finally in the intermittent streams of the arid regions (called "wadis," "oueds," or "enneris" in the Sahara).

The bedload of these rivers includes all grain sizes from fine flour sand (coarse silt, see Fig. 12) on up to the coarsest gravel particles 2.5 mm to 25 cm in diameter. Blocks over 25 cm in diameter occur, if at all, in the steepest headwater reaches of vigorous mountain streams. The majority of exposed blocks this size are non-fluvial in origin. They are either core-stones (*Grund-blöcke*: boulders spared by subsurface chemical weathering in a present or past moist warm decomposition sphere) or blocks that have been stripped of their weathering rind (mostly in the Pleistocene after deep weathering in Tertiary or pre-Tertiary times), such as those in the block-fields (*Blockmeere*) of the German Mittelgebirge. Rock which is very susceptible to block dis-integration may form block-strewn peaks (*Blockgipfel*) and block talus slopes (*Blockhalden*) of similar appearance both in the desert and in the subpolar environment. Unsorted moraine deposits may also be rich in large boulders, and ice-rafting on rivers may occasionally transport isolated blocks.

The larger discoidal pebbles of a gravel bed, especially those between 5 and 25 cm in diameter, lie like roofing-tiles in the gravel bed, with their flat sides facing the current and their long axes at right angles to it.

During stages of medium and low water in gravel-rich permanent streams, *continuous* through-transport of bedload is present only at the bottom of the main channel. When diving in such a river, for instance in the Danube near Vienna, one can clearly hear the constant, typical clattering of this conveyor belt of gravel on the floor of the streamline. But it is only during floods (which in desert streams, of course, occur very rarely, sometimes only at intervals of many years) that the bedload is worked through over most or all of the riverbed. Thus most bedload transport occurs *discontinuously* (Bremer, 1959).

Only during the "minor catastrophes" of such floods can the gravel floor of a river be activated right through to the underlying bedrock, and only then can the channel bed actively be attacked by bombardment and grinding, and can pools be scoured out by eddying. But where no ice rind is present to loosen the bedrock beneath the entire riverbed, down-cutting is restricted not only to occasional time periods, but also to individual pools that migrate downstream. Thus even in gravel-rich streams, vigorous down-cutting occurs only on unconsolidated sediments, where the ground beneath is already amenable to transport. On solid bedrock where no ice rind is present, vertical erosion occurs many times more slowly and irregularly.

The surface of such gravel beds is not flat, but displays a very consistant irregularity, produced by long gently arched bars of sand or gravel. At high water, bedload material is removed from the narrow upstream sides and deposited to the lee, so that the bars migrate downstream. In Spitzbergen along a little stream having a gravel bed only 70-80 m wide we found that gravel bars were shifted six m downstream in 24 hours during only moderately high water. At low water, the anastomosing branches of the *braided stream* (*Pendelfluss*) flowed in channels between the gravel bars, which then lay inactive. Only with the next flood were the bars once more attacked and moved downstream. At the same time, the main current tended to migrate from one side of its gravel bed to the other, in cycles of larger amplitude than those of the shifting bars.

These *transitional deposits* (*Durchgangsaufschüttungen*) accompany the entire course of such gravel-rich streams. Even in their upper reaches, periods of bedload inactivity alternate with *minor catastrophes* when the bedload is mobilized, permitting active erosion of the bedrock beneath. No river can be neatly divided into an upper section of active erosion, a middle section of passive transportation, and a lower section of deposition. Rather the case is simply that minor catastrophes with bedrock attack occur more frequently in the steeper slopes of the upper reaches. With decreasing gradient and increasing gravel load (concurrently becoming finer grained), erosion of the stream bed occurs more rarely and in ever more widely separated pools, while the quiet periods when the transitional deposits are in repose become longer and more frequent, until eventually the sediments are permanently deposited in the delta area. Yet even in the upper and middle reaches of a river, low-gradient wide valleys with lower reach characteristics may occur in areas of soft bedrock, while in the lower reaches one may find steeper narrows (water gaps) where the bedrock is subject to more intense erosion. The deepest pools of the Danube are found in its lower course at the Banat Gap, and reach below the level of the Black Sea.

1.8.6 PERMANENT DEPOSITION

Small permanent deposits occur in continental areas, especially in lowlands where relief-building has cut off transitional deposits from the transportation process. Thus the moraines and gravel fields of the Würm have remained largely unmodified, as have the Pleistocene periglacial gravel terraces, loess blankets, and cold phase solifluction blankets.

Such permanent deposition occurs primarily in terminal lakes and areas of continental subsidence such as tectonic grabens (e.g., the Upper Rhine Graben) or deep basins (e.g., the Vienna Basin or the Upper and Lower Hungarian Basins). The major depository of permanent sedimentation, however, is just offshore, particularly in the deep-sea trenches and in offshore areas of marine subsidence. Examples of filled depressions are the Hungarian Basins and the Po Plain of N Italy, where sediments have accumulated many km deep. Such sediment bodies may later be raised to great heights through eustatic changes in world sea level or through epeirogenic uplift. Young Miocene sediments have been raised to heights of over 2000 m in the Cilician Taurus of Turkey, while Mesozoic sediments reach heights of over 3000 m in the highest peaks of the limestone Alps. The majority of these sediments, altered diagenetically if at all, form the geologic superstructure, an important part of the lithosphere. This sediment sphere could be regarded as the most comprehensive companion of the relief sphere, for sedimentary rock is the grand and final product of all the numerous and intricate transformations occurring in the relief sphere as a consequence of the primal light-to-heat conversion. The chain of transformational processes, which we have divided here into six stable and six mobile elements, constitutes the entire relief-forming mechanism which, since the dawn of the geomorphic era, has led through such multitudinous changes to produce the continental relief which exists around us today.

1.9 The Climatically Controlled Variants of Today's Relief-Forming Mechanisms, Making Up the Modern Morphoclimatic Zones

In considering the stable and mobile elements of the relief-forming mechanism (Sections 1.6-1.8), it has become clear that each element changes according to climatic zone. This is even more true of the relief-forming mechanism as a whole, for its various elements overlap and affect each other.

Not only do the individual elements vary with climate, but also the different mechanisms accentuate different elements. Thus each type of relief-forming mechanism leads to a very different relief product.

This is true even from a purely quantitative viewpoint. Endogenous conditions (petrovariance and epeirovariance) being equal, the controlling mechanism of one climatic area will remove more material per unit time than that of another climatic region. Such differences have often been calculated from measurements of the total load-carrying capacity of similarly sized river systems of different climatic regions. It has, however, been customary in such studies to extrapolate from measurements taken at normal water levels, a practice which excludes the exceptional cases, the minor catastrophes, which may sometimes be morphologically the most effective. But even were it possible to circumvent this problem, one would still have only cumulative values which even out all highs and lows and thus say little about the details of relief-development, the actual subject of geomorphology. Such studies can only yield the very general conclusion that erosion is more effective in some climatic zones than in others.

Far more important are the qualitative differences in the appearance of the relief, the quite different landform styles of different climatic regions. These are revealed by the exact study of all elements of the relief-forming mechanism and of their effective capacity individually and as a whole. Such a study is executed by exact geomorphologic analysis of convenient natural test sites, and by comparison between the different morphoclimatic zones. It is particularly important that one distinguish rigorously between features produced by today's climatic and developmental conditions and features produced by former processes, relict features which still participate in the total relief about us today.

If one considers only those relief-forming mechanisms at work today and only those relief products being created under Holocene conditions, then it will be seen that these remain uniform over large areas, but change profoundly along wider or narrower transition zones. Areas governed by similar relief-forming mechanisms are called *morphoclimatic zones*, depicted in Fig. 13 (see endpaper). In places these zones stray outside the boundaries of the climatic zones as set up by current meteorological data: clearly the relief-forming mechanisms depend on climatic effects which cannot be measured by the usual atmospheric data. This is not to be wondered at, considering all the above-described energy conversions which, via the various elements, lead to the relief-building mechanisms as a whole.

2

Climatic Geomorphology

2.1 The Significance of the Morphoclimatic Zones

Figure 13 (endpaper) shows the morphoclimatic boundaries defining areas within which the active relief-forming processes dependent on today's climate behave uniformly.

Since Thorbecke (1927), morphoclimatic classifications have assumed that the regions where these processes dominate (regions whose variants had at that time scarcely been studied), and even the relief assemblages of the total relief (which must include many relict forms), could simply be equated with the standard continental climatic zones themselves.

The presentation here is instead based entirely on the analysis of the active relief-forming mechanisms themselves. These are related to their climatic primogenitor through a chain of processes involving much alteration and selection. The final product is often distributed differently from the various interrelated processes that created it. Since Büdel (1948a), systematic investigations have increasingly convinced us of this, and have led to the present model presented in Fig. 13. Wilhelmy's presentation (1974, 1975), is also in line with this model.

In Fig. 13 we consider only those processes pertaining to the lowlands and to medium elevations, covering the major land areas of the continents and of the human habitat. The high mountains are all fairly similar above the snow line in all climatic zones, and have been expressly omitted.

In this zonation we are concerned primarily with representing core areas: zonal boundaries are not always clear, for reasons which will be readily understood.

2.1.1 Boundaries of Geographic Elements

The first of these reasons is implicit in the nature of all boundaries of geographic elements, even of landscape units. Thus Gradmann (1931, p. 41) rightly says that "the Black Forest has so characteristic an appearance, that it has been regarded as a unit since the earliest of times." The entire Black Forest differs markedly both culturally and physico-geographically from its surroundings, yet were a hundred geographers given the task of drawing it on a map of 1:100,000 or 1:200,000, not one of the resulting figures would correspond exactly with another. Laid on top of one another their fringes would form an approximate line on the W side, but be up to several km wide on the E side. Except for coasts, nature knows no sharp boundaries in the relief. Nevertheless, areas such as the Black Forest, the Po plain, the Appalachians, or the Tarim Basin, can and must be clearly distinguished for the sake of scientific classification.

2.1.2 THE LITTLE HILL NEAR VORONEZH

The second reason is that it is harder to draw geographic boundaries than boundaries between climatic zones or between natural plant communities. This seems obvious with regard to meteorological data, yet different authors have subjectively constructed varying pictures of the terrestrial climatic regions. Vegetational zones are also easier to distinguish: tundra, taiga, mixed oak forest, parkland, open steppe, herbaceous steppe, desert steppe, desert savanna, tropical scrub savanna, etc., are characterized by density and occurrence of plant species hereditarily bound to certain ecotopes. Where distributions overlap, simple statistics can determine exactly where, for instance, the taiga ends and mixed oak forest begins. Here again, core areas of plant communities are more important than precise delineations, but even the most stubborn of pedants could find satisfaction on the latter point.

With climatic soil types it is far more difficult to draw boundaries, for podzols, lessivés, brown forest soils, gray-brown podzolic soils, pseudo-gleys, and chernozems are not distinct pedogenetic types, but abstract types created (for good reasons) by science. Russian soil scientists such as Glinka (1914) made a major contribution to science when, at the beginning of this century, they founded the study of climatic soils, and this contribution is clearly reflected in the various languages of the scientific community (as, German, English, French, and Spanish). If Kubiena (1953) was able to distinguish fifty climatic soils subtypes in Central Europe alone, one can imagine what hybrids must occur there. The Russian mapping of the major soil types in Russia is therefore all the more valuable. Anyone who has seen such a comprehensive map knows how these types interdigitate. A. Penck was present at the first soil science congress in Russia where these epoch-making results were unveiled, and often related how, after they had been presented, a speaker rose and said: "I know of a small hill near Voronezh, on whose NE side are podzols and lessivés, on whose W and E side are brown forest soils, and on whose SW side are chernozems, all within one square kilometer.

If all these can coexist within such a small space, then the entire zonal division of the Russian soils you have presented must be false.'' (Voronezh lies on the Upper Don, 480 km S of Moscow, in the transitional area between wooded and open steppe.) At once another speaker stood up, saying in effect: ''If our scientists had not expended enormous energies on this general mapping of the climatic soils, then the differences on that hill would have remained inexplicable, had you noticed them at all. As it is, your observation does not disprove soil zonation, but rather substantiates its dependence on climate, for you found the major soils zones reproduced there in microcosm, according to microclimate.'' After this, ''Voronezh'' was long a stock term among German geomorphologists for the narrow-minded view that rejects such basic organizing of the geofactors, and thereby ignores the foundations for all scientific observation of the earth.

It is not uncommon for specialists (in contrast to generalists) to lean toward this mode of thought.

2.1.3 Delineating Process Areas

Soils, however widely they may vary, still have clearly distinguishable profiles, while statistics on type frequencies may help provide boundaries with a reassuringly firm basis. But to establish morphoclimatic zones we must segregate highly complex chains of processes which cannot be assessed by direct comparison. The predominance of a particular relief-forming mechanism must be inferred from the presence of its active products, i.e., characteristic relief assemblages. Though these are especially marked and recognizable in their core areas, an indirect mode of thought is required to recreate the chain of events responsible, and until these chains of events have been studied, it is impossible to distinguish morphoclimatic zones. For without this dynamic basis, the omnipresent misleading exceptions were weighted equally with those key relief elements which were unmistakably produced solely by the local set of relief-forming mechanisms. For this reason static-statistical methods based on maps and aerial photographs cannot always be used, especially as petrovariance may further complicate the picture.

2.1.4 The Difficulty of Genetic Delineation

A further problem is that it is very difficult to determine exactly which relief elements in a particular area were actually produced by current relief-forming mechanisms and which were inherited from long-vanished mechanisms. This is the problem of correctly associating generative processes and product features, for once imprinted in the solid crust, features often survive long beyond the era in which they were created. Hence the distinction between dynamic and active geomorphology.

2.1.5 INEQUALITIES BETWEEN MORPHOCLIMATIC ZONES

Certain inequalities therefore arise among the morphoclimatic zones as shown in Fig. 13. Some zones are being actively formed today (and of course, it was on the basis of the currently active relief-forming mechanisms that the boundaries were drawn), and their major visible features are the product of current mechanisms. These are primarily four zones: the glacial zone, the polar zone of excessive valley-cutting, the zone of excessive planation (with its two major variants, the peritropical zone and the smaller inner tropical zone of partial planation), and finally the coastal forms, which are largely independent of climate.

The relief-forming mechanisms of the glacial zone are more easily studied in the areas of former glaciation than under today's glaciers or inland ice sheets, as will be discussed in the next edition. Those morphoclimatic zones whose total relief is a mosaic of previous relief generations will be discussed in detail in Chapter 3. These include the zone of retarded valley-cutting in the region of the ectropic relief-covering soils and the zone of mixed relief development in the etesian subtropics. The very name of the latter indicates its mixture of recent and past influences.

Chapter 2 will largely discuss those morphoclimatic zones whose relief is solely or predominantly being created by current relief-forming mechanisms. These are the zone of excessive valley-cutting and the zone of excessive planation, along with the latter's variant, the inner tropical zone of partial planation. To this we will append a discussion of the tropical and ectropical arid regions, whose basic features are inherited from extinct (largely humid-tropical) mechanisms. Such relict features are particularly well-preserved in arid regions, for here the more recent processes (largely of the Late Pliocene, Pleistocene, and Holocene) modify the basic relief without effecting any basic change in the total relief style.

Relief-forming processes are at work today in other zones where they are not important to the total relief style. But it saves much repetition if their modest contribution in such areas is considered in connection with an analysis of the older relief generations which occupy the lion's share of the total relief. It is therefore of practical advantage to discuss the zone of retarded valley-cutting and of mixed relief development in terms of climato-genetic geo-morphology in Chapter 3.

The most important reason for this sequence of discussion, however, is that it is only when one knows which relief features are being created in the most active zones of today that one can then recognize their relicts in the other zones, and thus know to which current features the survivors of the older relief generations correspond.

2.2 The Polar Zone of Excessive Valley-Cutting (the Active Periglacial Region)

2.2.1 THE FROST DEBRIS ZONE OF SE SPITZBERGEN (RESULTS OF THE STAUFERLAND EXPEDITION). PROBLEMS AND SELECTION OF STUDY AREA

In the unglaciated areas of Central Europe and of the ectropics, Passarge (1914), Salomon (1916), and workers of the last three decades have isolated the landforms produced by the climatically controlled periglacial processes of the Pleistocene cold phases. We have, moreover, learned that these forms constitute so much of the relief around us that simply stripping off the plant cover would reveal that the major ectropic cultural centers of the world inhabit a landscape of 95% of which is a relict ice age relief.

The processes which created this relief are now long gone and can no longer be observed and measured. Only their product remains, so that we know *what* was produced, but not *how*. This much alone is sure: all elements of the periglacial relief-forming mechanism differed from those of Central Europe today, both individually and in total overall effect, for together, they worked far more ruthlessly and effectively per unit time than any ectropic mechanism today.

Some things could be more or less deduced, but even investigations of subpolar patterned ground or of ice wedge development did little more than construct theoretical models. The simplest thing to explain was fluvial lateral erosion, which was clearly very efficient then and created the wide valley floors that characterize even the smallest non-glacial mid-latitude river[9] (see Photo. 10). Asymmetric upper valley reaches could also be explained (Büdel, 1944, Photo. 53). But beyond that, uncertainty set in. What above all defied explanation was the extraordinary down-cutting capacity of non-glacial cold stage rivers. Even the smallest rivers were able to cut down their valley floors over a breadth of many hundred m, the most extreme case being the Kinzig River (below Gelnhausen, E of Frankfurt), whose valley floor is over two km wide. Such valley floors consist of a thin one to four m layer of loamy alluvium (*Auelehm*) overlying far thicker deposits of coarse sands and gravels which often contain Ice Age faunal remains (mammoths, reindeer, and paleolithic artifacts). These "modern" valley floors are in fact Würmian *Nie-*

[9] The term "non-glacial" is being used here to describe rivers which are not fed by glaciers.

derterrassen or "Lower Terraces." Their wide, even surfaces, often gently convex in cross-section, also indicate this origin. All except the largest of today's rivers continue to follow the talwegs of the old surfaces, meandering freely between the low gravel fans and covering only a portion of the valley floor's width.

Longitudinally the valley floors climb smoothly without waterfalls or sharp breaks into the heart of the Central European Mittelgebirge. Usually they are accompanied by terraces of older cold stages, particularly of the Riss Glacial, which may further broaden the valley considerably, the Riss terraces being 15-30 m high and up to 100 m wide.

Over the ages these valleys floors and their terraces have proven themselves particularly suited for settlement and for communication routes. Providing easy access into the mountains and highlands of Central Europe, they have contributed greatly to the spread of the mid-latitude old-world civilizations. Widespread loess blankets and a soil skeleton loosened by Pleistocene perma-frost have further provided Europe with fertile farmland from the Neolithic to the present.

The Riss terrace stands about 15-30 m above the Würmian Lower Terrace. As will be explained later, the corresponding lowering of the valley floor took place during the roughly 30,000 years of the Early Würm Glacial (ca. 60,000 to 30,000 B.P.), indicating down-cutting at the rate of 0.5-1 m per thousand years. As in all drastic climatic changes, down-cutting must have been most rapid at the very beginning of the new cold stage, perhaps proceeding at a rate of two or even three m per thousand years. Similar averages are produced when one examines the Pleistocene as a whole. The valleys dissecting the relict planation surfaces of the European Mittelgebirge today were formed almost exclusively during the Pleistocene. In the areas of little uplift, as in

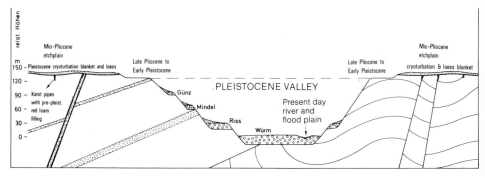

Fig. 14. Typical valley profile in the Pleistocene periglacial region of Central Europe. Plain and terrace sequence are roughly that of the Main valley below Würzburg. Underlying rock approxi-mately that of the lower Saar valley between the Lorraine limestone plateau (left) and the Hunsrück (right).

the Swabian and Franconian Gäuland, they were cut ca. 150 m deep. In areas of greater uplift, some were cut over 300 m deep, as in the lower Moselle, the Black Forest valleys, and the higher valleys of the Erzgebirge, while the Styrian Randgebirge was dissected to even greater depths. Assuming that the Pleistocene had four to five early glacial stages of highly effective erosional mechanisms, and that these stages lasted around 30,000 years each, the down-cutting must have taken place during the 150,000 years of all the early cold stages put together. (These early periods of down-cutting were separated both by interglacial interruptions, when erosional conditions resembled those of the Holocene, and by the brief glacial maxima, when the valley floors, later converted into terraces, were widened by predominating lateral erosion.) This again suggests down-cutting at the rate of one to three m per thousand years, occurring in the earlier cold stages over a much wider floor (see Fig. 14). No theory of down-cutting had so far come close to explaining this phenomenon. A major question of the Stauferland Expedition, therefore, was: how could the cold stage rivers cut down so far over so wide an area, far into the mountains, without breaks in profile?

The answer could be found only in an area having a climate similar to that of the Early Würm: a wet, cold climate with permafrost, but without glacial covering. Furthermore, the Holocene features had to be clearly distinguishable from the older ones, in order to perform a qualitatively and quantitatively exact relief analysis of the Holocene features. Finally, the endogenous setting (epeirovariance and petrovariance) of the comparative polar area had to correspond as nearly as possible to a well-studied area of the mid-latitudes, such as Franconia or Hesse.

The expedition was particularly interested in the *kind* and *efficiency* of periglacial valley-cutting. Today's polar periglacial regions were mostly covered by continental ice masses during the Würm. While the mid-latitude periglacial valleys were produced over a period of at least 150,000 years, then, polar periglacial valley-cutting has been limited almost exclusively to the brief 10,000 years of the Holocene. We therefore had to choose an area where the periglacial rivers had already managed to produce a relief clearly showing this unique type of valley-cutting.

Two additional conditions had to be fulfilled. Firstly, the area could not consist of rapidly uplifted crystalline uplands, for in such an area, thick, swift valley glaciers would have created a glacial relief which would have remained firmly imprinted on the land long after the melting of the ice, as in the mountains of W Spitzbergen, in the marginal mountains of Greenland, in Ellesmere Land, in Axel-Heiberg Land, and in the E of Baffin Land. Our task was to find an area where thin Ice Age sheets had moved slowly with little erosion over broad relict surfaces. With little glacial relief, the Holocene valley-cutting processes would be more visible, having developed more freely. Secondly, we had to make sure that the rocks of the investigation area were not metamorphic, so as to ensure full-strength penetration of mechanical frost weathering. The area also had to be one which had undergone some Holocene uplift, for no valley-cutting will occur without uplift, regardless of climate. The W Canadian archipelago was subject to very little glacial erosion and had a suitable rock substrate, but here Holocene uplift was lacking (Pissart, 1979). Where the necessary conditions were coupled with Recent uplift, however, typical periglacial valley-cutting processes could be expected to have produced their characteristic relief type even during the brief post-Pleistocene period.

The "polar zone of extreme valley-cutting" shown in Fig. 13 does not display the relict glacial relief which still exists here today. As with the other morphoclimatic zones, we have defined this zone according to active relief-forming processes and mature relief produced. These can only be recognized in small areas where the favorable conditions described above are present.

The large islands to either side of Freeman Sound in SE Spitzbergen, namely Barents Island and Edge Island, were chosen as particularly suitable. Together they have an area of 6640 km², or roughly that of the Black Forest. They rise from the Barents Sea Shelf, a flat surface broken only by the very shallow indentations of a former subaerial drainage system (Ahlmann, 1931). Most of the shelf was exposed during the cold stages and in pre-glacial times, and it acquired its subaerial features at that time. It was then covered by the Late Glacial marine transgression, which left only a few very narrow, flat fringes of coastal lowland exposed (Fig. 15). From these coastal fringes the steep cliffs of the two island blocks rise to relict surfaces at 250-500 m altitude (600 m altitude where covered by ice). SW to NE the snowline sinks from ca. 500 m to ca. 300 m, approaching sea level on the small island of Kvit Oya (located far to the NE at 80°4'-80°14' lat. N.), and reaching sea level only as far N as Franz Josef Land. The low island block of SE Spitzbergen thus has relatively little ice cover considering its high latitude (between 77°15' and 80°38' lat. N.), and 55% of the area of Barents and Edge Islands is glacier free.

The geologic structure resembles that of the Keuperland of Swabia and Franconia. Subhorizontal Mesozoic limestones, marly limestones, shales, clayey sandstones, and arkoses (Wirthmann, 1964; Nagy, oral communication) are shot through with basaltic sills, much as in the Vogelsberg and Rhön Mountains. These sills tend to follow the bedding planes dipping at up to 1°, or occasionally follow steeper faults.

The height of the snow line on the S and W of the island is due to the Gulf Stream, which brushes the W coast of Spitzbergen (the Norwegian Svalbard Archipelago). Though the current has cooled here, it keeps the coast free of ice nearly all year round. A second N branch of the Gulf Stream reaches into the S Barents Sea, sweeps round the N end of Scandinavia, and keeps Murmansk open for many winters. It is met on the E side of the Svalbard Ar-

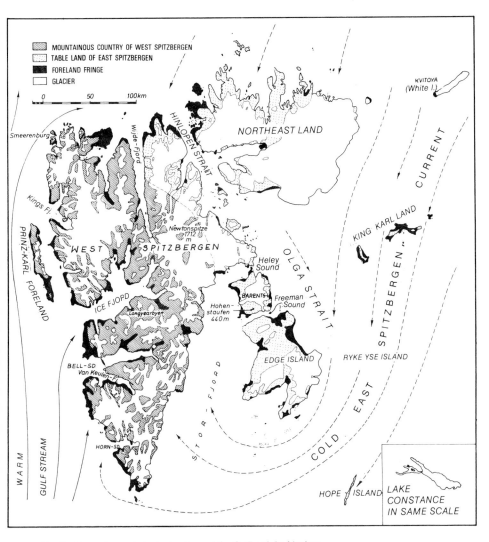

MOUNTAINOUS COUNTRY OF WEST SPITZBERGEN
TABLE LAND OF EAST SPITZBERGEN
FORELAND FRINGE
GLACIER

Fig. 15. General morphologic structure of the Svalbard Archipelago.

chipelago by the cold polar E Spitzbergen Current (Fig. 15), which even in summer may often fill Olga Strait as far as Hope Island with dense fields of pack ice. The pack ice then joins with the local ice of Stor Fjord, which likewise survives many summers, making the islands difficult or impossible to reach even with reinforced ships. The Stauferland Expedition (conducted in the summers of 1959, 1960, and 1967) had luck each time. In the first two years we were able to reach Kvit Oya without obstruction, but in 1967 Olga

Strait was completely filled with dense pack ice, which wind and tidal currents (the latter running at up to 2.5 m per second) brought in large masses through Freeman Sound westwards, on one occasion endangering our ship.

Politically, the islands were easy of access. The Svalbard Declaration, signed in 1920 by Germany and other countries, granted Norway the entire island except for two small areas on the W coast, and gave all signatory countries the right to conduct research there.[10]

An important prerequisite for our work was that we be able to distinguish the relict features of an older relief generation from the set of features being shaped by the relief-forming mechanisms of the periglacial Holocene climate. Here again the island presented a convenient ''natural test site,'' for the remnants of the pre-glacial Tertiary relief generation were readily identified. It was particularly convenient that the entire Svalbard Archipelago had been covered during the Würm by a continuous sheet of continental ice, for this meant that all older relief features had been remodeled uniformly. Since we managed to determine the extent as well as the phases and termination date of this glaciation, we were able to obtain a clear picture of the parent relief as it existed 10,000 to 12,000 years ago at the beginning of the Holocene. It was then easy to distinguish the features created in the Holocene from the parent relief, permitting an exact qualitative and quantitative relief analysis of the modern features.

2.2.2 THE PRE-GLACIAL LANDFORMS

It was shown that the Pleistocene glaciation (as in Scandinavia and the Alps) merely spread a veil of new relief forms over the framework of the pre-glacial relief. The pre-Pleistocene relief generations were studied by Wirth-

[10] We owe a great debt of thanks to many Norwegian persons and institutions who granted us their kind help. Preeminent among these is the ''Norsk-Polarinstitutt'' and its director Tore Gjelsvik; further, the various *Sysselmänner* (governors) of Svalbard who were in office during our expeditions and who granted us unceasing support; also the Jacobsen line in Tromso and the captains, officers, and crews of the ships ''Norsel'' (1959 and 1960) and ''Norwag'' (1967) chartered there. These experienced seamen can justifiably be regarded as the descendants of the ancient Vikings, as can our superb Swedish helicopter pilot. Furthermore, I thank my scientific colleagues. Twenty people (other than myself), representing four nations (Germany, Norway, Switzerland, and Czechoslovakia), exhibited unflaggingly cooperative and selfless teamwork. Many of these individuals have since achieved high academic rank. Their names, in alphabetical order, are: A. Fugel, G. Furrer, U. Glaser, W. Hofmann, J. Kvitkovič, G. Nagel, I. Nagy, G. Philippi, H.-G. Preuss, A. Semmel, H. Späth, G. Stäblein, D. Tannheuser, H.-G. Wagner, O. Weise, H. Wennrich, G. Widgand, F. Wilhelm, and A. Wirthmann. Of these only the last accompanied me on all three expeditions. Tannheuser and Kvitkovič have accompanied Glaser on subsequent expeditions. I stress once again that all of us are extremely grateful to the German Research Council (*Deutsche Forschungsgemeinschaft*) and especially to the Ministerialrat E. Gentz, who was at that time the Council's representative for Geography. Ninety percent of the necessary funds for the three expeditions are obtained from this source.

mann (1964), who, with the aid of his colleagues (U. Glaser in particular), successfully unraveled the story. His conclusions on Edge Island, reproduced below with a few additions of my own, far outstrip any previous research, and consist of the following points.

Firstly: the Barents Sea Shelf, whose subaerial creation was first plausibly suggested by Nansen (1922) and corroborated by Ahlmann (1931), was recognized by Wirthmann as a relict etchplain with shallow drainage lines resembling wash depressions. Subsequent attack (at lower eustatic or higher isostatic level) by marine abrasion or glacial areal erosion played little if any role in its development. The etchplain surrounds the Svalbard Archipelago on all sides: even in the W, where the continental slope lies closest to the island group, an etchplain shelf separates the two by 40-80 km. This old etchplain forms typical triangular reentrants (*Dreiecksbuchten*) reaching into the higher island body. The largest of these is Stor Fjord, which separates the massif of SE Svalbard from the main island of W Spitzbergen.

This surface grades gently up to the coastal lowlands of Edge and Barents Islands, emerging from the sea to meet the cliffs of the island massif. That the shelf surface and the foreland surface form a single unit has been confirmed by Wilhelm's soundings (unpublished) from the 1960 expedition. The forelands are not on headlands, as would be the case with shore platforms, but run between headlands right into the island body (see Fig. 16), just like triangular reentrants on active etchplains of the seasonal tropics.

Secondly: the massifs are crowned by a gently rolling relict planation surface at 250-450 m. A fairly even basal plain is surmounted by higher knolls and ridges separated by old dells. Seemingly unrelated to these dells are two younger valley networks, consisting of broad, U-shaped, clearly Pleistocene valleys running far into the island's interior, and short, steep, V-shaped gullies in the Holocene cliffs. The island heights thus represent another etchplain relief, which, judging by its shape and independence from geologic structure (apart from some accordances between harder beds and the etchplain margins), from the younger drainage network, and from the Holocene cliffs, surely "derives from before the Würm glaciation, though its origins must be far older than that" (Wirthmann, 1964, p. 16). These conclusions are corroborated by Barents Island's best preserved etchplain remnant, Hohenstaufen, 440 m high, on which was located one of my main test sites. This is a nearly flat etchplain over one km² large, cutting evenly across gently undulating thin-bedded arkoses with intercalated basalt dykes.

Thirdly: Wirthmann found that the gently rolling relict surfaces of Edge Island rise from E to W, the knoll summits rising from 250 to ca. 450 m. A characteristic relict surface occurs again W of Stor Fjord (on the E side of W Spitzbergen) at 640 m (Aghard Mountain), again on a horizontal Mesozoic substructure. From here a gipfelflur with broad plateau remnants cuts at 600-800 m W across a syncline filled with Cretaceous, Paleocene, and Eocene

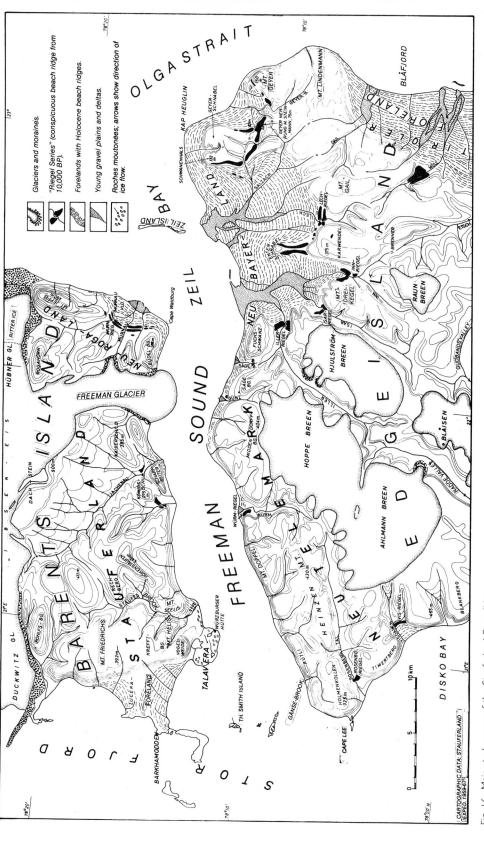

Fig. 16. Main study areas of the Stauferland Expedition.

sediments. (The W limb of the syncline, highly metamorphosed and later uplifted, overlays the crystalline rocks of the sharp crest of W Spitzbergen, for which the archipelago is named.) Fossils in the Lower Tertiary sediments cut by the relict surface include such cold-sensitive species of today's mixed oak forest as *Sequoia, Taxodia*, hasel, walnut, cyprus, and magnolia. These indicate a warm-temperate to monsoonal-subtropical climate, such as that found today around Shanghai (31° lat. N), or Savannah, Georgia. Such frosts as occur in these places do not affect the chemical weathering of the lowland relief-forming mechanisms.

The very long periods of time necessary to warp these beds and then plane them off to a level surface imply a climate in which chemical weathering permitted the development of broad etchplains, for such planation could not have occurred in any other way. In this conclusion we go further than Wirthmann.

The lower etchplain of the shelf and coastal lowland fringe must be younger where it intersects the island massif. In the W there is still "a major escarpment about 400 m high between the old highland and the flat bottom of Stor Fjord with its outrunners, Tjuv Fjord and the Disko Bay Foreland. Due to the original tilt of the upland, however, the foreland embayments on the E side are only about 200 m below the highland" (Wirthmann, 1964, p. 23). His further conclusion is very important, namely, that the Oligo-Miocene upper surface, whose shallow dells are independent of today's drainage, must have remained in a state of at least traditionally continued development (*traditionale Weiterbildung*: a concept introduced later in Bremer, 1971 and Büdel, 1971)[11] into the Late Tertiary, or "roughly to the Earliest or Early Pleistocene" (Wirthmann, 1964, p. 23).

The enormous Barents Sea surface (ca. 1000 × 1000 km) could not be wholly a product of the subsequent time period (approximately the Early to Mid-Pleistocene), especially as it indicates a climate with predominantly chemical weathering. The situation can be explained as follows: the E dipping relict planation surface of the islands of SE Spitzbergen (and the even lower surface on the small island group of King Karl Land to the E) must at the time of formation have joined the etchplain of the Barents Sea Shelf around 50 km SE of King Karl Land and about 100 km E of Edge Island (see Fig. 15). Only thus can we interpret Wirthmann's opinion (1964, p. 23), so advanced for its time, that "valley-cutting on the high relief occurred during the creation of the Barents Sea Shelf, when the escarpment was being formed between the first relict surface, which was undergoing inactivation, and the second relict surface, which, divided from the first, was being cut down." This was a revolutionary thought at the time, though stimulated by the dawning explanation of tropical etchplains.

---

[11] Traditional development may briefly be defined as relief development which is controlled and adapted to the pre-existing relief. For a detailed discussion of this, see Section, 2.3.9.

Bremer (1971, 1973b) showed that when part of an etchplain is raised (here the Upper Tertiary uplift of the W block of Spitzbergen above the old Barents Sea Shelf), the surface continues to form on the lower portion as before. The upper portion, however, upon which only traditional planation continues, is attacked by triangular reentrants like those along the edges of the SE Spitzbergen islands. Bremer's most important discovery (1971, 1973c) was that such triangular reentrants grow only until they have formed a steep *etchplain escarpment (Rumpfstufe)* which is stable and cannot retreat further. In the tropics such steep etchplain escarpments cannot wear back as in the ectropics, and help to interrupt erosion and preserve the periphery of the upper etchplain (see Fig. 49).

Clearly this discovery throws a very interesting light on the development of the Barents Sea Shelf (including Franz Josef Land) and on the creation of other shelf seas and their islands, especially the much-debated Norwegian "strandflat" or shore platform.

2.2.3 THE WÜRM ICE SHEET ON AND AROUND SVALBARD

This part of the Stauferland Expedition's program was investigated largely by Büdel (1960, 1961a, 1968a and b) and Glaser (1968). The ice was thickest and most fast-moving on the lower parts of the archipelago, and hence imprinted these areas most strongly. At the same time the amount of isostatic depression and uplift could only be investigated along the coast. To understand the effects of the Pleistocene glaciers, therefore, the most important areas of study were the sea cliffs and the foreland fringes. The work could be divided into two portions: first, study of the direct traces of continental ice and postglacial rebound, and second, study of the features inherited from these processes.

The suggestion had been made quite early (Blüthgen, 1942b), and then simultaneously by ourselves and others (Corbel, 1965, and Feyling-Hanssen, 1965b), that the entire N Barents Sea Shelf had been covered by extensive Würm glaciers. Our investigations were able to elaborate on this. We found evidence of very heavy glaciation, whose location moved from the Early to the High Glacial periods, and covered not only all of the Svalbard Archipelago, but also much of the adjoining Barents Sea Shelf E beyond King Karl Land and Kvit Oya, perhaps even to Victoria Island. It may have been at least temporarily connected with the large ice cap over Franz Josef Land, for today only 70 km of shallow sea divide Victoria Island from the W cape of the Franz Joseph Archipelago. On the other hand, the passage between Franz Joseph Land and Novaya Zemlya, which today is 390 km wide and over 200 km deep, almost certainly remained free of continental ice (Glaser, oral communication), even though the former ice cap of Novaya Zemlya extended far S into areas of heavy snowfall. It reached the Ural ice in the SE and crossed

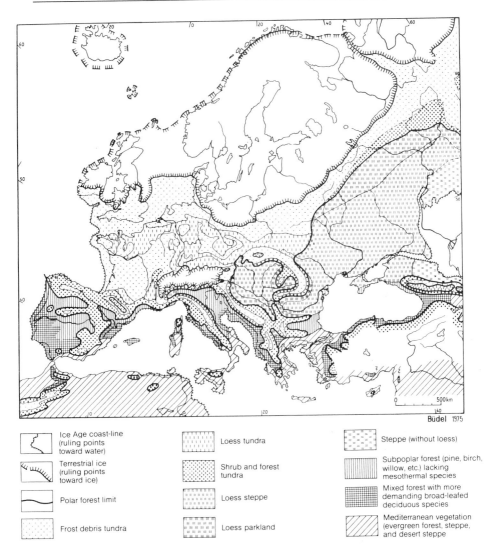

Fig. 17. Climatic zones and vegetational cover in Europe during the Würm cold stage. The Mediterranean vegetation in the S and SE included much more forest during the cool and humid early glacial. In the dry-cold Glacial Maximum steppe types expanded at the expense of the forest belt, which in the mountains was constricted both from above and from below.

Kolguev Island to the Kanin Peninsula on the SW, where it contacted the moraines of the Scandinavian ice sheet along a line ca. 120 km long (Rudovitz, 1947).

The old etchplain of the Barents Sea Shelf was thus not completely covered by Würm continental ice, but was surrounded by a ring of ice caps to the N, E, and S. Bear Island bore its own small cap, for though it shows complete glacial reworking, it lacks the foreign erratics which would otherwise have been introduced from, e.g., the Svalbard Archipelago. Taking Valentin's results (1957) into consideration, it seems that an unbroken ice barrier almost 4000 km long, nearly 2000 km wide, and up to 3000 m high extended off the entire NW corner of Europe from Svalbard (the N end of the Novaya Zemlya ice cap) down to S Wales and S Ireland (see Fig. 17). Low pressure cells today usually enter the continent through this area, bringing rain and ameliorating the winter climate. The significance of this for the Würm climate of Europe and for the differences between the Early Würm, when this ice barrier was still rudimentary, and the Würm Maximum, when the barrier was fully developed, will be discussed in Chapter 3.

The expedition dealt with Svalbard, the barrier's northernmost extreme. A first major discovery was that this possessed two centers, much like the N American ice sheets. The development of the Scandinavian ice sheet during the Würm will be discussed in Section 3.3.2 and Fig. 76.

Our investigation showed that the glaciation of the Svalbard Archipelago and Barents Sea Shelf resembled that of N America, though with characteristic differences (see Fig. 15). We found two centers of glaciation. One, which we called the W ice, formed around the mountains of W Spitzbergen. This had the character of an Alpine type *ice-field* (or *network of transection glaciers*), and resembled the model of the Ice Age Alps even more than does the current ice sheet on W Spitzbergen. The subglacial erosional forms on Spitzbergen have been destroyed more extensively than in the Alps by vigorous Holocene reworking: slope dissection (see Sections 2.2.13 and 2.2.15) has attacked the cirque headwalls, and the glacial U-shaped valleys can be recognized only by their straightness. The second glaciation center, the E ice, formed a completely separate sheet of purely continental ice without nunataks on the Barents Sea Shelf around the small island group of King Karl Land. This area must have been higher in elevation then than now, for post-glacial rebound here has not yet ceased, and, as everywhere else, the sea sank during the Glacial Maximum to 110 m below its present level. The highest peak of King Karl Land, around 350 m high, barely reaches snow line today. Most of the surrounding shelf must have been laid dry by the time that the pre-Würm snow line dropped below sea level, so that an independent center of glaciation formed on the flatland, much like the great Laurentian and Keewatin ice sheets in Canada. In Canada it was the lowland ice shield (not the one originating in the high mountains of the Cordillera) which produced the broad ice stream that filled the mid-continental depression of the Great Lakes and

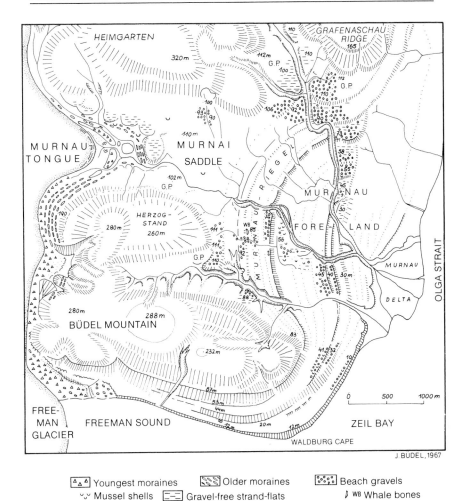

Youngest moraines   Older moraines   Beach gravels
Mussel shells   Gravel-free strand-flats   WB Whale bones

Fig. 18. Moraines and raised beaches in New Roga Land (SE corner of Barents Island, SE Spitzbergen; see Figs. 15 and 16). G.P.: Grenzplatte (highest post-glacial marine stand at around 11,000 B.P.). The S portion of New Roga Land was at that time an island, the Murnau Saddle an arm of the sea. Marine clays deposited there were lifted above sea level by the post-glacial rebound and subsequently plowed into moraines by the Murnau tongue of Freeman Glacier.

the Mississippi Basin down to the Ohio River. The Würm glaciation of the Svalbard Archipelago resembles this in miniature. The main center during the Glacial Maximum lay to the E on King Karl Land, as is shown by the westward orientation of glacial striations and roches moutonnées around Freeman Sound and on the Freeman Sound forelands (Büdel, 1960, 1968b, and Fig. 16), and by the distribution of foreign erratics. These cannot be discussed in greater detail here, but were the subject of much of the work by Büdel and Späth.

The final proof of a major glaciation center from which the E half of Svalbard was covered is that the highest marine terraces of unquestionably Holocene age on the entire archipelago are found on King Karl Land at 145 m, the highest outlying beach ridges even reaching 170 m (Büdel, 1960, p. 6). The next highest beach terrace, on the E side of Barents Island, is the so-called *Grenzplatte* (''Border Platform''), and is only 110-112 m high (see Fig. 18 of New Rogaland). On the NE corner of Edge Island the corresponding terraces are 85-95 m high. According to Glaser (1968), the highest beach terraces on the W side of Barents and Edge Islands do not exceed 85 m in height, while on the W side of Stor Fjord they reach only 60 m. The highest terraces then rise again to 150 m toward the center of W Spitzbergen, even reaching 156 m in the highland area of the W ice on parts of Prince Karl Foreland (Glaser, oral communication).

It has lately been questioned whether the present uplift in Scandinavia and Spitzbergen is isostatic or tectonic. If tectonic, then it is no indication of the thickness of the former ice covering. We do not adhere to this theory. In Scandinavia the area of uplift corresponds with admirable precision to the area covered by the Würm continental ice, while the isobases of uplift coincide so perfectly with the Würm ice thickness as deduced from other clues, that a direct relationship must be inferred. Given a 300-400 m indentation in 30 km of crustal thickness, and given corresponding post-glacial rebound, the stresses placed on the crust are such that latent tensions will be released, causing it to vibrate. Observations along the edges of these plates have shown such secondary movements. The region around the edge of the Svalbard continental ice, particularly at the edge of the continental block around W Spitzbergen and on Prince Karl Foreland, may also have been subject to such motion.

In Scandinavia the thickest ice stood 3000 m high over the Baltic Sea trough. Here the post-glacial rebound was greatest, and hence the previous indentation deepest. Over 300 m uplift occurred there, or about one-ninth the thickness of the ice cap, whose specific gravity is much lower. There is no reason to assume that the similarly constructed crust of the Barents Sea Shelf should have different elasticity. If, then, the ice cap was nine times as thick as the isostatic movements it occasioned, we may reason that its two peaks stood 1300-1500 m above King Karl Land in the E, and at most 1300 m above W Spitzbergen. This is corroborated by the upper limit of ice scour.

The E and W ice masses met in Stor Fjord, from which one ice sheet spread S and another N through Hinlopen Strait. Heley Sound and Freeman Sound were filled with westward moving ice from the E ice mass, as is shown by the direction of glacial striations and by the distribution of erratics.

This means that the upland surfaces of Barents and Edge Islands were covered only by thin, barely moving ice caps which had little erosional effect on the relict surfaces, and even protected them against the frost attacks of the atmospheric climate. The pre-glacial network of gentle swales was barely altered during the Glacial Maximum, as its radial layout was discordant to the flow of the thin, slowly creeping E ice. The swales were not deepened into U-shaped valleys until the two Late Würm reglaciations when each island formed its own centrifugally flowing glaciation center. Thorough investigation on various levels enabled us to reconstruct the overall glacial relief assemblage in detail.

This could be done even more precisely for the Early Holocene relief assemblage formed directly after the disappearance of the ice. After around 12,000 B.P. the land began to rise isostatically, at first rapidly, then more slowly, and a tendency to sink back again has recently been noted. The shore below the marine terraces became covered with marine clays and clayey sands, richly larded with C-14 datable mussels, whalebone, and driftwood. Today the marine clays still cover most of the foreland and extend into the lower reaches of the glacial trough valleys. At the same time the marine regression left clear terrace structures behind it. At about 10,000 B.P. a temporary halt both in isostatic uplift and eustatic marine transgressions created a particularly marked terrace which lies like a bar or barrier (*Riegel*) across many of the valley mouths. This landform, which we have named the "Barrier Series" or "Riegel Series," forms a broad terrace surrounding the edges of Freeman Sound at a height of 70 to slightly over 80 m (Figs. 16 and 18).

2.2.4 Convenient Natural Test Sites for Identifying Relief Features of Holocene Origin

The above observations on the overall relief history point the way toward a clear reconstruction of the main Holocene processes and features, using the following clues as a starting point.

Most of the elevated relict planation surfaces on Barents and Edge Islands were preserved under their Pleistocene ice cover, and suffered little erosion even in the frosty climate of the Holocene. One might even say that traditional planation continued on them. Only along the periphery of the relict surfaces could valleys cut back somewhat or the upper slope rims retreat, and even this occurred very slowly where sharp slope rims impeded erosion.

Traces of Pleistocene ice scour are best preserved where scouring was most intense, i.e., in the depths of those fjords, such as Freeman Sound, which paralleled the direction of ice-flow. Ice-scoured surfaces (*Rundhöckerfluren*) veneered with ground-moraine are widespread under the Holocene marine clays of the coastal lowlands, and are best preserved where the present elevations are lowest.

On the stage set between these two assemblages of inherited features, that is, in the area between the elevated pre-Pleistocene relict surfaces and the lines of vigorous ice scour along the fjord depths, the overall relief of the emergent valleys and slopes was reworked by chains of processes governed by the rigorous Holocene climate. Knowing the parent relief, we were able to pinpoint the morphologic effects produced here during the known span of Holocene times. To start with we chose the high relict surfaces which have been traditionally reworked by Holocene cryoturbation.

2.2.5 HOLOCENE CRYOTURBATION PROCESSES ON THE RELICT SURFACES AND COASTAL LOWLANDS

One of the most striking features of unglaciated flat ground in polar regions (the frost debris and tundra zones), is the *frost-patterned ground* produced by *cryoturbation* (frost mixing and sorting). The commonest form of patterned ground, distinguished by Troll (1944), is *frost-structured ground (Frost-strukturboden)*, where networks of sorted stone rings and polygons surround cores of fine-grained earth. This develops only where frost shattering of the local rock produces coarse as well as fine material, yielding silts rather than clays. The rings are about 0.5-2 m in diameter, uniting in honeycomb-like patterns. Where the local rock breaks down into fine material only (as with clays, shales, and marly shales), *frost-textured ground (Frosttexturboden)* is produced, consisting of fine networks of cracks enclosing square to rectangular areas which are usually 0.1 to 0.8 m in diameter.

Both types of patterning had so far been investigated chiefly on the marine clays and moraines of the coastal lowlands. Furrer (1959), Pissart (1964, 1966), Washburn (1967, 1973), and Journaux (1972) have tried to explain the development of this phenomenon by experiments and orientational measurements *in situ*. What processes might lead to these quantifiable aspects of the structures, however, remained for the most part a mystery. We therefore considered it our task to elucidate the active physical processes responsible for these structures by excavating them on the relict surface, where they have surely been longest subject to the present frost climate. We also dug deeper trenches into the permafrost layer to investigate its developmental history as well, and to estimate the erosional effectiveness of the present frost patterns, for this determines the amount of traditional planation occurring on the relict surface.

2.2.6 THE PEDOSPHERE AND DECOMPOSITION SPHERE
IN THE PERMAFROST LAYER

The profile of the pedosphere and decomposition sphere in SE Spitzbergen is shown in Fig. 19.

In the coal pits of W Spitzbergen the permafrost layer is around 350 m thick. Beneath this, in the *frost-free layer (Niefrostbereich)*, Corbel (1957) found relict karst features from a pre-Pleistocene warm period, further proof that a warm climate was present in Spitzbergen before the Ice Age. The boundary between the permafrost layer and the frost-free layer is formed today by the lower limit of the decomposition sphere. This boundary, formed where the permanent influence of the severe winter cold penetrating from above is balanced by the heat emanating from the earth's interior, reached its present approximate position in the Holocene after the Würm ice had disappeared. Continental ice sheets are accompanied at most by thin layers of permafrost along their margins; underneath them the ground is protected from the winter cold by the ice, which at the base of the glacier is usually at the pressure-melting point, even in polar latitudes. The exceptional thickness of the permafrost layer in E Siberia (maximally 1300 m thick) is due not only to the extremely severe winter cold there today, but also to the fact that during the Pleistocene Ice Age (with its even more rigorous climate), protective sheets of continental ice were largely absent.

In non-polar latitudes, ground and rock below ca. 12 m are unaffected by annual variations in air temperature; this is likewise true of the permafrost layer in Svalbard. The depth of this layer is marked by the lower boundary of the ice wedges, the largest of which reach down as far as 8-10 m (30 m in E Siberia). Below this lies the *isothermal permafrost layer*, where no temperature fluctuations occur. The lower boundary of this layer, at the frost-free zone, is at 0°C, while the upper limit next to the ice wedge zone is at a temperature of $-6°C$ to $-10°C$. This approximates the average annual surface temperature, namely, $-5.8°$ to $-10°C$ (Knothe, 1931; Dege, 1960; Wagner, 1965).

Above this lies the ice wedge layer, whose wedges are formed by episodic deep frosts, that is, by frosts reaching to unusual depths in uncommonly severe winters with little snow. Freezing is accompanied by cycles of contraction and cracking, the cracks being filled immediately with needle ice. Once filled, the cracks can no longer close, even when the temperatures rise to nearly zero in the following summer.

Over the ice wedge layer lies the ice rind, which we will discuss in detail in Section 2.2.17. This layer forms the region penetrated by annual deep frosts with accompanying temperatures and volume fluctuations. It is completely shot through with ice-filled clefts, which fragment the bedrock utterly.

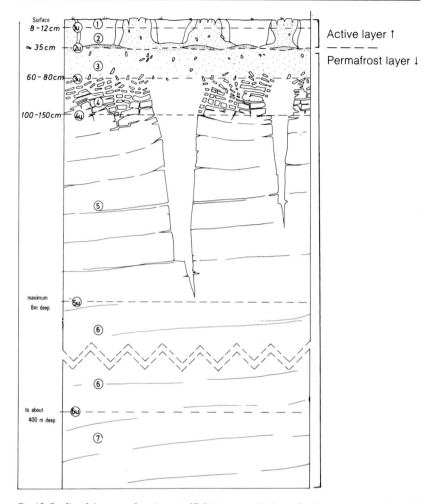

Fig. 19. Profile of the permafrost layer in SE Spitzbergen. (1) Area of midsummer desiccation; (1b) its lower limit. (2) Active layer with frost structures; (2b) its lower limit. (3) Relict active layer from the post-glacial climatic optimum, containing scattered pine pollen, now permanently frozen; (3b) lower limit. (4) Ice rind, the layer of periodic (annual) temperature and volume fluctuations; fragments of bedrock in largely continuous matrix of ground ice; (4b) lower limit. (5) Ice wedge layer with sporadic temperature and volume fluctuations; (5b) lower limit (the largest ice wedges in SE Spitzbergen reach eight m deep). (6) Isothermal permafrost layer, reaching from the lower edge of the ice wedge layer to around 400 m depth (as found in coal pits of W Spitzbergen); (6b) lower limit. (7) Frost-free ground in limestone areas, with karst features below 400 m.

The upper 20-25 cm of the permafrost layer contain the relict active layer, which has less ice. This zone, which thawed out in the summers of the post-glacial climatic optimum (around 6000-500 B.C.), belongs today to the de-composition sphere.

This is in turn overlain by a yearly active layer, averaging 30 cm thick, with extreme values from 18 to 80 cm. This layer forms the pedosphere, and consists of a frost-debris blanket.

2.2.7. THE FROST DEBRIS ZONE AND THE MOSS TUNDRA

Like the vertical succession of climatic, soil, and vegetational zones in mountains such as the Alps, a horizontal succession of these zones may be observed with increasing proximity to the poles as the annual temperature drops and the growing season shortens. Roughly speaking, a climb of about 1000 m in the Alps is accompanied by about the same temperature drop as is a journey from there 1000 km toward the N pole. Above the forest line lies the tree line, beyond which lies the limit of the elfin forest (in the Alps) or forest tundra (in Lapland). Beyond this lies tundra (in the Alps the turf region, or *Mattenregion*), whose hardiest variant, the moss tundra (Büdel 1960), reaches the islands of SE Spitzbergen. Moss tundra covers everything except bare rock (a polished rock surface or roche moutonnée) with a plant cover which is fairly continuous, though thin and broken in places by frost-patterning (forming moss rings or mud-pits). Moss tundra provides the main feeding grounds for wild reindeer (*Rangifer tarandus*), of which as many as 1000-1500 head may still exist on the islands of SE Spitzbergen.

The frost debris zone is clearly demarcated by a sharp boundary often involving only 10-30 m difference in height. Like the desert steppe, the unprotected frost debris is sparsely dotted with plants which are undemanding, uncompetetive, and highly resistant to cold and wind. The altitudinal boundary between the moss tundra and the frost debris zone (the Alpine *Felsschuttzone*, *regio alpina superior*, or Scandinavian *Fjellmark*) lies, according to Hofmann (1968), at about 200 m in the W portions of Barents and Edge Islands; this can rise to 250 m in wind-sheltered valleys, with outposts up to 300 m. In the E part of the islands, the limit sinks to about 100 m. This accords with the snow line, which descends in the same direction from nearly 500 m to about 400 m. The frost debris zone, both in the E and the W of the island, stays within an altitudinal zone about 300 m wide, covering the entirety of the unglaciated old etchplains which are of so much interest to us.

2.2.8 THE DEVELOPMENT OF PATTERNED GROUND

On the Hohenstaufen etchplain (located 440 m above sea level on the SW corner of Barents Island in the so-called Stauferland), we accurately marked and surveyed a nearly flat test site 16 ha large (see Fig. 20). Here a series

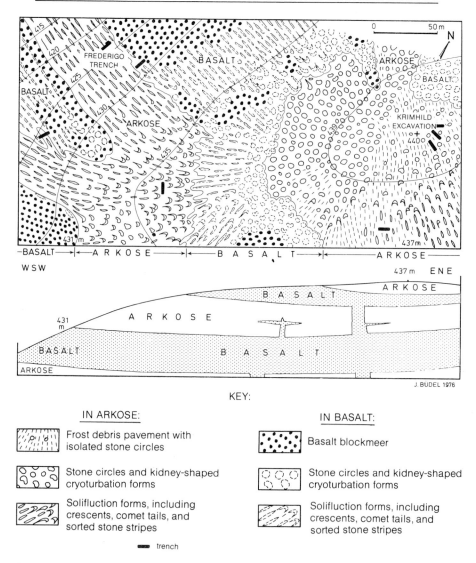

KEY:

IN ARKOSE:

Frost debris pavement with isolated stone circles

Stone circles and kidney-shaped cryoturbation forms

Solifluction forms, including crescents, comet tails, and sorted stone stripes

IN BASALT:

Basalt blockmeer

Stone circles and kidney-shaped cryoturbation forms

Solifluction forms, including crescents, comet tails, and sorted stone stripes

trench

Fig. 20. Frost structures on the Hohenstaufen relict planation surface (440 m altitude, SW part of Barents Island; see Figs. 15 and 16). *Above:* part of a precisely measured and mapped test site (etchplain and gently convex upper slope). Each symbol represents eight to ten corresponding forms in nature. *Below:* subsurface profile along the lower edge of the plan. As soon as the slope exceeds 2-3° the lithic boundaries at the surface become blurred by solifluction. Exceptions are found at the heads of basalt blockmeere, around which solifluction flows (see left).

of highly differing, gently dipping lithic units (largely basalt and dolerite sills alternating with Upper Triassic thinly bedded arkose) are beveled by the flat surface (see Fig. 20 and Photo. 1).

The entire surface and its surroundings are covered with a great variety of patterns ranging from textured ground on clay and marl outcrops through frost pavement with sparse structure-like turbations to block fields (*Blockdecken*). The latter occur on rocks such as quartzite or thick basalts which yield no fine material at all when attacked by frost. Most of the patterns between these extremes are found on rocks where frost weathering produces both coarse matter and fine silty material, e.g., marly limestones, arkoses and most of the slaggy-tufaceous basalts. Here we find the classic annular *sorted stone circles* and *stone polygons*. Proceeding on the geomorphological principle that a phenomenon should be investigated in its purest form, we chose for our first excavation (Krimhild, see Fig. 20) an area in which these annular forms were best developed. Later excavations showed that all other variants could easily be derived from the classic annular form.

Sorted circles are about 0.5-2 m in diameter. Their centers are formed by *fine earth cores* (*Feinerdekerne*) of gray or brown silt and sand and a few more or less vertical stones. The core is surrounded by a 1-2 decimeter broad *fine pebble mantle* (*Feinkiesmantel*), whose components are usually 0.8-2 cm, at most 2.5 cm large. The area between circles is occupied by *coarse rubble beds* (*Grobschuttbeete*), typically containing subangular stones of at most 2.5-15 cm length. These are clearly oriented tangentially to the circle (Furrer, 1959 and 1969), surrounding it in a ring 5-12 cm wide. It is these upright tangential stones which first convey the full impression of the annular forms (see Photo. 2).

Where frost weathering causes the local rock to produce much fine material (as with slaggy-tufaceous basalt), the fine earth cores are large and closely spaced, so that the coarse rubble beds are restricted to small triangular areas between them. The polygonal honeycomb pattern is best revealed here, and this is also the best area in which to observe how the fine earth cores and the fine pebble mantles are completely free of vegetation. What little there is of mosses, lichens, and hardy angiosperms such as the arctic poppy, is limited to the coarse rubble beds, a sure sign that these remain fairly quiet, while the fine earth cores are continually in motion.

It is on such polygonally patterned ground that the impression of acres of evenly distributed structures is most striking. On arkose, on the other hand, which disintegrates into coarse platelets, the circle foci are often several m apart, so that the circles occupy less area than the coarse rubble beds. Here we find both large and small circles, some even "embryonic," with only their fine pebble mantle visible as a small spot in the coarse rubble. The fine earth core may barely show at the surface, or may be found only by excavation, but it is always present. The distribution of circles is not very regular on such

1. Etchplain on Hohenstaufen (440 m altitude, Barents Island, SE Spitzbergen) with patterned ground. View S over the pack ice of Freeman Sound (five km wide) to the relict planation surface on Edge Island. (Photo: Büdel, August 2, 1959.)

2. Patterned ground (cryoturbation-sorted circles) on the Hohenstaufen surface. View ESE over Freeman Sound to Edge Island. On the structures, see Figs. 20, 21, and 24. (Photo: Büdel, July 31, 1959.)

substrates, and a transition often occurs to a continuous stone pavement having but a few circles with small fine earth cores 10-30 m apart.

In areas of midsummer desiccation, both large and small circles display a network of polygons about a hand's breadth in size, formed by desiccation cracks in the upper 8-10 cm of their fine earth cores.

So much for the external picture. Our excavations, penetrating (for the first time, as far as we know) to the permafrost layer, showed that in the cool summer of 1959 the permafrost table on the Hohenstaufen lay around 30 cm below the surface, sinking to 40 cm below the surface in the warmer summers of 1960 and 1967. At all times it was 8-16 cm deeper under the fine earth cores than under the coarse rubble beds. Since a constant film of water circulated just above the permafrost, being unable to trickle further downwards, we required a sturdy motorized pump in addition to powerful digging equipment (a Swedish ''Kobra''), as the hole would otherwise have filled with water. Our first discovery was that the structures of the active layer do not continue into the permafrost, a radical change setting in at the permafrost's upper boundary.

The structures are thus limited exclusively to the active layer. According to our investigations, they are created and developed by a series of hitherto undifferentiated processes. It therefore fell to us to identify these processes and clarify the manner in which they interact both spatially and through the annual cycle, not only in the present, but also through the history of their Holocene development. This then enabled us to estimate the role played by each of these subprocesses in the entire picture.

Proceeding polewards from the tropics, the role of chemical weathering decreases, partly due to falling temperature, partly to increasing phase changes (freezing and thawing) of water. Yet chemical activity is not entirely lacking even in the frost debris zone. Our maximum temperature measured in the air next to the ground was $+11.0°C$ on a sunny July day in 1960, at which time we measured a temperature of $+20.5°C$ on the surface of shaley frost debris lying on a SW facing slope. The platelets here were covered by a thin reddish-brown varnish. Mechanical weathering caused by frost is still the most effective agent here. Frequent freezing and thawing can only reduce stone as far as coarse silt (20-200 μm); medium and fine silts (2-20 μm) or clays (<2 μm) are rarely found in the fine matter of the frost debris zone, forming but a few percent of its components.

More chemical weathering occurs in the moss tundra, where incipient soils tend toward nanopodzols (Kubiena, 1953). Semmel (1969) and Nagel (oral communication) dug up basalt blocks showing slight traces of chemical attack. Even on the plant-covered moss tundra, Semmel (1969) found no more than 8½% clay fraction in the soil, and he suggests that this clay fraction, derived from plagioclase and olivine, may have formed during the post-glacial climatic optimum. Other clay components may derive from clayey parent rock, or from the Holocene marine clays on the coastal lowlands, for no frost-weathering occurs in the sea. Clay minerals of all origins form but a small component of the fine earth cores, and are found largely in the so-called mud-pits of the moss tundra.

Frost shattering is the most powerful mechanical weathering agent on earth. Water penetrates in liquid form during the summer and in vapor form at all times of year into the finest crevices, and surrounds the individual mineral grains. Repeated warming and cooling may loosen these grains, due to differing coefficients of expansion. When the water freezes it increases in volume by around one-ninth, creating pressures in nearly closed cracks of up to 2000 bar. No other form of mechanical weathering produces such a drastic effect. Each new freezing cycle renews the attack. As Troll (1944, 1947, 1973a) has pointed out, this effect is most marked with frequent cycles of freezing and thawing.

It was on this basis that our investigations began. We found that in the subpolar climate of alternating freezing and thawing, the upper part of the active layer goes through an annual cycle of reactions resulting from thawing and refreezing, from hydration and dehydration, from moistening, flushing, and desiccation, as well as from the various processes of sorting; these lead to the striking patterning of the ground. We were able to distinguish no less than sixteen physical component processes of cryoturbation, whose chronological and spatial interactions constantly keep the formation of patterned ground in progress. These sixteen component process are listed here (see also Fig. 21).

[1] *Snow cover and snow melt* provide most of the water, for it does not rain every summer. The snow thaws, refreezes, percolates down, flushes, and rises again through capillary action, keeping the mechanisms of ground patterning in progress. Unlike the mid-latitudes where thaws occur throughout the winter, the snow blanket of SE Spitzbergen accumulates steadily from around Christmas-time onward and melts all at once in June. The melt water has almost no impurities, and is chemically fairly inactive.

[2] *Upward freezing* is an important member of these processes, whose basic principles, first described by Hamberg in 1916, are shown in Fig. 22 in five phases. I. Two stones stand more or less upright near the surface of a fine earth core, the left stone being about three times as long as the right (unfrozen initial condition). II. The ground freezes in winter, and frost heaving (see [6] below) raises both stones equally. (Frozen ground is shaded, height of frost heaving slightly exaggerated.) III. Thawing works down almost to the lower end of the small stone, with concomitant resettling of the fine earth. The stones do not settle back, as their lower portions are still frozen fast, and they protrude slightly above ground level. Both are equally visible above ground, at most 10% of the smaller stone being exposed. IV. The ground has

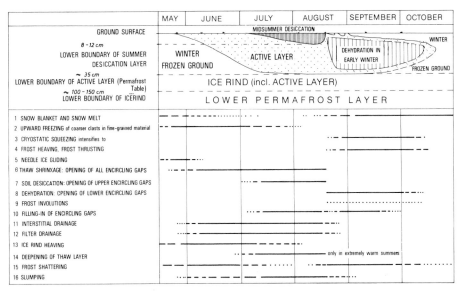

	MAY	JUNE	JULY	AUGUST	SEPTEMBER	OCTOBER

GROUND SURFACE
8 – 12 cm
LOWER BOUNDARY OF SUMMER DESICCATION LAYER
~ 35 cm
LOWER BOUNDARY OF ACTIVE LAYER (Permafrost Table)
~ 100 – 150 cm
LOWER BOUNDARY OF ICERIND

MIDSUMMER DESICCATION
WINTER FROZEN GROUND — ACTIVE LAYER — DEHYDRATION IN EARLY WINTER — WINTER FROZEN GROUND
ICE RIND (incl. ACTIVE LAYER)
LOWER PERMAFROST LAYER

1 SNOW BLANKET AND SNOW MELT
2 UPWARD FREEZING of coarser clasts in fine-grained material
3 CRYOSTATIC SQUEEZING intensifies to
4 FROST HEAVING, FROST THRUSTING
5 NEEDLE ICE GLIDING
6 THAW SHRINXAGE: OPENING OF ALL ENCIRCLING GAPS
7 SOIL DESICCATION: OPENING OF UPPER ENCIRCLING GAPS
8 DEHYDRATION: OPENING OF LOWER ENCIRCLING GAPS
9 FROST INVOLUTIONS
10 FILLING-IN OF ENCIRCLING GAPS
11 INTERSTITIAL DRAINAGE
12 FILTER DRAINAGE
13 ICE RIND HEAVING
14 DEEPENING OF THAW LAYER — only in extremely warm summers
15 FROST SHATTERING
16 SLUMPING

Fig. 21. Annual cycles of the component processes involved in cryoturbation (frost-patterned ground in the active layer).

thawed down to the lower end of the larger stone, and has settled accordingly. The smaller stone, no longer frozen into place, has likewise settled, but the larger stone has not, its lower end still being cemented by ice. Its upper end protrudes even further, approximately three times as much as the small stone, or barely 10% of its own height. V. Thawing has reached the base of the fine earth. Settling has compensated for the total frost heaving, and the fine earth has returned to the initial level of phase I. The two stones now protrude by the amounts attained in phases III and IV. In a few years they migrate fully above the surface and fall over. Frost-gliding (see [5] below) then removes them to the edge of the fine earth core.

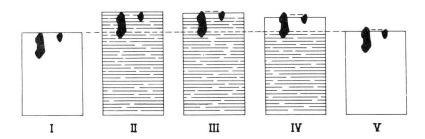

I   II   III   IV   V

Fig. 22. Upward freezing of coarser clasts in fine matrix. Large stones migrate more rapidly than small stones (explanation in text). After Hamberg, 1916.

In all fine earth cores, upward freezing causes the stones to migrate in yearly stages to the surface, larger stones traveling faster than small ones. One hundred thirty-two steel nails pounded in (each 22.5 cm long and 7.5 cm thick) were raised by frost heaving an average of 2.7 cm per year, with a maximum of 8.4 cm per year. Stones are seldom this long, but averages of 5 mm per year for small stones to 12 mm per year for large ones are surely normal. In this fashion the fine earth cores quickly purify themselves of coarse material, producing their most striking feature. The few stones they contain are continually supplied on a small scale from below by the processes described in [13], [14], [15], and [16]. If a stone 10 cm long migrates at a rate of eight mm per year, it would require 50 years to travel from the base of a fine earth core to the surface 40 cm above. Stones one cm long would require 500 years. But these are minimal values. As long as filter drainage (see [12] below) provides for substantial lateral growth of the fine earth core, it will continue to expand sideways, taking up stones along its sides as well as from its base, so that some stones may make the journey through the fine earth core several times. The cycle ends only when the coarse rubble beds between the sorted circles have been washed free of fine earth. The slightly subangular stones of the coarse rubble beds lie dry and loosely piled, having little fine earth in their interstices. This is a gradually attained mature condition, just as is the final purified and stone-free state of the fine earth core.

The patterned ground structures of the Hohenstaufen represent a climax stage achieved during the 2500 years since the post-glacial climatic optimum. This agrees well with results obtained in other ways (see Section 2.2.10). Wirthmann (1962) likewise describes the majority of the sorted stone circles on the relict surface Edge Island as fully stabilized climax forms. Particularly where working rims on such relict surfaces have interrupted denudation and induced desiccation, such patterned ground forms an almost motionless protective cover, preventing any further removal of the substrate.

[3] *Cryostatic squeezing*. The initial areas to freeze in the first winter frosts (end of August to early September) are the surfaces of the fine earth cores. This process draws water from below toward the frost table with 20-30 times more energy than toward the unfrozen ground. In the ensuing hydration, water molecules become lodged in and on the crystal lattices of the individual minerals. The fine earth beneath the frost table becomes waterlogged, even becoming somewhat plastic in the clayier soil of the moss tundra. Pressure from the frozen soil above can cause it to flow plastically, e.g., to ooze up along the periphery of the frozen fine earth core. According to Semmel (1969), this process, called cryostatic squeezing, is very important in the moss tundra of Spitzbergen and Lapland. Many researchers have suggested that this process is responsible for the relict involuted soils of the Pleistocene periglacial region in Central Europe. In the frost debris zone of Spitzbergen cryostatic

squeezing plays no great role, and we were unable to find any traces of recent involuted soils.

[4] *Frost heaving and frost thrusting* are considerably more important in the frost debris zone than cryostatic squeezing. During the early winter, as the fine earth cores freeze down to the base of the active layer, their water content, already much greater than that of their surroundings, expands by around 10%, so that the fine earth cores swell up above the surrounding ground. This leads to frost-gliding (see [5] below).

The expansion of the fine earth cores in early winter also acts laterally to produce *frost thrusting*. This occurs primarily in the later stages of freezing, when the upper portion of the fine earth core has frozen solid, and yields less to pressure from below than the surrounding coarse rubble (which, having large interstices and no cementing ice matrix, can yield). In this manner, a silt core will gradually armor itself with a carapace of stones pressed flat along its margins. This is how the cores become ringed with tangentially oriented fine and coarse frost debris, the latter, of course, being particularly striking to the eye.

[5] *Pipkrake or needle-ice gliding.* When the upper surface of a fine earth core freezes and heaves, material is sloughed off toward its edges. Meltwater runoff in spring also carries material off the domed surface. But the most effective agent for this transport is *pipkrake gliding* or *needle-ice gliding* (*Kammeisgleiten*), which occurs during the rapid freeze and thaw sequences of spring and fall. Stones of all sizes are lifted by ice needles for as much as a few cm perpendicular to the domed surface of the core, i.e., slightly diagonally. After the core surface has melted back, the stones resettle vertically, so that each pipkrake frost causes the stones to inch a little further toward the core perimeter. In time, even the largest stones wander in this way out to the surrounding ring of rubble. Here they rest a while, until further frost thrusting forces them into tangential-vertical alignment. This is also aided by the opening of encircling gaps (see [6], [7], and [8]).

In summer the domed surface settles and becomes concave, yet a reverse movement toward the center does not occur. The pipkrake effect is lacking, and meltwater is no longer present at the surface, but circulates lower down in the active layer.

Pipkrake gliding is even more important in the frost zone of the tropical mountains. Troll (1944, 1959, 1973b) rightly draws a basic distinction between the annual cryoturbation and solifluction cycles in the subpolar region and daily cycles in tropical or even subtropical highlands, where needle ice generally creates smaller forms and effects a more superficial movement. Noticeable smoothing of the slope is brought about on rocks which are susceptible to frost-weathering. The total erosional efficiency over Holocene times remains less in such highlands than in the polar periglacial region.

[6] *Thaw shrinkage, first opening of the encircling gaps.* During the spring and summer thaws the fine earth core shrinks laterally and vertically by around 10%, producing shrinkage gaps around the core next to the rubble armor. This is the first stage in forming the encircling gaps (*Randspalten*).

[7] *Soil desiccation, further opening of encircling gaps (upper portion).* The upper portions of the encircling gaps open wider during the summer as the upper 8-12 cm of the soil dry out. This desiccation produces the minute polygons mentioned above. The desiccation fissures, especially the encircling gaps, which may gape as much as two cm wide at the top, trap tiny pebbles that keep them open through the winter.

[8] *Dehydration, further opening of the encircling gaps (lower portion).* When the fall frosts set in, water rises to the freezing table, draining the lower silt layers even of capillary water. This process, named dehydration (*Dehydratation*) by Schenk (1955), causes marked shrinking in the lower portions of the fine earth core, widening the shrinkage and desiccation gaps there.

[9] *Frost involutions.* The thorough kneading of involuted soils is generally attributed to pressure from new ground ice forming in fall, which affects the unfrozen layers simultaneously from above and below (see Fig. 21, upper right, for September and October). Relict involuted soils in Central Europe are found primarily in loess and similarly fine-grained material. As these materials are lacking in Spitzbergen, we saw very little involution there.

[10] *Filling in of the encircling gaps.* The encircling gaps, which annually gape up to two cm wide from top to bottom, take up all manner of smaller stones fed into them by pipkrake gliding. Over the years these gravel-filled fissures grow by repeated frost thrusting, thaw shrinkage, desiccation, and dehydration, producing the fine pebble mantles (*Kleinkiesmäntel*). These envelop the silt cores in bands 5-12 cm wide, clearly defined both with respect to the cores and to the surrounding coarse rubble beds. Frost-thrusting in the following winter not only orients the larger stones tangentially around the fine pebble mantles, but also tends to heave the latter into low embankments, forming well-marked stone circles around the fine earth cores.

[11] *Interstitial drainage (Drainagespülung).* Like simple frost pavements, the coarse rubble beds between the fine earth cores consist of medium-coarse to coarse frost debris. In the summer this lies loosely piled, dry, and free of any fine earth matrix; the stones, occasionally flecked with moss or lichen, may easily be picked up, as the interstices are filled with air. During the summer the sinking permafrost table is covered with an active water film. This is little affected by molecular forces in the rubble beds, and runs freely, though the slope on the relict surface is rarely over 1°. The water is almost

pure H_2O. Even in the moss tundra, lack of root or bacterial respiration means that the CO_2 content of the soil air (according to measurements by Nagel, oral communication) is only about a hundredth of that in mid-latitude soils, and about one thousandth of that in tropical soils. The dissolved fraction in the water film running along the base of the rubble beds in summer is very small. The suspended load consists not of clays, but almost entirely of silt prepared by frost weathering. The fact that interstitial drainage is not a more effective erosional agent on the relict surface will be explained next.

[12] *Filter drainage (Filterspülung)*. We saw above that the permafrost table in summer is 8-16 cm lower under the fine earth cores than under the surrounding coarse rubble beds. Instead of following the gentle overall slope of the relict surface, therefore, drainage water generally runs down the much greater slope into the silt core. It streams gently through the lower part of the core, becoming filtered of its silty suspended matter. The water film flowing onward is purified, while the core is by this means continually provided with new silt. In this way the rubble beds are cleansed of the fine matter produced there every winter, and the already existing silt cores become enlarged. Drainage and filtering perform little noticeable erosion, instead feeding the patterned ground or cryoturbation sheet, helping this sheet conserve the relict surface.

[13] *Ice rind or permafrost heaving*. In winter, particularly in severe winters, pressures build up below the permafrost table. These remain latent during the winter ice-bonding, but in summer, when the active layer encroaches from above, this ice-bonding is loosened, and small boils and welts pop up on the permafrost surface, much like frost-heaving in a street. Such permafrost heaving feeds new material to the active layer.

[14] *Deepening of the active layer*. As frost heaving pushes up from below, it meets a second process encroaching from above: in very warm summers the active layer works 5-10 cm deeper into the ground. It eats into the upper portions of the permafrost layer, bringing this temporarily into the cryoturbation cycle. New material, mostly fine-grained, is activated and fed into the fine earth cores by filter drainage.

[15] *Frost-shattering* produces debris and leads to slumping.

[16] *Slumping*. Every winter silty material is produced in the coarse rubble beds and is carried to the fine earth cores by filter drainage, causing the rubble beds to slump in gradual increments. Upward freezing in the fine earth core carries larger stones rapidly to the surface, where pipkrake gliding removes them to the edges of the stone circle. The material lost in the rubble beds is thus more or less compensated for. These sixteen processes in combination bring about very little removal of material from the relict surface.

2.2.9 CONTINUED GROWTH OF RELICT SURFACES: TRADITIONAL PLANATION, NOT CRYOPLANATION

The sixteen processes described in the last section have brought the patterned ground to its present state, usually a climax state, which protects the relict surfaces against severe erosion.

As Wirthmann (1964) made clear, the dissected relict etchplains of SE Spitzbergen are products of a warm Tertiary climate. At the onset of the Pleistocene their much broader remains were covered by a flat, supercooled, and nearly motionless Pleistocene continental ice sheet, which at times and places may even have taken the form of an ice skin frozen fast to the underlying ground. This ice did little to affect the land surface below it; indeed, it protected more than attacked it. At any rate, the overall *flat nature of the relict forms was fully preserved*. This flattened form is *permanently stamped* into the bedrock, and forces all newly arising relief-forming mechanisms (especially those incapable of planation) to adapt themselves to the given relief and to maintain the given forms. Following Bremer (1965c) and Büdel (1971) we call this process *traditional development* of surfaces (for a detailed discussion, see Section 2.3.9). The important point is that these newly arising processes *could never create* such flat surfaces by themselves, but rather *had to adapt* to the given relict surface.

This continued to hold true even after the continental ice sheets disappeared and a Holocene periglacial climate took over. Again, the new relief-forming mechanisms could not create the relict surfaces, but since the latter were inherited and unavoidably present, the new processes were forced to adapt to them. Indeed, their climax form, the patterned ground sheet, even protected the surface.

Anyone who has not personally studied the relief-forming mechanisms of tropical planation, and observed the extraordinary durability of their product through repeated climatic changes, will have difficulty realizing how such a plain, once created, will force all other mechanisms, even under quite different exogenetic conditions, to adapt to a regimen of traditional planation.

The relief-forming mechanisms of cryoturbation were by no means capable of creating such etchplains. In today's periglacial, frost debris, and tundra zones, we found not the slightest evidence of any etchplain creation such as the "cryoplanation" or "altiplanation" proposed by some authors. The examples quoted by Demek (most recently in 1972) and Karrasch (1972) give no data on the physical mechanisms which would make planation possible under periglacial conditions. All proponents of this theory, moreover, overlook any developmental approach to such supposed cryoplanation terraces, and thus risk attributing the creation of very ancient flat relief forms to much weaker current processes which had to adapt to the situation in which they found themselves, and were at best capable only of traditionally expanding the surface slightly (see Fig. 51).

Vigorous cryoturbation took place in Central Europe not only on the old etchplains and etchplain stairways of the first relief generation, but also on the surfaces (Broad Terraces or *Breitterrassen*, pediments, and glacis) and older cold stage terraces of the Günz, Mindel, and Riss I terraces (see Figs. 14 and 62). But even on younger terraces this cryoturbation found rock surfaces already created by fluvial lateral sapping. The steeper slopes between terraces were at the same time leveled and carved back by solifluction and slope wash, the oldest slopes being most affected. Any relict gravel left on the level surfaces from the time of the terrace's creation has long since been taken up by cryoturbation, and has disappeared. Cryoturbation, however, did not create the level bedrock surfaces, as Demek (1964, 1969, 1972) and Karrasch (1972) assume when calling them cryoplanation terraces (Demek in connection with active terraces in Siberia, and Karrasch with respect largely to relict terraces in Central Europe). The chain of processes involved in cryoturbation, according to our analysis above, contains no mechanism capable of creating such slope terraces or etchplains, particularly in such a short time and at such a height above the valley floor (the local base level of erosion).

Solifluction and the much more powerful processes replacing solifluction on steep slopes can accomplish much leveling and backwearing of slopes, and can create working rims and piedmont angles. Inselbergs resting on relict surfaces in Central Europe may have been lowered by solifluction and slope wash during the repeated periglacial climates of the Pleistocene ice ages; a ridge relief inherited from tropical times may become evened out. But the processes involved are types of traditional development, for which I prefer not to use the terms "cryoplanation" or "altiplanation," as these are fraught with too many other meanings. For the lowering of inselbergs implies a pre-existing basal surface, or in the case of the relict ridge relief, a pre-existing general level. Any pre-Würm moraine landcape shows how effectively periglacial processes have eradicated the closed depressions of the fresh subglacial relief, converting the whole into a fluvial relief. But this was brought about in unconsolidated deposits by valley-cutting, not by planation. Even under such favorable conditions, it would take interminably long to level out the entire relief. To assume that this could occur on solid rock, without even undertaking an exact relief analysis or citing the mechanisms supposedly responsible, seems to me an unconscious relapse into the long-refuted peneplain theory of Davis (see footnote 17, p. 98). In treating the seasonal tropics, we will come to know the only specific relief-forming mechanism which is *a priori* clearly capable of forming such etchplains. It seems unacceptable to ascribe to other, quite different mechanisms, the ability to form etchplain surfaces merely because they are found acting on the stage set by etchplains. Etchplains formed during tropoid paleo-earth times are found as plateaus over the entire earth, and their very existence forces more recent relief-forming mechanisms to adapt themselves to a type of traditional planation.

Our following discussions will show quite clearly that both the active polar and the relict mid-latitude periglacial regions bring about the exact opposite of planation, causing excessive valley-cutting instead. Working inward from the peripheries of relict surfaces, this valley-cutting dissects and destroys them.

2.2.10 THE RELICT ACTIVE LAYER
OF THE POST-GLACIAL CLIMATIC OPTIMUM

Our two conclusions on the patterned ground sheet of the relict surface, namely, that it is a climax form and that it protects against erosion, are supported by the existence of an underlying relict active layer from the post-glacial climatic optimum. This climatic optimum, according to foraminiferal analyses of Holocene marine clays by Feyling-Hanssen (1955a and b) and Feyling-Hanssen and Olsson (1960), lasted on Spitzbergen from about 6000 to 500 B.C.

On flat surfaces and gentle slopes, our excavations invariably revealed in the permafrost an upper zone 35-50 cm thick, sharply divided from the active layer of today. The almost pure, unstructured fine material of this zone (see Figs. 23 and 24) contains thin, nearly horizontal ice laminae, and is marked by stained horizons: at first yellowish-brown, it turns blue-gray, then grades below into a dark, highly humous horizon containing pine pollen.[12] Radiocarbon dating of this horizon has yielded an age of 3000 to 3100 years, thus putting it at the end of the post-glacial climatic optimum. All levels of the relict active layer still display a green-tinged rusty brown coloring.

The entire relict active layer must therefore have formed during a climatic optimum on a tundra having greater chemical weathering. Though patterned ground may have existed during the colder first half of the Holocene, what form it took can no longer be reconstructed. The subsequent climatic optimum created the tundra soil whose lower portions are still preserved today. The few (mostly fine) clasts found in it are in the vertical position typical of upward freezing. Increased upward freezing at the end of the climatic optimum may have created a frost pavement, which at first would have been rather uniform. In the thinner active layer of today, gradual segregation has produced

[12] The bogs on nearby Bear Island have revealed no traces of pine. The pollen must therefore have been blown from N Norway, about 800 km away. Although N Norway has only a few relict stands of pine now, this species may have grown thickly there at that time. We may assume that the post-glacial climatic optimum had stronger S winds.

Fig. 23. Patterned ground and ice rind on Hohenstaufen, altitude 440 m. Parent rock: Triassic thin-bedded arkose. Krimhild I Trench, July 15, 1967 (for location see Fig. 20). (1) Permafrost table. (2) Coarse rubble beds in cryoturbation patterned ground. (3) Fine pebble mantle around the fine earth core. (4) Fine earth core. (5) Relict active layer from the post-glacial optimum, upper gray horizon. Much fine material, little coarse matter, thin ice lenses. (6) Lower greenish-brown horizon of same. Much humus, scattered pine pollen, more coarse matter, larger ice masses (shown in black). (7) Ice rind, highly fragmented bedrock in ground ice matrix (black). Some fine material in upper part. (8) Transition from ice rind to undisturbed bedrock.

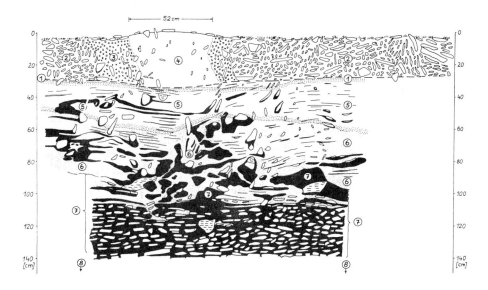

Fig. 24. Similar to Fig. 23, but from the deeper trench of Krimhild II. August 2, 1967. Numbers 1-7 as in Fig. 23. (8) At 140 cm the bedrock had still not been reached, indicating an unusually thick ice rind.

the fine earth cores and frost patterning. The fact that a tundra soil from the climatic optimum is still extensively preserved below the surface shows what little erosion has taken place in the ensuing two-and-a-half thousand years, and shows how the climax forms produced here during that time have inhibited erosion.

2.2.11 THE ICE RIND

In the profiles shown in Figs. 19, 23, and 24, the first traces of bedrock appear about 60-80 cm below the surface (i.e., 30-50 cm below the permafrost table) and are cut above by the relict active layer of the post-glacial climatic optimum. These bedrock traces form an intermediate zone of shattered clasts in an ice matrix. Forty to sixty percent of the zone may be ice, even 100% in areas of continuous ice sheets. In the trench wall ice masses 0.25 to 0.5 m^2 in size are not uncommon, though these usually contain pieces of shattered bedrock. Figures 23 and 24 show the ice in black. The fragmentation of the bedrock is even greater than indicated there, for the larger pieces are shot through with ice-filled haircracks. Below the ice-rich layer (which is usually 50-100 cm thick), the bedrock becomes more continuous, and is only broken at intervals by ice wedges (see next section). We have named the continuous ice-rich layer the *ice rind (Eisrinde)*.

The ice rind is the lowest zone reached annually by the deepest and coldest frosts. The mean January temperature in the area of Freeman Sound is $-18°C$ and that of Hohenstaufen is around $-20°C$, while the coldest winters may reach below $-40°C$. At such temperatures, frozen rock contracts like any other solid. An ice column one m long will shrink by ca. 0.5 mm between 0° and $-10°$, ca. 1 mm between 0° and $-20°$, and ca. 1.5 mm between 0° and $-30°$.

Frozen rock behaves similarly. Every winter the severest frosts tear the ground open to depths of 1-1.5 m and inrushing air immediately fills the clefts with needle ice. When the frozen ground warms up at the end of the winter, the cracks can no longer close. New fissures are filled with ice every winter, quickly fragmenting the bedrock. This greatly facilitates summer erosion on the slopes, and, as we will see, is particularly beneficial to fluvial erosion.

Excavations on all relief types of Stauferland showed that the ice rind is a *general phenomenon characteristic of the upper surface of the permafrost.* On relict surfaces (and on the very gentle grades of the adjacent upper slopes) the only permafrost layer overlying the ice rind is the relict active layer, whose lower ice content suggests that most of the underlying ice rind originated in the colder first half of the Holocene (12,000 to 6000 B.C.). Where the relict active layer is absent, as on all steeper slopes (where it has been removed during the last 2500 years) or on the forelands (most of which were first exposed after the climatic optimum), the ice rind lies *directly* beneath the thin Recent active layer. On the flatter coasts where the sea ice freezes to the sea floor in winter, the ice rind extends at least six m out from shore, even during the summer. We will see later that it is also found under all river beds. Its distribution is only interrupted in those few areas of outcropping hard and widely jointed basalt, but even here rock-splitting ice may occur at depth. Where columnar basalt has been beveled by former continental ice, columns lifted up to 50 cm above the ice-scoured surface (Büdel, 1960, Figs. 25 and 32) show that an ice plug of this size has formed in a transverse joint.

The ice rind in all these features was a *hitherto unknown phenomenon*. It also throws new light on the Siberian stone ice (*tjäle* in Norwegian terminology) which, though often interrupted by deposits of sand and gravel, may be up to 30 m thick. Tjäle formed along the same principles as the ice rind of SE Spitzbergen, though it reaches such great thickness only in depositional areas such as those I saw along the Riss terraces of the Lena and Aldan Rivers.[13] It is characteristic that the well-preserved mammoths shown to us were all located in thick tjäle (ice rind) in these Riss age depositional terraces (see Fig. 38 and Photos. 13 and 14).

2.2.12 ICE WEDGES,
ICE WEDGE POLYGONS, and PINGOS

While the ice rind is formed by periodic deep frosts, the far deeper reaching ice wedges are due to episodic frosts in exceptionally cold and snow-free winters, presumably occurring, as in Central Europe, only a few times a century. The longest ice wedges in SE Spitzbergen reach about eight m deep. As their tops are sharply cut by the relict active layer (where present), they must have developed chiefly before the post-glacial optimum. The Siberian ice wedges, formed over practically the whole of the Quaternary, reach a maximum of 30 m deep. Naturally, during the million years or so of the Quaternary, they have experienced far more and far colder winters. The interior of E Siberia today has mean January temperatures of $-50°C$, and an average extreme temperature $-63°C$. The absolutes are lower today, and must have been far more so during the Pleistocene cold stages.

After the Würm ice left SE Spitzbergen, the first deep frosts pierced far into the unprotected rock, wherever this was loose enough, tearing fissures averaging 0.5-1 cm wide, spaced at intervals of 10-12 m. As in any contracting medium, these fissures formed hexagonal patterns, producing *ice wedge, tundra*, or *taimyr polygons* 10-12 m in diameter. These are best developed

[13] This was on an expedition of the Russian Academy of Sciences in the summer of 1969, led by Gerasimov, Markov, Katasonov, Solovev, Velitschko, Giterman, and others, to all of whom I am very thankful.

on the soft marine clays of the forelands.[14] The next severe winter, occurring possibly after years or decades, attacked these lines of weakness, adding new fissures to the old ones, which had stayed open due to their filling of ice needles. Each ice wedge thus consists of many more or less vertical *ice laminae* 0.2-1 cm wide. These laminae are separated by thin ice walls containing fine dust particles wafted in each time the gap is opened. Internally these laminae consist of horizontal ice needles, whose entrapped air bubbles make it possible to distinguish the laminae by color (see Photo. 3 of a Siberian ice wedge, and also Fig. 25). Each lamina corresponds to one winter of deep frost. Since the upper part of an ice wedge experiences more deep frosts, it has more laminae, is broader, and puts greater pressure on the surrounding

3. Ice-wedge from the six-m terrace of the Lena, 150 km N of Yakutsk. 2000-3000 years old. Compare Fig. 38. (Photo: Büdel, July 29, 1969.)

[14] Taimyr polygons are absent on very hard, widely jointed basalts. No other rock is so resistant to all forms of frost-shattering. On marl and marly shales we made the unprecedented discovery of beautifully formed polygons on a slope of 31°. Usually slopes of over 25° erode too rapidly for ice wedges to form in the substrate.

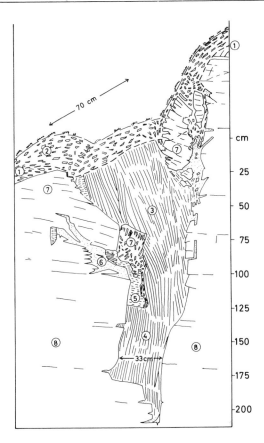

Fig. 25. Typical ice wedge in SE Spitzbergen (slope 31°, 185 m altitude, on the "Vogelweide," SW corner of Barents Island; see Fig. 16). (1) Permafrost table. (2) Active layer. (3) Upper portion of ice wedge with criss-crossing laminae. (4) Lower portion of ice wedge, laminae nearly parallel. (5) Thin ice laminae, parallel to plane of drawing. (6) Thicker ice rind veins. (7) Particularly fragmented portion of bedrock in the ice rind layer. (8) Deeper, less disturbed beds of marly shale in the permafrost layers.

soil, causing the soil to turn up. The laminae near the top often meet each other and the wedge border at acute angles. Lower down the number of laminae drops (as, with increasing depth, fewer frosts penetrate so far), and run parallel to each other and to the wedge border. Here the surrounding rock can better resist the disruptive effect of the growing ice wedge. In horizontally bedded rocks, the pressure of the rock changes with each bedding plane, so that the ice wedge does not taper gradually, but in spurts. We may observe this in two ice wedges, each two m long, one an active wedge from Spitzbergen (Fig. 25), and one a cast from the Muschelkalk near Würzburg, filled with red loam and sand (Photo. 4). The very tip of a wedge often consists of only

4. Ice-wedge cast in the Muschelkalk of the Riss Terrace of the Main River near Karlstadt, Franconia. Filling: Eemian sand and loam. (Photo: Büdel, June 1955.)

two or three laminae, showing how few deep frosts were able to penetrate this far during the 12,000 years of the Holocene.

As our second example shows, ice wedge casts (loess or loam wedges) are easily recognized by their characteristic shape. They are important evidence of former permafrost conditions: in stratified loess deposits of the mid-latitudes one may often find several vertically distributed generations of ice wedges from the various cold stages.

The existence of a former ice rind in the former periglacial region of the mid-latitudes has been overlooked till now, for it disappeared at the end of the Würm cold stage, allowing the overlying active layer to collapse by the corresponding amount. Yet one widespread feature clearly shows that the ice rind was once widely distributed, namely, a striking unconformity between the remains of the former patterned ground, which are often well preserved, and the nearly undisturbed rock beneath (see Photo. 5)

On gentle slopes, solifluction in the area of downslope tipping and pseudostratification (see Fig. 67, layers 5 and 6) has overthrust soil sheets along shear planes for up to 10 m. These sheets sometimes even seem to have been moved upslope or over bedrock knobs, indicating a particularly thick ice rind which vanished during the late glacial stages of permafrost melting, causing slumping of the overthrust sheets of earth.

Relict pingos are further evidence of a former permafrost regime, though so far their origin has not been clearly explained. According to F. Müller (1959), Wiegand (1965), Washburn (1973), and others, they form in incipient permafrost, when frost-resistant, waterlogged dome or column-shaped bodies (the pre-existing form seems a rather arbitrary assumption) are surrounded by a frost table encroaching gradually from above. Such permafrost-surrounded, unfrozen waterlogged bodies are called *talik* in Russian. Subjected to pressure from all sides (eventually even from below) they theoretically become round, with diameters of a few to several hundred m at their base. During this process the overlying layers may be heaved or torn up, and material thrown up explosively from depth. The frozen superstrata build up into an encircling rim wall (often with radial dilation cracks) around the rapidly freezing pingo core, while the central crater-like depression over the ice core may soon fill in with sediment. On Spitzbergen we found fresh pingos only in young fluvial deposits; closer study of these was undertaken largely by Wirthmann (1962). Material brought up from depth by pingos served to prove the existence of marine clay pockets underneath the river gravels. Melting of a covered-up pingo ice core at the end of a cold stage creates a circular, rimmed lake, which may quickly silt up.

2.2.13 DENUDATION PROCESSES ON A SIMPLE CONVEX-CONCAVE SLOPE: SOLIFLUCTION, SLOPE WASH, RILL WASH, SLOPE DISSECTION, AND AUGMENTED SOLIFLUCTION

The relatively intact relict planation surfaces of the island heights are separated from the forelands, whose glacially worked rock surfaces were the scene of Holocene marine transgressions and terrace building, by the slopes and valleys which are the main setting for Holocene erosion and relief-cutting. To measure the extent to which these processes have taken place, an exact relief analysis is necessary. We therefore begin by discussing the slopes, starting with the simple convex-concave slope (see the profile in Fig. 26).

The ice rind, which completely covers the permafrost layer, prepares the bedrock for all forms of erosion. Areal denudation processes take place in the active layer, working with an energy that far outstrips all equivalent processes on slopes of the ectropics and of the dry and semi-dry tropical mountain ranges.

Descending from the relict planation surfaces to the convex upper slopes via gradients of under 2° to circa 25°, we find the patterning of the active layer with all its sixteen component processes maintained. Interstitial and filter drainage, of course, increase rapidly, and gravity causes downslope displacement of the active layer as a whole. This is called *solifluction* (gel-isolifluction or cryosolifluction) or *soil creep*. This is accompanied by surface wash, a phenomenon which has been inadequately studied, as it occurs only

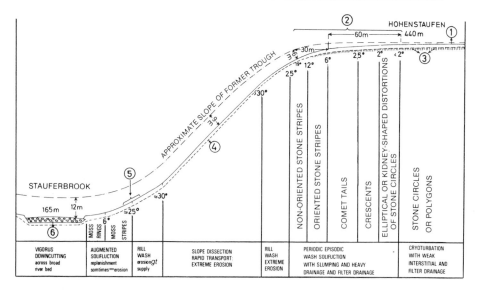

Fig. 26. Erosional processes on a convex-concave slope in SE Spitzbergen. (1) Traditional planation of relict surface. Sorted stone circles (annular frost patterns) still undisturbed at 0-1.8° slope. (2) Increasing erosion on the upper convex slope. (3) Complete elimination of the remnant active layer from the post-glacial climatic optimum at slightly over 12° slope. (4) Lower boundary of active layer (permafrost table). (5) Upper limit of moss tundra at slightly over 200 m. (6) Braided gravel bed of today's river.

once or twice a decade in the rare arctic summer rains.[15] The total movement involved, however, equals that occasioned by slow and steady solifluction. The two processes are closely linked, for the fine material preferentially carried downhill by slope wash is reactivated by the constant turnover of solifluction. These rare but highly effective periods of slope wash are an example of minor catastrophes whose effect adds up over time. We were able to measure the effects of one slope wash event, when material was washed from a snow-free upper slope onto a snow bank which still remained below. After a single day of rain in July 1967 we measured a fan of slope wash debris 30 m wide, over 50 m long, and up to 22 cm thick. The contents consisted of coarse sand and shale platelets up to 1.2 cm wide. Similar masses of slope debris were washed down from the other snow-free slopes around. When such minor catastrophes are separated by intervals of several years, most of the down-washed masses are reincorporated into the patterned solifluction sheet. Where relict sheets of slope debris have been found between loess or

[15] Only in the summer of 1967 was the Stauferland Expedition rained upon. The rains lasted three days and occasioned massive surface wash and soil slippage. Thiedig and Lehmann (1973) experienced the like in W Spitzbergen. In the summers of 1959 and 1960 our only precipitation was in the form of snow.

solifluction deposits in Central Europe, many observers have tried to interpret them as traces of an interstadial or even of an interglacial.

On all slopes of less than 25° solifluction creates *sorted stone stripes* which follow the slope exactly, giving them a hachured appearance. Where the transition from the relict surface to the upper slope is gentle, sorted stone rings may grade into sorted stone stripes in various ways. The simplest, but by no means most common way is for the circles to get drawn out into parallel stripes. This has been described with reference to needle ice solifluction by Troll (1944) in the High Andes of Bolivia, and we found it on the Hohenstaufen in basalt (see Fig. 27A and B, where the appropriate gradients are shown). On rock with platy disintegration, where interstitial and filter drainage are stronger, the transition occurs via a richer sequence of stages. On slopes of up to 1.5-1.8° the stone circles remain unchanged. At 2° kidney-shaped distortions occur, which grade into crescents by about 2.5°. In both forms the convex side is uphill (see Fig. 27C and Photo. 6). The horns of the crescents then distend into comet tails, which straggle out to increasing lengths. This type dominates until about 6° (see also the overview in Fig. 26). The fact that the convex side of the crescent is uphill indicates increased interstitial and filter drainage. The core of the crescent is formed of fine material, which is now better fed by increased filtering. In the early winter freezing the water-logged material heaves up, so that the surrounding embankments of fine pebbles and coarse rubble are particularly marked. On slopes of 2.5° we found that from the summer of 1959 to the summer of 1960 (neither of which had any rain) an average displacement in the silt stripes of 1.1 cm took place, in the fine pebbles of 0.9 cm, and in the coarse rubble of 0.7 cm. In rainy years

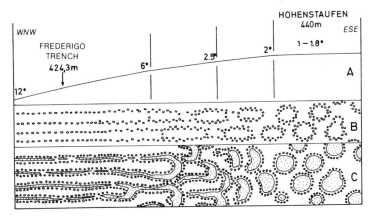

Fig. 27. Transition from cryoturbation to solifluction patterns on gently convex upper slopes. (A) Slope gradients. (B) Distortion of sorted stone circles via ellipses into sorted stripes with weak interstitial and filter drainage. (C) Stronger interstitial and filter drainage, creation of intermediate shapes: kidney forms, crescents, and comet tails.

6. Solifluction sheet on upper slope of Hohenstaufen, inclination 3°–5°. Underlying rock: thin-bedded arkose. Patterns: crescents and comet tails. (Photo: Büdel, July 31, 1959.)

the movement must certainly be greater. The silt stripes, which collect more moisture, move faster than the dry rubble beds, creating the surface pattern of the comet tails. We measured the displacement with steel nails 22 cm long. Not only did their numbered heads move downhill, but the nails also became tilted, showing that movement is faster at the surface of the solifluction sheet than at depth.

By a 6° gradient, the comet tails become very long, and by 11° the oriented stone stripes are fully developed (see Photo. 7, and Figs. 28 and 29). Excavation showed that the silt stripes are actually little silt walls, running through the coarse rubble at intervals of 0.5 to 1 m. Like the sorted circles, they are accompanied by fine pebble mantles. Below the walls, the permafrost table sinks by 1-3 cm, directing the filter drainage toward the walls. This feeds the silt walls from below, causing them to be broader at their base, while upward freezing cleanses them of rubble. In the stated year the silt walls crept an average of two cm downhill, while their pebble mantles crept 1.7 cm, and the rubble beds 1.5 cm. Until 12° the solifluction debris, whose

5. Profile of remnant patterned ground sheet (at Kleinrinderfeld, S of Würzburg, Franconia). Above: Holocene gray-brown podzolic soil on thin sheet of loess, overlying a contorted cryoturbation layer. Beneath this, silt-rich residuum of melted ice rind (20 cm thick). Clearly marked boundary between this and the bedded Triassic Muschelkalk (Quaderkalk), containing pre-glacial karst pipes filled with red loam. (Photo: Büdel, May 1, 1958.)

7. Solifluction sheet on upper slope (11° gradient) of Hohenstaufen, on thin-bedded arkose. Structures: oriented stone stripes at intervals of 0.5-1m. Site: upper left of Fig. 20. Surface pattern for cross-sections in Figs. 28 and 29. (Photo: Büdel, August 7, 1960.)

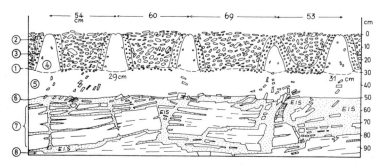

Fig. 28. Cross-section of oriented stone stripes on 11° slope (Frederigo Trench on the NW slope of Hohenstaufen, August 9, 1960; view uphill). (1) Permafrost table. (2) Coarse rubble bed (frost pavement). (3) Fine pebble mantle around fine earth stripes. (4) Fine earth stripes (silt walls). (5) Silt rich relict active layer from the post-glacial climatic optimum. (6) Boundary between the relict active layer and the ice rind. (7) Ice rind with solid ice masses up to 1/8 m in volume. (8) Zone of fragmented bedrock (here thin-bedded arkose) beneath the ice rind.

particles are aligned with the slope and tilt slightly upwards (see Fig. 29), overlies thin remnants of the relict active layer, a clear sign that at this inclination erosion during the last two-and-a-half thousand years has remained slight.

Between 12° and 14°, however, conditions change on all forms of bedrock. Instead of oriented stripes, we now have *non-oriented stone stripes*, with gravity outweighing all sorting agents. Most of the solifluction sheet here consists of a vaguely striped mixture of coarse and fine. This increasingly forces the interstitial and filter drainage to the surface, making surface wash

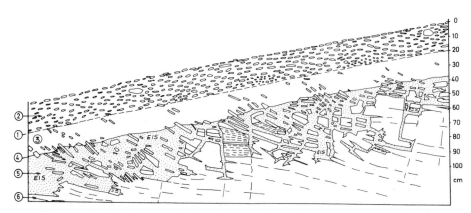

Fig. 29. Same cross-section as Fig. 28 (Frederigo Trench, view parallel to slope). (1) Permafrost table. (2) Coarse rubble bed (frost pavement), with oriented stones. (3) Relict active layer. (4) Boundary between relict active layer and ice rind. (5) Ice rind. (6) Upper zone of fragmentation (bedrock here thin-bedded arkose).

particularly effective, especially in rainy years. For this reason, all remnants of the relict active layer are absent downslope of this gradient. At 17°, solifluction is even more effective: the fine earth stripes move at 2.7 cm/yr (maximum 4.6 cm/yr), mixed stripes at 2.2 cm/yr, and the coarse rubble at 1.9 cm/yr. The steepest slope still covered by a continuous solifluction sheet (about 25°) had even higher values during the same year, the maximum here being 9.9 cm/yr.

On the steepest mid-sections of convex-concave slopes, at gradients exceeding 25°, the erosional processes change again. Slope wash reinforced from above becomes dominant, cutting rills or runnels into the increasingly blurred silt stripes. The rills run in parallel downhill between 25° and 30°, cutting through the entire active layer. Solifluction is thus replaced by the most vigorous process of areal slope erosion known: *rill wash*. As Mortensen (1930) has already emphasized, these runnels do not cut into the bedrock. The striped appearance of the slope is thus retained.

Where mid-slopes reach gradients over 30°, neighboring runnels unite to form small, steep erosional systems. The collection runnels cut deeper into the ground, so that rill wash turns into *slope dissection*. This is not annual, but takes place intermittently in minor catastrophes.It is clearly promoted by loosening of the bedrock through the ice rind.

In the simple profile (see Fig. 26), the steep mid-slope is followed by the concave footslope. Here fine material from slope dissection is temporarily deposited. It does not remain here, however, but is taken up by a new type of solifluction and moved onward. Due to the increased supply of water and the higher mobility of the fine material, the *mobile debris sheet* moves faster in this zone of *augmented solifluction (Zufuhrsolifluktion)* than at the same gradient on the upper slope. Moreover, footslopes are often located in the moss tundra zone, and therefore experience more chemical weathering and clay mineral formation. Solifluction movements on footslopes at 4-5° gradient take place at an average rate of 1.2 cm/yr, with extremes of 2.5 cm/yr, which is more than we could measure on inclinations of 11° on the upper slope.

Frost patterning reappears on these footslopes, but is of a different kind from that on the upper slopes. It is a form of surficially textured ground, based on a network of cracks which are limited to the surface of a silt-rich sheet of augmented solifluction and which more or less migrate downhill along with it. The crevices are often filled with fine and coarse stones, echoes of the patterned ground on the relict planation surface. With the better climatic and soil conditions on the footslopes, dense *moss cushions* grow on these stone-filled cracks. On grades slightly over ca. 6° these form *moss stripes*, which, carried by solifluction, follow the slope exactly, recreating a striped pattern. At grades less than around 6° we find *moss circles* or *mud-pits* whose barren silt cores are slightly arched in winter and early summer by frost heaving (and according to Semmel, 1969, by cryostatic squeezing). The superficial and short-lived nature of these circles is shown by the fact that on slopes of up to 6° they are rarely stretched and broken into stripes, but retain their round shape despite the rapid solifluction here.

Excavations have given us a look at the developmental stages of such a frost debris sheet; we have measured its rate of solifluid movement across a slope, and we have observed the spatial distribution of the various forms. Only on this basis could we come to understand how these overlapping processes interact in time and space, an insight which could not be obtained by measurement alone.

2.2.14 Holocene Erosion on Convex-Concave Slopes

The simple slopes still show traces of the old glacial U-shaped valleys, particularly in such features as the steepness of the mid-slopes (see Fig. 26). Two fixpoints can inform us as to the amount of subsequent erosion and transformation. Firstly, little erosion has occurred on the relict surface and on the upper slopes down to about 12°, for part of the relict active layer from the post-glacial climatic optimum still remains here. The gentle convexity of the working rim is probably also inherited from the Würmian subglacial relief. Secondly, we may reasonably estimate that during the Holocene the bed of the Staufer Brook (the local base level in Fig. 26) has been lowered around 12 m below the Würmian U-shaped valley. Since the footslope with its augmented solifluction grades smoothly into the 70-80 m wide gravel bed of the Staufer Brook, it is clear that denudation on the mid-slopes and footslopes must have kept pace with this down-cutting. This seems reasonable in light of the efficiency of augmented solifluction and especially of rill wash and slope dissection on the mid-slopes.

2.2.15 The Tripartite Slope:
Erosional Processes and Their Efficiency

The simple frost slope is easily transformed in the larger and deeper valleys into a tripartite slope. In the simple concave-convex slope, the former U-shaped valley still shows through, only the middle and lower slopes being subject to more Holocene erosion. Intensified erosion leads to a tripartite slope. With greater steepness, slope dissection expands greatly in area: the V-shaped gullies cut deeply and their tributary runnels reach back to the upper edge of the slope, forming a sharp working rim where they intersect the relict surface. In contrast to the convex-concave slope, then, erosion on the relict surface is completely interrupted along the edge. The periphery of the relict surface dries out, intensifying the climax nature of its cryoturbation sheet as described by Wirthmann (1964, 1973).

Below this working rim is the deeply runneled upper slope (Figs. 30, 31,

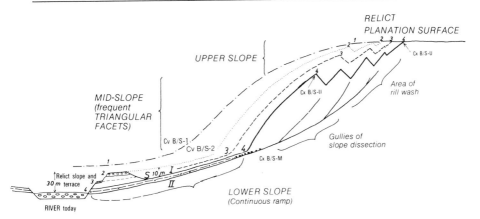

Fig. 30. Profile of a tripartite frost slope in SE Spitsbergen. Greatest erosion by rill wash gullies and slope dissection on the upper slopes; weakest erosion on the footslopes with their augmented solifluction (I) across the slightly abraded bedrock with ice rind (II). Cv B/S, 1-4: stages through time of the concave break in slope (piedmont angle) at the foot of the mid-slope. Cx B/S-M, 1-4: stages through time of the convex break in slope (working rim) at the upper edge of the triangular facets. Cx B/S-U, 1-4: stages in the convex break in slope (working rim) at the upper edge of the upper slope. S: Zone of subsequent drainage at the upper edge of the relict slopes and terraces.

33). Its runnels unite on the mid-slopes into deep, widely spaced, V-shaped gullies, between which the remaining segments of the mid-slope still suggest the former shape of the glacial trough. Due to their appearance in bird's-eye view (see Fig. 31), we named these ungullied mid-slope segments *triangular slope facets (Dreieckshängen)*. Their apices are not supplied with water or fine material from the heavily dissected upper slopes. Gossmann (1970) reck-

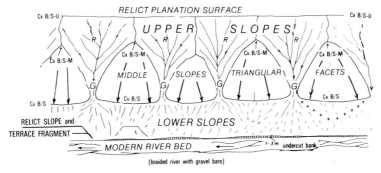

Fig. 31. Tripartite frost slope in SE Spitzbergen, bird's-eye view. R: rill wash on higher parts of the upper slopes. G: gullies of slope dissection lower on the upper slopes, between the midslope triangular facets. Cv B/S, Cx B/S-M, Cx B/S-U, see Fig. 30. Line of crosses on lower slope right: line of theoretical Cv B/S if it were caused solely by lateral erosion of the alluvial fans issuing from gullies G. Independent erosion, however, does take place on the footslopes (see Figs. 30 and 32).

oned that this is one reason why their broad bases have no gently concave transition, but intersect the footslopes at a sharp piedmont angle. The frost footslopes then run asymptotically into the broad gravel bed of the valley floor. The steeply dissected upper slope and the deep erosional gullies on the mid-slopes may reach inclinations of 40°, while the triangular segments have gradients of 19-26°. The broadly sweeping lower slopes fall from 11-13° (even as much as 15°) in their upper sections down to 2-3° at the edge of the riverbed.

Mass wasting on the upper and mid-slopes is easily explained. Far harder to explain is the erosion on the gentle lower slopes, and particularly the straight, continuous, and marked piedmont angle between their upper edges and the triangular facets. The deeper gullies (G in Fig. 31) which separate the triangular facets also end at this same piedmont angle. The extensive material washed off the upper slopes is carried across the mid-slopes through these gullies alone. It then spreads across the footslopes in alluvial fans so shallow as to be hardly distinguishable from the slopes on which they lie. This is the key to the problem of the lower slopes.

In fact the alluvial fans are an important factor in creating the lower slopes, but they do not form them simply by lateral erosion in the sense of Wissmann (1951). If this were so, then the line of the piedmont angles would zig-zag, as shown by the crosses in Fig. 31 (right), instead of being straight. That this edge is usually straight is because some material is areally removed (through slope wash and solifluidal movements) from the surface of the steep but smooth, undissected triangular facets. This material is blocked by the large alluvial fans issuing from the V-shaped gullies and merges with the fans, causing the feet of the triangular facets to recede along a straight line.[16]

The material brought down from above does not simply accumulate on the footslopes, but is immediately subjected to augmented solifluction, which spreads the material thinly across the entire lower slope. Aided by the ice rind effect, solifluction areally wears down the bedrock of these footslopes, and is therefore a major factor in forming the wide and continuous footramps found in so many valleys. This form of erosion is by no means fully autonomous, for it is also regulated by vertical erosion of the valley floor below the footslope. From the piedmont angle on down, the footslopes are governed entirely by the river level, a fact which shows how powerful fluvial erosion is here.

To demonstrate this, we dug a trench in a particularly broad and gentle footslope (525 m wide) on the S flank of the Isar Valley (Fig. 32) at its lowest margin just above the riverbed. The gradient here ranged from 11° to 3-5° right above the undercut river bank, here barely two m high. The active layer

[16] Another factor in maintaining this active piedmont angle may be Gossmann's footslope effect (1970), whose importance I have at times overestimated (Büdel, 1970a). Where the triangular facets are somewhat dissected, a bipartite slope is formed (Wirthmann, 1964).

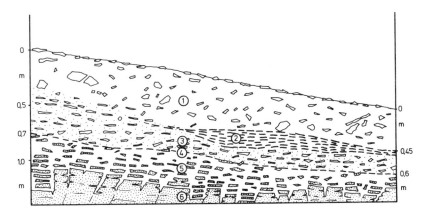

Fig. 32. Augmented solifluction on frost footslope. (Isar Valley in New Bayer Land, Edge Island, SE Spitzbergen; see Fig. 16. Excavation August 30, 1967.) (1) Upper layer of augmented solifluction. Fine matter with allochthonous clasts, movement partly turbulent, partly along shear planes. (2) Semi-autochthonous frost-loosened debris, moved in frozen condition along shear planes together with (3) lower layer of augmented solifluction. Purely autochthonous material, clearly oriented along shear planes. (4) Permafrost table. (5) Ice rind. (6) Upper fragmentation zone of bedrock (hard thin-bedded marly limestone).

was 73-80 cm deep, the thickest layer found in SE Spitzbergen. At only 52 m above sea level, a relict active layer was of course absent, the modern active layer lying directly above a well-developed ice rind. This contained thick ice lamellae and ice-filled vertical joints, tearing the bedrock of marly limestone into thin platy fragments (see Fig. 32)

The augmented solifluction sheet had two clear layers. The lower part consisted exclusively of fragments of the underlying blue-gray to ocher-yellow marly limestone. These fragments were embedded in fine material, but moved along well-defined shear planes, exactly as seen at the base of many relict solifluction sheets in Central Europe (where it was formerly called pseudo-stratification).

Clearly differentiated from this was an upper layer of finer material. Some of its coarser clasts derived from the underlying material, but worn fragments from the higher slopes were numerous. Any fragments over two cm long lay flat with their long axes pointing parallel to the slope, exactly as in the corresponding levels of the relict solifluction sheets in Central Europe. This orientation shows that periodic movement along shear planes still occurs here. The smaller fragments were not oriented, showing the presence of periodic turbulent movement. The planar movement presumably occurs in early summer, as the permafrost table gradually sinks; the turbulent movement occurs in midsummer.

During the main melting season, these gentle slopes are deluged with fine material, through which the coarser clasts then rise by upward freezing. This movement keeps the plant cover discontinuous despite the low altitude: moss stripes and moss rings are absent, and other structures are barely visible.

The Isar Valley which we have been discussing is typical of the larger valleys in SE Spitzbergen. Viewed as a whole, it is clear that the piedmont angles and triangular facets have not receded 525 m during the Holocene. The cross-profile instead reflects the shape of the former shallow glacial trough whose simple concave-convex profile is still preserved in the smaller valleys. Two major areas of erosion have developed in the larger valleys during the Holocene. One, on the heavily dissected and steepened upper slopes, created the sharp working rim along the relict surface. This area also includes the large V-shaped gullies of the mid-slopes. The second area is the broad erosional area of solifluction on the lower slopes, which is in phase with fluvial down-cutting, and which, as described above, also sharpens the piedmont angle.[17] The triangular facets remain untouched between these two areas of vigorous erosion, and retain the resemblance to the old U-shaped valley wall. Though very steep, they are but little worn, because they are quickly cut off from supply (especially of water) from the upper slopes. The effectiveness of erosion on the footslopes, on the other hand, is shown by the fact that their margins often bear remnants of higher, older footslopes. These remnants, whose lower edges may be lined with river boulders, carry relict solifluction sheets which formerly moved toward the river (see Figs. 30, 31, 33, and Photos. 8 and 9).

Given the former U-shaped valley profile, these tripartite slopes must be considered a type of traditional remodeling of the pre-existing relief. The processes acting here today could never have created such decisive forms by themselves.

2.2.16 FLUVIAL LATERAL EROSION

Erosion on all parts of both simple and tripartite slopes, but most especially on footslopes, annually supplies the rivers of the frost debris zone with more material than is received by the rivers of any other climatic region on earth. This includes not only fine matter, to be removed in suspension, but also much coarse, highly erosive bed-load material.

But this supply of erosive material only becomes morphologically effective

[17] The very vigorous erosion here is a type of slope development closely linked to the processes on the higher slopes and to the excessive valley-cutting on the footslopes; it is completely integrated into the system of slope development described above. It therefore seems illogical to me to describe the footslope, whose development is governed exclusively by valley-cutting, and which may be up to 15° steep at its upper edge, as a product of so-called frost planation phenomena (cryoplanation, pediplanation, etc.), and to lump this in one category with the creation of huge etchplains in the seasonal tropics. In relief-forming mechanisms, in the shape and expanse of the product, these two phenomena are completely unrelated and cannot be compared.

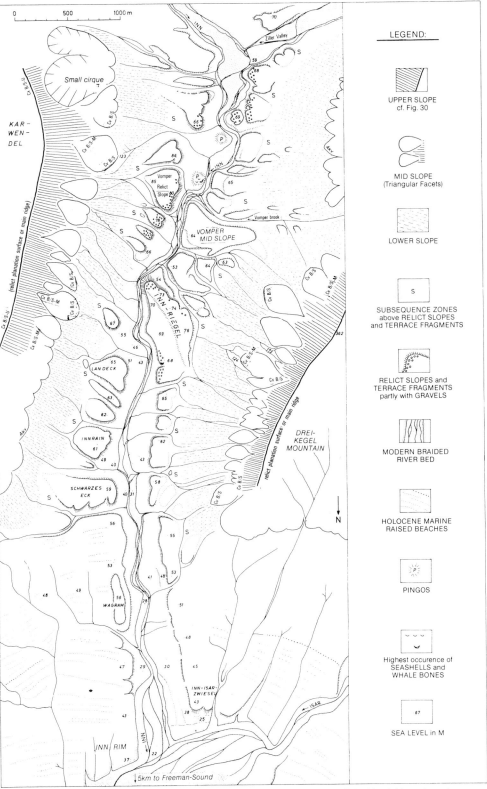

LEGEND:

UPPER SLOPE cf. Fig. 30

MID SLOPE (Triangular Facets)

LOWER SLOPE

S — SUBSEQUENCE ZONES above RELICT SLOPES and TERRACE FRAGMENTS

RELICT SLOPES and TERRACE FRAGMENTS partly with GRAVELS

MODERN BRAIDED RIVER BED

HOLOCENE MARINE RAISED BEACHES

PINGOS

Highest occurence of **SEASHELLS and WHALE BONES**

SEA LEVEL in M

Fig. 33. Valley with tripartite frost slopes. (Inn Valley in New Bayer Land, Edge Island. Map oriented S showing area between 78° 6′ and 78° 9′ lat. N.) After photo by Büdel and Späth, 1960 and 1967. For explanation see also Figs. 30 and 31.

8. Early Holocene marine level (10,000 B.P.) is marked by the prominent "Barrier" or "Riegel Series" (see Section 2.3.3), which continues far inland as river terraces along the Inn. Ice wedge polygons on the terrace surfaces. New Bayer Land, Edge Island, view looking S (see Fig. 33). (Photo: Büdel, August 9, 1967.)

9. The highest terraces of the Riegel Series in the Inn Valley. View E over the "Vomper relict slope" to the Karwendel Mountains (see Fig. 33). Below this terrace the Inn has cut down 30 m over a wide bed during the last 10,000 years. Above, tripartite frost slope. New Bayer Land, Edge Island. (Photo: Büdel, August 9, 1967.)

when the early summer meltwater permits rapid fluvial transport of the coarse matter. With plant cover nearly absent above 200 m, rain and meltwater are directed along the sorted strips and runnels along the shortest way to the riverbed. Direct evaporation at mean summer temperatures of barely over 0°C is slight, plant transpiration is minimal, and percolation into the frozen ground completely absent. In warm summers (when the active layer may work 15 cm further down than in cool summers), additional meltwater may even be set free here, contributing to an extremely high runoff. Furthermore, most of the snow blanket which collected between December and June melts during the first early summer warm spell, with results resembling the minor catastrophes of the sporadic midsummer rains: the entire valley floor fills to the brim with a gray-brown turbulent flood of shallow rushing water. Even when such a flood was abating, we found that large gravel bars were displaced as much as six m in 24 hours. At such times the river braids continually, forming straight banks without meander loops. Anastomosing channels define fan-like gravel bars, which in some places take up nearly the entire riverbed (compare Büdel, 1944). These bars shift rapidly, and wander with the line of maximum velocity from one side of the riverbed to the other. Riverbed and valley floor become one and the same, but the river is nowhere deep, and a pool one m deep (found below a gravel bank) is a rarity. Instead of the cross-bedding typical of meandering streams, this produces parallel bedding, such as is found in the cold stage valley floors and terrace gravels of Central Europe (see Photos. 10 and 11).

10. The Tauber Valley above Mergentheim in N Württemberg, Germany. Above: Gäu surface (etchplain, last active at the end of the Middle Pliocene). Below: broad, gently arched Würmian gravel floor. The present river is only active in a very narrow band at the left edge of the picture. (Photo: Büdel, June 1963.)

11. Advent Valley near Longyearbyen (W Spitzbergen). Floodwater in early summer snow melt. (Photo: Büdel, June 1960.)

12. Staufer Brook (Barents Island, view looking up valley). Water level in late summer: the few stream channels on the broad gravel bed are fed only by the last remaining snow patches. During the spring melting season the bed is completely filled; in winter it lies dry. (Photo: Büdel, August 23, 1959.)

The braiding caused by the excess of gravel leads to heavy lateral erosion. Streamlets only six km long may have gravel beds 70-80 m wide, while rivers only 12 km long may have beds 100 to 200 m wide, just like the relict Würmian floodplains (*Wiesentalsohlen*) of the Central European rivers. In slip-off areas the footslopes meet the riverbed at banks only a few dm high, while in undercut areas these banks may become 2-5 m high. In the course of a few years, as the braid loops work downstream and turn the former undercut edge into a slip-off slope, the banks are rapidly worn down by solifluction and surface wash. This continual leveling (which we were able to study quite well from 1959 and 1960 to 1967) of the laterally eroded banks again proves that the gentle footslopes are capable of cutting down at the same rate as the river.

The question yet remains, however, as to what mechanism enables the river to cut down so rapidly over so wide a bed.

2.2.17 THE ICE RIND EFFECT:
THE DRIVING FORCE OF VERTICAL EROSION

That these rivers also erode *downward* very rapidly is an ineluctable conclusion, given the fact that they have all kept pace with and cut down parallel to the post-glacial uplift.[18] Down-cutting did not occur linearly in narrow gorges, but across wide gravel beds, while the longitudinal profiles of the riverbeds run without steps deep into the island block. Only an exceptionally effective relief-forming mechanism could have made this possible.

The answer lies in the ice rind. We have already seen how this uppermost zone of permafrost underlies the entire land. On gentle seacoasts it even extends out under the sea floor for the distance to which the winter sea ice, freezing to the sea bottom, temporarily lends the ocean certain characteristics of solid land.

The permafrost layer accordingly even underlies the gravel beds of the braided streams, due to the unusual regimen of all non-glacial rivers. These are periodic, like rivers of the semiarid regions, but for quite different reasons and with far better conditions for erosion. After the floods of the main melting season, the rivers dwindle rapidly to a few channels winding through a dry gravel bed. From mid-July to the end of August these narrow channels are fed only by the few surviving snow patches and by the thin active layer (see Photo. 12). Any supply from ground water or from springs fed by ground water is absent, for here the ground is frozen. The first frosts of fall (which

---

[18] The only exception is Staufer Brook, which crosses a particularly hard, thick, and sparsely jointed basalt layer just above its mouth. This was formerly a subglacial discordant junction, and even today the stream overcomes the basalt only by means of a waterfall 17 m high, the only waterfall in all of SE Spitzbergen. Upstream of the fall, the brook has managed in Holocene times to cut at least 12 m into the former glacial trough (see Fig. 26).

we experienced between August 23 and September 5) cut off this surficial supply of water, and the streams again shrink suddenly to thread-like rivulets. Throughout the winter the streambeds lie dry, and since the snow blanket remains thin for a long time, frosts penetrate the riverbed unhindered. Permafrost and ice rind consequently underlie these gravel beds, just as they underlie all other parts of the relief.

At midsummer low water we made eight profiles with over 80 one-m cores across the gravel bed, which at this time contained only a few shallow water channels. Below an active layer 28-66 cm thick, the cores consistently revealed a typical ice rind with pure ice between rock fragments. Figure 34 depicts two series of cores along the same profile, showing that between July 22 and August 11, 1967, the permafrost table under the Staufer Brook sank by several dm. On September 6, after the first frosts, when the stream had become quite small, we dug a trench across a gravel bank in the middle of the riverbed (see Photo. 12, center). We were naturally hampered by inrushing water, which we could only master with our strongest pump. Even this became inadequate at 107 cm, but by this time we had already penetrated 38 cm into the ice rind (which here sets in at 69 cm depth). At this level 40% of the ice rind was pure ice, which presumably increased with depth.

This is extremely important. Since the ice rind underlies all rivers, it breaks up the bedrock even here. Clearly, then, *mechanical down-cutting of the rivers is greatly facilitated, and is governed by a process totally different from that in any other climatic zone.* The crucial point is that down-cutting is facilitated equally across the entire gravel bed. Elsewhere rivers can cut down mechanically only where they carry much coarse gravel. Such a bed

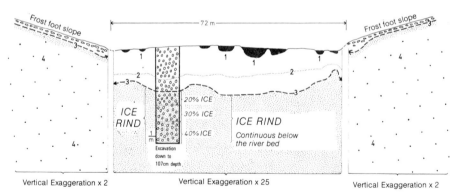

Vertical Exaggeration x 2   Vertical Exaggeration x 25   Vertical Exaggeration x 2

Fig. 34. Proof that the ice rind is continuous beneath the riverbed. Staufer Brook excavation, Sept. 5-6, 1967. (1) Remaining channels of Staufer Brook on gravel bed after the first frost. (2) Permafrost table on July 22, 1967. (3) Permafrost table on August 11, 1967 (by Sept. 5, 1967 it had sunk only a few cm more). FS: Footslope. Heavily dotted area: ice rind. (4) Lower permafrost layer.

load, however, can only gradually scour away solid rock by tenths and hundredths of mm, and not year round over the whole riverbed, but only during exceptional floods in pools reaching down through the gravel bed. The main work of any kind of vertical erosion consists of dissolving or breaking loose transportable fragments from the bedrock. *In periglacial rivers, this work has already been performed by the ice rind.* The river need perform no mechanical work, but only melt the underlying ice rind and incorporate the prepared rock fragments into its gravel bed without further ado. Since most or all of the gravel bed is worked through annually by the melting season floods and even more effectively by episodic minor catastrophes once or twice a decade, this uptake of material through melting of the ice rind is not limited to individual pools. Fluvial down-cutting across the entire riverbed, therefore, is made possible by the water temperature, which according to our measurements may reach $+5.8°C$ in midsummer, but whose maxima are presumably still higher.[19]

Summing up, the ice rind effect, which uniquely promotes both vertical and lateral erosion, is the driving force causing the periglacial frost debris zone to be morphodynamically the zone of *excessive valley-cutting* (Büdel, 1969b).

2.2.18 The Amount of Holocene Vertical Erosion, Measured by Exact Relief Analysis

The last link in the relief analysis presented here is the evidence for how rapidly this Holocene vertical erosion occurred. This can be demonstrated with the help of isostatic rebound. The phases of isostatic uplift coinciding with eustatic rises in sea level are marked by beach ridges, which can be radiocarbon dated by their fossil contents (here whalebone and driftwood). During the early Holocene, a prolonged period when the two movements balanced each other created the raised beach of the Riegel Series, dated at 10,000 years B.P. This continues inland along all the larger rivers in the form

---

[19] The continuous ice rind beneath the riverbeds and its driving role in vertical erosion have since been confirmed by Semmel (1975), using the same methods inland of Ice Fjord in W Spitzbergen. The terrain here does not consist of ice-free tableland, but of high mountains, heavily glaciated even today. The major rivers are all glacially fed, so that their beds are not as dry in winter as those of the periglacial rivers in SE Spitzbergen. In the Würm Ice Age, thick, steep, and highly erosive valley glaciers radically reshaped the valley beds (which were mostly of hard, crystalline rock), and in many cases scooped them out to below present sea level. SE Spitzbergen, on the other hand, consists of a soft, relatively low Mesozoic plateau. This was covered by a very slow-moving ice cap which left little glacial imprint. Here we were able to show both quantitatively and comparably that the ice rind-controlled down-cutting of the brief Holocene epoch achieved the same effect as was brought about in the periglacial mid-latitudes by the combined Early Pleistocene Ice Ages, which lasted thirty times as long. This was yet another ground for choosing SE Spitzbergen as a model of current periglacial excessive valley-cutting.

of marked river terraces and forms a single unit with the fragmentary relict footslopes. In Freeman Sound, the center of our work area, the Riegel Series averages 80 m high. On the Inn River (the largest river flowing into the Sound from Edge Island) the terraces can be traced up to 12 km inland (see Fig. 33 and Photos. 8 and 9), where they become part of a relict footslope 30 m high lined with scattered fluvial boulders along its lower margin and dropping steeply to a 90 m wide gravel riverbed. During an intermediate stage this riverbed was even as much as 800 m wide. This means that in 10,000 years, 30 m of down-cutting have taken place over a broad bed, or an average of 3 m per thousand years.[20] Other parts of our work area yielded values of 1.5 to 2 m per 100 years.

This agrees fairly well, as already mentioned (Section 2.2.1), with the value obtained for the relict periglacial valleys of Central Europe by measuring the distance between the Riss and Würm terraces. This difference averages 15-30 m throughout. Since other clues indicate that erosion took place over an extremely wide riverbed (the Würm valley floor of the Kinzig River, below Gelnhausen E of Frankfurt, is over two km broad) during the approximately 30,000 to 35,000 years of the wet and cold Early Würm Glacial, we derive here down-cutting rates of a comparable magnitude, namely, just below 0.5 to almost 1 m per thousand years.

In SE Spitzbergen, our small but ideal demonstration ground for current periglacial processes, Holocene down-cutting took place along the bottom of previously formed Late Würm glacial troughs. The coastal lowlands onto which these opened were still submerged just after the ice melting, and pockets of Early Holocene marine clays may even be found in deeply scooped sections of the valleys. The question now arises as to what role Holocene uplift played for simultaneous fluvial down-cutting.

Clearly uplift was an important cause of down-cutting (nowhere can any form of denudation occur without previous uplift above sea level). But how erosion occurred and what effect it had on the relief are stamped by climate-specific features. In other climates, narrow ravines would have extended inland from the coast; growing by headward erosion, they would have been blocked by every harder bed of rock, and would have formed steep valleys leading in steps up to the relict surface.[21] In actuality however, we have an entirely different picture, one that could be produced only by the ice rind effect. Five features distinguish this form of valley-cutting from all others:

---

[20] It is meaningless to calculate this value in terms of single years, as down-cutting does not occur in regular yearly increments, but rather during minor catastrophes, separated by lengthy intervals. This example shows once more how very misleading it can be to use even the most accurate data on the events of one year to extrapolate values for thousands of years, such as for the entire Holocene or Quaternary.

[21] Headward erosion, formerly much overestimated in geomorphological and geological literature, is quite limited in effectiveness, as Bremer (1967a) has demonstrated.

[1] First is the *vertical amount of down-cutting*, which in extreme cases is even greater than three m per thousand years, for the highest marine stand in the Freeman Sound area is marked by the Grenzplatte (see Fig. 18), which lies at 110 m, 30 m higher than the Riegel Series. This rapid erosion was not in steep and narrow ravines, but across gravel beds which, even in the short non-glacial rivers of SE Spitzbergen (the longest of which is exactly 25 km) can be several hundred m wide. Of all the morphoclimatic zones of the earth, this is a maximum value.

[2] These broad valley floors *climb evenly, without steps*, and at moderate gradients into the upper reaches of the rivers, so that valleys maintain their width far into the center of the island block.

[3] The valleys *cut across the most varied of rock types smoothly* and without steps. They run over basalts overlain by pockets of marine clay in their lower reaches, over the whole sequence of Mesozoic sediments, and over basalts again in their upper reaches. Naturally the valleys narrow where they cross more resistant rocks, and broaden at less resistant ones. In longitudinal profile, however, narrower places show only very slight steepening. Nowhere are steps formed.

[4] Rivers were able to *erode laterally and vertically at the same pace across their entire bed*. Yet this did not produce high banks, for vigorous erosion on the adjoining footslopes wore these down with the gravel bed, so that the one runs straight into the other. Both lateral and vertical erosion here can only be explained by the ice rind effect.

[5] That *uplift stimulates down-cutting* is important for small ravines along cliff coasts.[22] But where larger rivers flow over wide flat forelands into the sea, this impetus is *far less*. The marine regression exposed surfaces hardly steeper than the river valleys themselves, so that erosion was barely accelerated, even in terms of vertical effectiveness, let alone in terms of breadth and lack of steps.

The effect of all these influences has caused us to classify the active periglacial area (pedologically known as the frost debris zone) morphodynamically as the *zone of excessive valley-cutting* (see also Büdel, 1948b, 1969b).

2.2.19 RELICT EXCESSIVE VALLEY-CUTTING
IN THE MID-LATITUDES

These observations apply far beyond the narrow periglacial belt of today, for during the Pleistocene cold stages, a basically analogous climate covered

---

[22] Wirthmann (1964, 1968) has described the Holocene development of ravines along coastal cliffs. Such ravines are actually enlarged collection rills of slope dissection which has cut back more vigorously.

the entire mid-latitudes (outside the glaciers) down to 45° latitude, and covered a far greater area than did the ice sheets. In Central Europe the Würm zone of tundra and frost debris extended as far S as the French Plateau Centrale, the Alps, and the Carpathians (see Fig. 13). In the Mediterranean countries it occurred above forest line, which rose from a few hundred m at the S edge of the Alps to around 2000 m in the Anti-Atlas.

Periglacial traces in Europe have been fairly thoroughly investigated since around 1935. Remnant ice wedge polygons, cryoturbation and solifluction sheets, blockmeere, simple and composite loess sheets, dells with their characteristic asymmetry, and above all, the non-glacial river terraces in the Mittelgebirge, all crowd the pages of the relevent journals in yearly growing quantities. In all this mass of material, four points are most important.

Firstly, the valleys of the Central European Mittelgebirge, whose broad, usually steep-walled, box-shaped forms are sunk 100-300 m into the Tertiary relict planation surface, are all products of the Pleistocene and particularly of the Early Pleistocene cold stages. This represents an extraordinary amount of erosion in a geologically extremely short-time period. Where tectonic uplift coincided with downwarping of the base level of erosion, as in the Upper Rhine Graben, rivers such as the Lower Neckar and Middle Main even cut down twice; during their second down-cutting cycle they could not always relocate the former valley course, and cut directly through bedrock.

The second point is the related fact that the broad "modern" valley floors (*Wiesentalsohlen*), leading without steps right into the heart of the European mountains, are simply non-glacial Würmian Lower Terraces. The floors are much wider than their rivers, especially in the case of smaller and medium-sized rivers such as the Upper Maas, Upper Moselle, Upper Weser, Neckar, Main, the Saxonian and Franconian Saale, the Upper Elbe, Vltava, Morava, Raab, and their tributaries (e.g., the Kinzig; see Sections 2.2.1 and 2.2.18). *These typical flat-floored or flood-plain valleys (Sohlentäler) are the most important relief feature distinguishing the unglaciated ectropic mountains from the mountains of the tropics.* They have contributed decisively to the development of transportation routes and to the settlement of the ectropic highlands. Below a thin covering of Holocene loamy and sandy alluvium, their floors consist of several m of coarse sands and gravels, often containing remains of Pleistocene fauna, especially mammoth and reindeer. During Pleistocene times, their gravels were quickly and regularly worked through. In a Würm spring, even the charming Tauber Valley of Photo. 10 resembled today's rivers in Spitzbergen (Photo. 11) in their active spring flood stage. In late summer, the rivers of Spitzbergen dwindle to a few channels in the gravel bed (see Photo. 12 and Fig. 34), only to become fully activated again during the next spring.

In a relict valley such as the Tauber (Photo. 10), the gravel beds are now fully stabilized; yet beneath their covering of loamy alluvium they still betray their cold stage origin in such features as their gentle convexity and their structure of gravel bars and partial gravel fans (see Büdel, 1944).

Our third point has been discussed before (Büdel, 1936, 1944, and Section 2.2.1). Just as the box-shaped valleys (*Kastentalsohlen*) with their sharp piedmont angles have almost completely retained their cold stage shape (only the largest rivers such as the Rhine and the Middle Weser have reworked much of their inherited floors in the Holocene), so the rest of the Central European relief in the Pleistocene periglacial areas retains about 95% of the shape stamped upon it by the last cold stage. All slopes of up to 27-30° contain remnants of the Pleistocene frost debris sheets under thin blankets of Holocene soil. These frost debris sheets lie unmoved today, but they indicate that the entire Holocene has not sufficed to remove the cold stage pedosphere, let alone the bedrock beneath. All mid-latitude centers of civilization, including those of N America, W, E, and Central Europe, N China, and Japan, developed in a relict Pleistocene relief. With its fertile loess sheets, preparatory frost-loosening of other soils, and broad, unstepped box-shaped valleys giving access to the mountains, this relief type has significantly contributed to the spread of these civilizations. One need only imagine the landscape minus its plant cover to see a relief, 95% of which still looks as though the cold stage climate had only just vanished.

This leads to the fourth point, which motivated the Stauferland Expedition. We knew quite well *what* aspects of mid-latitude relief-formation were controlled by climate, and we knew that thanks to the weakness of Holocene processes, most of the inherited landscape remained unchanged. But until recently we did not know *how* this happened, i.e., which individual mechanisms were responsible—for the respective processes are long gone, and can no longer be observed and measured. The feasibility of such observation and measurement in the highly comparable Recent periglacial region of E Spitzbergen is borne out by the above-described results of the Stauferland Expedition.

2.2.20 THE RECENT TUNDRA ZONE

We have already touched upon the Recent tundra zone in our example of the moss tundra of Spitzbergen. Much of Central Europe was once covered by broad tundras (mostly loess tundras) just S of the Pleistocene frost debris zone (see Fig. 17).

Present and past tundras, including the coldest moss tundras, summer-warm forest tundras, and loess tundras, must, on the basis of fluvial activity (though with some reservations concerning erosional processes), all be classified in the zone of excessive valley-cutting.

In today's arctic tundra on the N fringe of Eurasia and N America, solifluction and cryoturbation are *vegetation-bound* instead of free as in the frost debris zone (see Büdel, 1948b). Troll (1944) in particular has studied the numerous variations of bound soil movement and compared them with forms at the corresponding altitudinal zones in the high tropics, where daily needle-ice solifluction occurs.

The former mid-latitude tundra vegetation grew almost exclusively on thick layers of permafrost, such as are still present in continuous form under the Siberian continental tundra. In the oceanically influenced tundra on the N edge of Europe, the permafrost substrate is discontinuous both in time and space, and it is wholly lacking in the extremely maritime climate of Iceland.

Where a permafrost substrate is accompanied by an ice rind, erosion remains basically the same as in the frost debris zone, but the plant cover and its root network cover and hold the soil, inhibiting frost patterning and soil flow. The amount of chemical weathering also increases, leading at times to thin "nordic nanopodzols" (Kubiena's classification, 1953). This has been discussed by Semmel (1969), who compares the fully developed soils of the Lapland moss tundra with the far punier soils of the moss tundra in Spitzbergen. The soil-forming capabilities of the Lapland tundra show that soil movement here is weaker and more intermittent; yet even here, evidences of soil movement are widespread.

On nearly horizontal ground in Lapland, we find enlarged moss rings (mud-pits), leading in the old world tundras to the name *patchwork tundra (Fleckentundra)* (Büdel, 1948b, p. 41). The polygonal patchwork remains undisturbed to about 2-5° slope, depending on factors such as local rock, depth and grain size of the soil, plant cover, and presence of permafrost. Even temporary ground frost lingering into the growing season may affect the surface patterns.

On steeper slopes the plant cover is linearly torn. The striping does not follow the slope, however, as in the frost debris zone, but forms garlands running more or less horizontally, deviating by at most 30° from the contour lines, where the inclination or other microconditions change at an angle to the overall slope.

The horizontal stripes form *turf banks (Rasenwülste;* see Fig. 35), found also by Troll (1944) in the high Alps. These may form a number of kinds of *garlanded ground (Girlandenböden)*. Local influences, climatic fluctuations, and human interference may disturb the plant cover, leading to *turf slippage (Rasenabschälung;* Troll 1973b). To what extent the "Goletz Terraces" (described at high altitude above the Siberian and Mongolian taiga) constitute an extreme form of this phenomenon remains to be explained.

It is clear that bound solifluction on garlanded ground is far slower, less continuous, and weaker than free solifluction in the frost debris zone. This difference is intensified by the greatly diminished role of episodic slope wash on garlanded ground. Bound solifluction, furthermore, takes place over a far greater range of gradients, being normally present up to around 20°, sometimes

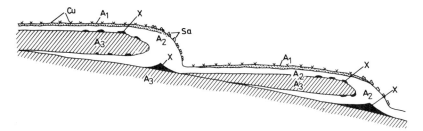

Fig. 35. Roughly horizontal solifluction terraces (garlands or turf banks) in the Alpine turf zone, after Troll (1944, Fig. 66), near the Trögeralm above the Glocknerhaus in the High Tauern, at 2300 m. Similar forms predominate in the tundra zone of Lapland. Soil horizons. (A1) Dark and humus-rich. (A2) Gray bleached sand. (A3) Stony reddish-brown weathered soil. (X) Chocolate brown raw humus-like soil. Plant cover: (Cu) *Carex curvula* turf with *Primula glutinosa* and *P. minima*. (Sa): Salicetum herbaceae with *Sibbaldia procumbens* and *Geum montanum*.

even 30°, without yielding to rill wash and slope dissection. The latter only become significant on cliff faces and talus slopes (see Fig. 36). In formerly glaciated Lapland, most of the steepest subglacial forms, e.g., cirque head walls, suffer as much slope dissection as the former ice fields of W Spitzbergen. Polished rock hummocks and roches moutonées on the flat areas are more commonly preserved than in Spitzbergen, where they occur almost exclusively on the freshly emerged forelands.

Despite reduced erosional efficiency in the tundra zone, the rivers erode vertically (where permafrost and an ice rind are present) about as much as

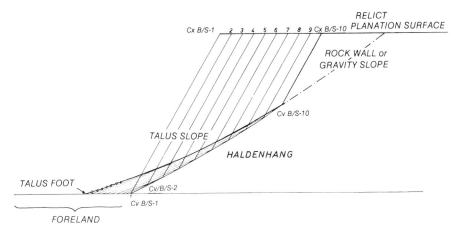

Fig. 36. Rock wall, talus slope, and haldenhang. The wall weathers actively and autonomously (without undercutting) by rockfall, forming a talus slope (coarsest fragments at the talus foot) with a bedrock haldenhang below. Cv B/S, 1-10, stages in the concave break in slope (piedmont angle) at the foot of the rock wall. Cx B/S, 1-10, stages in the convex break in slope (working rim) at the upper edge of the rock wall.

the rivers of the frost debris zone, though rivers of equal size have narrower gravel beds in the tundra. Where an ice rind effect is present, therefore, we may still assign the tundra to the zone of excessive valley-cutting. The Lapland tundra (which was glaciated during the Würm) accordingly lacks nearly all valley steps and waterfalls, while in the forested regions of the remaining Scandinavian highlands, where, as in the glaciated Alps, permafrost disappeared after glacial retreat, glacial valley steps and waterfalls are still well preserved. In Iceland (i.e., in a tundra without permafrost) the glacially formed waterfalls are equally well-marked. Iceland also has many large waterfalls created by young volcanism (see Photo. 35). The particularly striking waterfalls springing directly from the cliffs (Photo. 36) issue from networks of tubes where the last fluid streams of lava have left empty channels in the cooling basalt.[23]

2.2.21 THE TAIGA ZONE OF VALLEY-CUTTING

In highly oceanic Iceland, located in the forest tundra zone, permafrost is lacking, so that the landscape morphodynamically resembles the zone of retarded valley-cutting in the mid-latitude forest belt. In extremely continental E Siberia, on the other hand, both the permafrost and the ice rind reach far into the forest zone in thicknesses known nowhere else on earth (see Fig. 37). East of the Yenissei both of these underlie the entire taiga. In most of E Siberia, moreover, no continental ice interrupted subaerial valley-cutting. Down-cutting therefore continued here, as in Central Europe, throughout the entire Pleistocene, the major difference being that in Siberia the climate preserved and even added to the permafrost not only in the cold stages, but also during the warm stages or interglacials.

The conditions necessary for fluvial *down-cutting* were thus continually present, as the ice rind persisted here through the entire Quaternary right up to the present. Nevertheless, the individual cold stages are marked here by pronounced terraces along the large rivers. The reason for this appears to me to be that although the permafrost substrate remained the same during the cold and warm stages, the biosphere changed markedly. During the cold stages the taiga disappeared, giving way to polar tundra and to the S Siberian loess steppe. These met and interpenetrated along a broad front (Frenzel, 1968a, 1968b; Giterman et al., 1968), intensifying erosion. More material was fed to the rivers from all slopes, so that lateral erosion greatly increased during the cold stages.

Thanks to the invitation of the Russian Academy of Sciences, I was able

---

[23] The natural basalt bridge at Ofaera Waterfall may also result from partial destruction of such a tube, like the many natural bridges in karst.

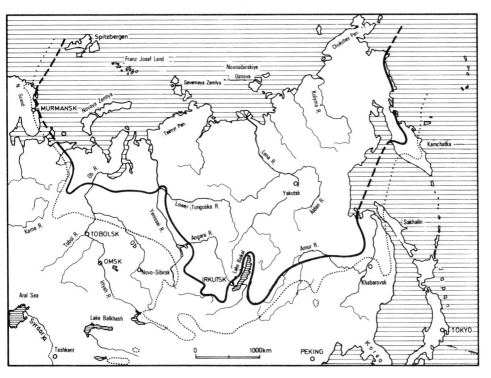

Fig. 37. Permafrost region in the Old World today. Thick line: distribution of continuous permafrost. Dotted line: distribution of discontinuous and annually varying permafrost.

in 1969 to participate in a field trip to the area of the Middle Lena and Aldan Rivers near Yakutsk. My numerous observations were collated into a composite profile of the terraces of the Aldan River (100-300 km above its junction with the Lena River, see Fig. 38). Here the Aldan is about 100 m above sea level, and averages 1.5 km in width. At the Lena-Aldan junction the Lena is about 4.5 km wide, reaching a width of 12 km in its lower reaches before splitting up at its delta.

The taiga forms a continuous forest cover 2000-2500 km wide over the entire Siberian lowland. With an area of about seven million km², it is the largest forest belt on earth, containing spruce, pine, larch, birch, willow, aspen, and alder in varying distributions, with wild currant common in the shrub story, and an extremely flowery herbaceous undergrowth in the glades. It is a silent forest: songbirds and many other small animals cannot withstand the Siberian winter. Yet there is, or was, a wealth of larger fauna, including lemmings, red squirrels, Siberian roe deer, red deer, reindeer, wolf, bear, lynx, and even some 700 Siberian tigers, which are probably closely related or identical to the large Ice-Age feline known as the cave lion.[24]

The E Siberian climate is so dry (the mean annual precipitation in Yakutsk equals 187-200 mm) that the taiga could not subsist on precipitation water alone. Were this the only water available, tundra and steppe would meet here today as in the past. The largest forest on earth owes its present existence to a combination of high summer temperatures and permafrost. A water film forms during the growing season above the impermeable permafrost layer. From early May, when the snow melts, to July, this water film drops as thawing works downwards. The active layer is 1-2 m thick, little wider than the thickest active layer in Spitzbergen. Thus both the taiga trees (most of which are shallow-rooted) and their undergrowth can be fed for most of the summer by the water film which moistens the active layer by capillary action. Thus the rule: no permafrost, no taiga.

Below the permafrost table we find a massive ice rind, often with inter-layered sands and clays. The ice rind can be up to 30 m thick, while the permafrost layer as a whole in the Yakutsk Basin can be over 500 m thick.

The older relief of this basin is dominated by broad, Later Tertiary surfaces, into which the terraced valleys of the major rivers are cut 100-110 m deep. The age of these terraces shows that all valley-cutting, as elsewhere in the mid-latitudes, occurred during the Pleistocene.

The profile shown in Fig. 38 shows the Aldan River at late summer low water. During the snow-melting in early May, when the lower reaches of the Lena are still ice-locked, the flood waters usually reach 8 m, at most 11 m above this level. The important question is whether the permafrost layer and accompanying ice rind continue underneath such large rivers. It is well known that this is not the case beneath the Yukon River in Alaska, located at the same geographic latitude. Appropriate cores have not been drilled in the beds of the Aldan and the Lena, but it is very possible that the permafrost and ice rind do continue underneath them, as indicated with a question mark in Fig. 38. Thus the permafrost in the ground of the Yakutsk Basin is distributed continuously at a given thickness, while it is discontinuous in the Yukon Basin, where, with some exceptions, it is not more than 8-10 m thick, and does not even reach as deep as the deepest river pools.[25]

The deepest pools of the lower Aldan are 20 m deep, and some in the Lena are slightly deeper. Every summer the warm water envelops the riverbed with a layer of thawed ground, but sand banks exposed for only a few winters

[24] Despite extremely cold winters (Yakutsk has a mean January temperature of $-43.5°C$, with extremes close to $-70°C$), the mean July temperature reaches $+19°$. On several days at the end of July 1969, the noon temperature reached $+37°C$, at which time the surface water of the Lena registered a temperature of $+20.5°C$.

[25] The permafrost layer is thinner in the Yukon Basin because the mean January temperatures there only drop to $-25°C$, and because it gets somewhat more snow than the Yakutsk Basin.

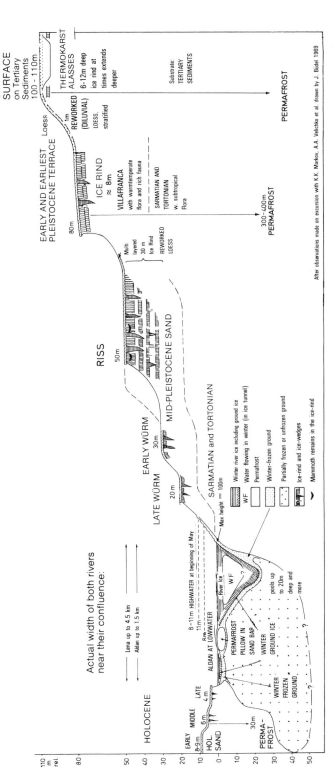

Fig. 38. Terraces of the Aldan River in E Siberia. Composite profile 100–300 km above its junction with the Lena River. Data on permafrost, river ice, ice rind, ice wedges, and thermokarst. River width reduced by one fourth. Ice rind (ground ice) is especially thick in the Riss depositional terrace. Winter ground ice penetrates even through the riverbed (permafrost may also underlay the riverbed at great depth), and shifting of the river across older Holocene terraces cuts into fresh permafrost (see profile left), producing the ice rind effect and making possible excessive valley-cutting even in the taiga.

contain pillows of permafrost. Both the Lena and the Aldan are covered every winter by an ice sheet 2-3 m thick, beneath which the water remains liquid only in the thread of maximum velocity. A thick coat of ground ice forms where movement is retarded along the edge and bottom of the river and at bridge pilings, so that in midwinter, the river, shrunken by the lack of ground water supply, flows through an icy tunnel. All subsidiary channels in the broad sand and gravel bed freeze completely. Frost also penetrates into the substrate surrounding this tunnel (shown as the winter frost area in Fig. 38), but to what extent this winter frost penetrates the unfrozen to partially frozen ground surrounding the river from last summer, has, as far as I have been able to ascertain, not been established by coring of the riverbed. It is quite probable that this layer of summer thaw may become completely frozen in winter around the smaller rivers. On the larger rivers, careful study of the Holocene terraces and their underlying permafrost has shown that the main river course shifts from side to side over hundreds or thousands of years, continually moving across areas of *permanently frozen ground. Once the river has abandoned such a terrace, permafrost immediately reenters the ground* (see Fig. 38, lower left). *In any case, the winter tunnel of ground ice around the main river course may also assume the function of preparatorily loosening the bedrock.* In this manner, ice rind loosening of the bedrock takes place in two ways beneath even the largest rivers.

The next two terraces above the high water level, dated by the Russian researchers to the Late and Early Würm, rise to the relative heights of 20 and 30 m. Within them the permafrost is already over 100 m thick, having a well-developed ice rind punctured by more recent ice wedges. Such ice wedges are also found in the Holocene terraces (see Photo. 3). This observation, as well as the fact that well-dated composite Holocene terraces lie far below the Würm terraces, proves that *even in the Holocene, intense vertical erosion over a broad surface has taken place with the help of the ice rind effect.* The colder winters and higher summer temperatures of the river water must have made this process even more vigorous in Siberia than in Svalbard.

A multilayered ice rind containing several generations of ice wedges is particularly well developed on the broad 50 m terraces, dated by Russian researchers to the Riss. We visited several places along the undercut bank of the Aldan where ice horizons separated by sand and clay layers are revealed to a thickness of up to 30 m. The largest such exposure was found at Momontova, 307 km above the junction of the Aldan and the Lena. Here the taiga grows on a thin 80-90 cm thick active layer of loamy loess. Below this the upper layer of the ice rind forms a steep face 5-6 m high (see Photo. 13). On July 31, 1969, the melting process at a noon temperature of $+36°$ was in full swing. One tree after another crashed over the retreating face of ice during the course of the summer, their trunks and root masses then slithering down a well-saturated mudflow sloping at about 20°.

13. Ice rind (ground ice) exposed six m deep (total thickness including interlayered sands is 30 m). Riss terrace of Aldan, relative height 50 m, 300 km above the junction with the Lena River (see Fig. 38). Above: taiga growing on 0.8-1 m thick frost-patterned active layer in reworked loess (*Fliessloess*). (Photo: Büdel, July 31, 1969.)

The entire 200 km represented by the composite profile in Fig. 38 are basically the same. Three of the famous sites where mammoths were found fully preserved in the ice rind are located along this stretch.

The next terrace at 80 m, probably representing an Early or Earliest Pleistocene erosional surface on Villafranche sediments, has an ice rind only 3-4 m thick, pierced in places (e.g., at Kuranack, 170 km above the mouth of the Aldan) by more recent ice wedges (see Photo. 14). The many *allases* (melt hollows of so-called thermokarst) on the broad, taiga-covered surface crowning this terrace sequence also show that the ice rind is merely 12-15 m thick. Why has the Riss terrace alone developed so thick an ice rind? This question is particularly important for the economic development of Siberia, as the width of this terrace would make it particularly suited for roads, pipelines, industrial centers, and settlements, were it not that all foundations are imperiled by the massive ice rind.

The answer seems to me to lie in the fact that it is this very Riss terrace, formed on a thick layer of Mid-Pleistocene sands, which bears the strongest traces of being a *depositional terrace*. During sedimentation, new ice rinds were constantly formed in vertical succession, while permafrost was preserved in the older ones. Only in such a continuously deposited body, where the permafrost grew upwards, does it seem to me that so many intact mammoths

14. Ice rind (ground ice) exposed 4.5 m thick, penetrated by a younger ice wedge. Located on Early Pleistocene terrace of the Aldan River, relative height 75-80 m; 170 km above junction of Aldan and Lena (see Fig. 38). Above: taiga growing on 0.8-1 m of frost-patterned active ground above reworked loess. (Photo: Büdel, July 30, 1969.)

could be preserved. The animals were surprised by high water, flooded with sediment, and frozen in, and were thus incorporated into the body of newly forming ice rind layers.[26]

As in Central Europe, high relict planation surfaces on anciently consolidated rock units dominate much of Central Siberia. In the Yakutsk Basin, the thick Tertiary deposits corresponding to the Tertiary erosion contain a rich Tortonian to Villafrancian warm-temperate fauna. Here again, valley-cutting only really set in with the first cold stages of the Pleistocene, when thick ice rinds and deep permafrost developed. Valleys were cut over 100 m

[26] The cause of this Riss deposition might have been a temporary tectonic downwarping of the Yakutsk Basin. Another explanation, however, is also possible: during the Riss Age the Verkhoyanski Mountains were particularly heavily glaciated. Ice tongues pushed forward nearly to the Lena just below the mouth of the Aldan, and extensive gravel plains and alluvial cones proceeding from these ice tongues may have caused deposition on the Lena and Aldan Rivers even upstream of the glaciers. The lowering of all altitudinal zones during the Riss further provided more bed load from the upper reaches.

deep, even in the fairly stable Yakutsk Basin. During the Holocene they have been cut down 30 m over floors many km wide. This is the same vertical amount which we find in SE Spitzbergen, even though the taiga-covered slopes provide less material. The fact that the particularly thick ice rind and the high summer temperatures of the river water nevertheless enable very vigorous fluvial down-cutting causes us to distinguish here a special *taiga zone of valley-cutting*.

2.2.22 THE CHIEF TEMPLATES OF
THE HOLOCENE RELIEF

It is the extreme polar and equatorial climates which have created the most effective relief-forming mechanisms. The polar zone of excessive valley-cutting (including the taiga and tundra valley-cutting zones) and the zone of excessive planation in the seasonal tropics (including the zone of partial planation in the perhumid tropics) are the two morphoclimatic zones whose relief-forming mechanisms are most effective. These are also the two zones whose pedospheres overlie the thickest decomposition spheres responsible for preparatorily loosening the bedrock and promoting erosion (see Fig. 9). True, these deep-reaching spheres are partly relict, but in both regions they are currently being extended, and are correspondingly widespread. In the intervening latitudes, as in the Pleistocene periglacial region, the remnant traces of former excessive valley-cutting are also widespread, but this cutting does not continue today. The relief relicts of the long Tertiary tropical climate have been bequeathed to us as highly resistant relict planation surfaces crowning many mid-latitude uplands. Remnants of the former decomposition sphere (which must have reached very deep then), or even of the pedosphere, are preserved only in a few favorable localities.

All remnants of this kind, however, indicate that the mid-latitudes have periodically experienced climates corresponding roughly to one or the other of today's two chief templates, the periglacial and the seasonally tropical climates. These have completely disappeared in the mid-latitudes today.

In our introductory examination of the morphoclimatic zones of the earth we began with the currently active pure types of these template climates; we proceed now from a consideration of the polar periglacial region directly to the seasonal tropics. After this we shall turn to the permanently humid tropics, then to the tropical and ectropical arid regions. The latter two zones, as we shall see, still display the major assemblages inherited from former humid periods, overprinted according to today's climate.

Older relief generations play an even greater role in the intervening mid-latitudes. Here the mechanisms operating during the Holocene were so weak that up to 95% of the relief confronting us today is inherited from a previous, clearly very different, very energetic relief generation. These regions will

therefore be treated in the following section as pattern zones for climato-
genetic geomorphology. The ectropic regions were covered in their virgin
Holocene state by continuous coniferous and deciduous broad-leafed forests
or by open and herbaceous steppes. Where not too steep, the slopes are
covered by stable, relief-covering soils. This entire zone is shown as the *zone
of retarded valley-cutting* in Fig. 13. Along with this we will also consider
the subtropics, where diverse influences overlap so variously as to necessitate
its being called the *zone of mixed relief-development*. This is especially true
of the *etesian region* on the W continental margins, separated from the humid
tropics by the great trade-wind deserts. The *monsoon region* on the E continental margins greatly resembles the seasonal tropics on which it borders.

2.3   The Peritropical Zone of Excessive Planation

2.3.1 THE ROLE OF TROPICAL RELIEF-DEVELOPMENT
FOR THE EARTH AS A WHOLE

Entering the tropics from the cooler latitudes, one enters a geomorphically
entirely new setting. Not only the climate, fauna, and flora, but also the
human race, speech, social structure, and cultural achievements in this region
are so completely foreign that one adjusts but slowly. This is perhaps especially true of the world of the relief. None of our customary interpretations
applies here, for we find no river terraces, gravel plains, or flood plains like
those of the ectropics; we find neither loess sheets nor solifluction blankets,
neither familiar soil types nor familiar relief features.

The soil is much deeper than in the ectropics, and is of a completely
different composition; it is rich in concretions and crusts (see Fig. 39) and
is inhabited by a totally different fauna, largely termites. The relief itself,
often so lushly overgrown that aerial photos show but little, does not display
the gradual transitions between broad ridges, gentle slopes, shallow dells, and
wide valleys familiar to us from the ectropics. Instead the forms are in all
ways more extreme. Barren rock inselbergs of all shapes and sizes thrust
precipitously out of wide erosional surfaces which, thickly covered with soil,
are often still forming. Where broad plains have been uplifted, their escarpment rims are slashed by rugged, stepped ravines. Steep dissection also characterizes the mountains. Even the occasional relief elements which are less
dissected are shaped quite differently from those of an ectropic hill country.
And wherever broad surfaces occur, every stream, from the smallest rivulet
to the largest river, flows through a completely unfamiliar and extraordinarily
flat concavity. The amount of water present varies dramatically between the
rainy and dry seasons.

Thus the initial impression gained by the geomorphologist upon first arrival

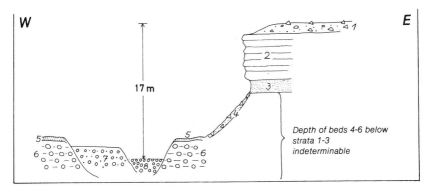

Fig. 39. Calcareous crusts in S Morocco. Thick dense crust above Earliest Pleistocene freshwater limestones. Thin recent crust. (1) One m of reddish fine material containing many angular fragments of the underlying calcareous crust, few rounded allochthonous pebbles. (2) Five m of very solid calcareous crust, clearly stratified. (3) One m of freshwater limestone (tufa) filled with plant stems. (4) Slope debris (fragments from 2 and 3 in fine material). (5) Thin younger calcareous crust. (6) Solidly cemented conglomerate of coarse, well-rounded allochthonous pebbles. (7) Well-rounded colorful allochthonous gravel in orange fine-grained matrix. (8) Well-rounded, colorful Recent wadi debris. (Bani Depression due W of Bou Izakarne; March 18, 1972; Büdel.)

in the tropics is that everything is completely foreign, necessitating that he abandon all accustomed perspectives and methods in favor of new ones. Much of the geomorphological work done in the tropics suffers from attempts to apply the methods used in the ectropics. This approach works quite well at altitudes above the forest line in the glacial and periglacial regions of the tropical highlands, and indeed, not a few "tropical geomorphologists" have withdrawn to this region.[27]

But the geomorphologist studying the relief history of the hot lowlands will quickly realize that the morphodynamics there are totally different, and will arrive by several means at a second conclusion: just as pre-Copernican man regarded the earth as the center of the universe, so the geomorphologist schooled in the mid-latitudes unconsciously assumes that the familiar relief assemblages found there dominate the entire earth.[28]

The science of geomorphology evolved almost entirely in the ectropics. Yet any equal-area world map shows that the morphoclimatic zone of the

[27] Faithfully following the verse once written by a German student (freely translated here by D. Busche): If lowland study is too much work/ Climb up and seek a glacial cirque,/ And never from your search refrain—/ You may at least find a moraine. (Originally: *Schaffst Du im Tiefland nicht das Wahre/ So steig hinauf und suche Kare/ Und lass die Hoffnung niemals schwinden/ Du könntest auch Moränen finden.*)

[28] The knowledge that today's world civilizations originated in the mid-latitudes, where most of the world's population lives, may have led us subconsciously to apply this "norm" to the world of nature.

tropics (including the warm arid regions) covers as much land as do the ectropic zones (see Fig. 13, endpaper).

Closer study of the tropical relief shows that with the exception of the highlands (see Fig. 13), tropical relief is controlled by far more uniform and continuous relief-forming mechanisms, namely, those responsible for creating the characteristic broad erosional surfaces (etchplains or *Rumpfflächen*) of the lowlands. Considering, moreover, that the high relict plains dominating the ectropics also derive from former tropical relief-forming mechanisms, it is plain that just as life evolved in the tropics (the ectropic biosphere being impoverished and clearly derivative), so the basic relief types originated in the tropics. The climate of the temperate latitudes has only lightly overprinted the basic relief, and this only during the Pleistocene. Without knowing both of the two clearly describable relief templates, that is, the tropical and subpolar morphogenetic extremes, their area of overlap in the mid-latitudes cannot possibly be explained, just as the existence of forest and steppe could not possibly be deduced from a mere study of wooded steppes.

Of the two basic relief templates, the variants of the tropical zone are more important for the earth as a whole than are those of the polar zone. The tropical relief dealt with in this section represents the major source of our interpretation of the total terrestrial relief.

2.3.2 THE CREATION OF ETCHPLAINS (PLANATION SURFACES, EROSIONAL PLAINS) BY THE MECHANISM OF DOUBLE PLANATION

The ability to create etchplains[29] on tectonically quiet or only weakly uplifted blocks is the dominant morphological characteristic of the seasonal tropics. This *excessive planation* occurs today where the rainy season is 6-9 months long, and continues weakly even where the rainy season is only 4-6 months long. A few features of this once almost worldwide zone have already been mentioned. We will now elaborate on them in two respects: firstly by discussing the morphodynamics, or relief-forming mechanisms, and secondly by discussing the forms produced. We therefore begin by considering the

[29] In English and French, a newer, more meaningful term for this characteristic assemblage of inselberg-studded *Rumpfflächen* has not been introduced. To some extent Davis's older concept of "peneplain" is still used, but this is based on the outdated concept of an *Endrumpf* or terminal torso, an old-age condition of all erosional processes. Equally inappropriate is the current term "pediplain," as this derives from the concept of a pediment, which belongs genetically to a completely different assemblage (see Section 2.6 below). Appropriate English terms for the German *Rumpffläche* (whose definition is very exact) are *planation surface* (or "truncated surface" according to some authors) or *erosional plain*. Even the German term *Abtragungsfläche* (denudational surface) can be used for *Rumpffläche*. Meanwhile, the term *etchplain* (cf. Fairbridge, 1968) can also be used in this general sense. (We have chosen to use the latter term here.—Trans.)

mechanism of double planation. The basic features of this have been described in earlier publications (Büdel, 1957a, 1957c, 1965), and subsequent work has refined our knowledge sufficiently that this planation region may now be considered to be as thoroughly investigated as the present periglacial zone of excessive valley-cutting.

2.3.2.1 The Basal Surface of Weathering; Basal Knobs and Core Stones; The Decomposition Sphere

In the humid ectropics, areal erosion involves only the upper surface of the thin relief-covering soils (*Ortsböden*). Once the relief-covering soil has developed a mature profile on any gentle slope, its base will show little further decomposition. This is shown by the fact that decomposition during the 10,000 years of the Holocene has not managed to eradicate the Pleistocene soil skeletons (cryoturbation, solifluction, or blockmeer skeletons in periglacial regions) on which the Holocene soil sheets were superimposed. Such skeletal remains are in fact common at the base of Holocene soil sheets, showing how little occurs here once a mature profile has developed.

The homogeneity of erosional processes is even more pronounced in the polar tundra and frost debris zones. Here areal slope erosion is not limited to the upper surface of the soil (as in the forested mid-latitudes), but uniformly attacks the entire active layer down to the permafrost table. This usually involves a layer less than 0.5-0.7 m thick.

In the humid and seasonal tropics, the basal and upper surfaces of the pedosphere are not only much further apart, but they also have the completely different functions of soil development and of soil erosion. It is from this phenomenon that we derive the term "mechanism of double planation."

The annual and daily temperature fluctuations, which are slight in any case, are completely evened out at the base of the thick lowland soils of the humid tropics. Water is always present here, supplied during the rainy seasons and lingering through all or most of the ensuing dry season. With the onset of the next rainy season, new moisture penetrates into the depths where the older water still remains in more or less bound form. Percolation is faster at the start of the rainy season because the ground is still torn by desiccation cracks and riddled by roots and animal tunnels. Small mammals, worms, termites, ants, mites, fungi, algae, and bacteria all play important roles here, the number of bacteria in tropical soils being estimated according to different authors at several hundred thousand to 20 million per gram of soil (Scheffer and Schachtschabel, 1973).

The excretions of these organisms provide tropical soil water with much of its chemical reactivity. As already mentioned, the aggressiveness of ground water in tropical soils is due to its greater acidity in comparison with ectropic soils. Carbonic acid, for example, in the soil air alone has been estimated by Nagel (oral communication) at 0.02-0.05% in the polar regions, 0.5% in winter to 2.5% in summer in the mid-latitudes, and 20% in the tropics. It is thus 20-40 times greater in the tropics than in the mid-latitudes, and 1000 times greater than in the polar regions (see also Gerstenhauer, 1972).

This CO_2 derives partly from precipitational water, up to 25% from root respiration, and over 50% from the respiration of the soil fauna. During the rainy season, when the tropical soils are poorly aerated, it is notable that the CO_2 content of the soil air increases with depth.

The living organisms in these soils supply far less organic acid than does dead organic matter (humus particles and mineralized humic matter). Barely if at all concentrated in a surficial A-horizon, this is usually distributed fairly evenly throughout the soil sheet. In dry weight, humus particles form 92-95% of the soil's total organic content. The soil water can thus be supplied equally at all levels with organic acids, largely humic and fulvic acid. The fairly constant rich supply of organic acids and warmth make the water percolating along the base of the soil particularly chemically aggressive toward the underlying parent rock. The soil base is therefore named the *basal surface of weathering* (*Verwitterungs-Basisfläche*; Büdel, 1957a, 1965). Both the German and the English term correspond to Mabbutt's "weathering front" (1961) and Ollier's "basal surface" (1960).

The intense chemical weathering taking place here through much or all of the year (not just in the rainy season) exceeds all chemical weathering found in any other climate. We therefore refer here to *intensive weathering (Intensiv-Verwitterung)*.

Four circumstances show how extremely corrosive this intensive weathering is. The first of these is the thickness of the humid tropical soils. These are nearly always at least 3 m thick, but are often twice this (6 m), and at times may reach 30 m.

The second is the fact that despite their thickness, these soils do not display the distinct horizons shown by mature ectropic soils. Once an ectropic soil reaches maturity, little basal decomposition continues. In the thick tropical soil, on the other hand, particularly in the latosols or red loams, the processes of reworking, eluviation, and clay mineral development occur constantly throughout the entire profile.

The third circumstance is that where soil directly overlies dense quartziferous, widely jointed rock (such as granite or quartz-porphyry), the transition at the basal surface of weathering from almost fresh, unaltered rock to completely altered, plastic kaolinitic clay occurs within a spheroid rind only 2-2.5 cm thick (see Ollier, 1965, especially Figs. I and VI). Were this process any less intense, the transition zone would certainly be thicker.

The fourth and final circumstance is that in many cases the humid-tropical weathered matter overlies a decomposition sphere many times thicker than itself. The weathered mantle is often 100-150 m, in exceptional cases 200 m thick. Precisely where or why such thicknesses occur has not yet been determined. Important factors may include the type of parent rock and the age or genetic development of the overlying soils (for age differences in soils, see Van Wambeke, 1962).

Three types of decomposition sphere may be distinguished. In the first type, the latosol extends downwards only along deep joints (*weathering along joints*, or *Kluftverwitterung*), while between these joints the fresh rock reaches to the basal surface of weathering. It is in these places that the above-mentioned abrupt transition from fresh rock to overlying clay soil takes place.

The second type of decomposition occurs in limestones, and consists of narrow *pipes* filled with red loam penetrating deep into the body of otherwise unaltered limestone. These pipes are on the average only 20-30 m deep, and often end in striking cul-de-sacs.

The third, deepest, and most widespread type of humid-tropical chemical decomposition is *grus development*, a form of partial decomposition, in which the dark components and feldspars in crystalline rock are altered into various grades of clay minerals, while the quartz grains remain preserved. Here the joint system and the structures in the rock remain recognizable. Such grus zones are often preserved underneath the relict etchplains of the mid-latitudes. They have been found, for instance, by Wurm (oral communication) in the Frankenwald and the Fichtelgebirge (both in N Bavaria) to depths of 70 m. They are also widespread in the Riesengebirge (between Silesia and Czechoslovakia) (Büdel, 1937), and depths of up to 20 m are exposed on the Felsberg in the Odenwald (see Fig. 68), where they contain unaltered floating core stones of granite. That castellated rocks and blockmeere may represent washed-out Tertiary tropicoid grus zones (transported by solifluction in the Pleistocene) was first shown by Hövermann (1950, 1953a) in the Harz Mountains (N Germany).

Obviously the basal weathering surface created by such attack is not a smooth surface, particularly on crystalline rocks. Between the joints (representing downward-extending fingers of red loam) the surface presents a knobby small-scale relief. We have decided to call the protrusions (usually 1-3 m high) *basal knobs (Grundhöcker)*.[30]

Sometimes, especially where the jointing system is very wide, intensive weathering works down around larger rock complexes, bypassing them and leaving them unaltered. These may form unweathered columns and pillars, which may pierce through the soil sheet nearly to the upper surface. Swarms of these have been called *penitent rocks (Büssersteine)* by Ackermann (1936, 1962). Where the stone complexes are broader and still soil-covered, one speaks of *rock pavements (isolierte Spülsockel*; Büdel, 1957a, 1965; see also

[30] The German term *Rundhöcker* is deliberately taken from *Grundhöcker* (roche moutonée), used in the terminology of glacial erosion (Büdel, 1957a, 1965), as the latter are inherited (e.g., in Sweden, Lapland, and Labrador) from a pre-glacial relief of basal knobs, whose red loam sheet was removed and whose form was then glacially scoured (see also Brochu, 1959).

Figs. 40 and 41), whose tips may often be exposed. Where larger rock complexes protrude above the soil surface, they often stand in characteristic shield fashion a few m above ground level. Such *shield inselbergs* are nearly always bare of soil, and can no longer be chemically weathered. Weathering continues only along their joints, breaking the shield inselbergs up into blocks (*block inselbergs* and *tors*).

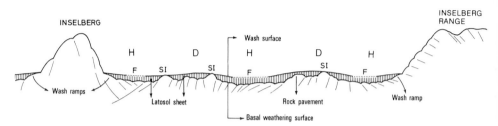

Fig. 40. Typical N-S cross-section through the Tamilnad Plain (E side of the Deccans), as an example of an active etchplain. H: Wash depressions. D: Wash divides. SI: Shield inselbergs. F: Fine sand in the rainy season riverbed.

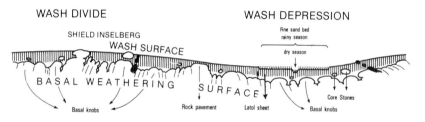

Fig. 41. Wash divide and wash depression (detail out of Fig. 40).

The irregular distribution of shield inselbergs, sometimes found isolated, sometimes in groups across the entire breadth of the etchplains, led Kayser (1949) to call them "azonal inselbergs." Thomas (1965) classifies features corresponding to our shield inselbergs (which he calls "domes")[31] as well as features similar to rock pavements (which he calls "ruwares") into a theoretical developmental sequence of inselbergs (which he calls "bornhardts"). Lack of soil cover is a characteristic which shield inselbergs share with many higher inselbergs, in accordance with the rule formulated by Bremer (1971) of "divergent weathering" (called "differentiated weathering" by Bremer in earlier publications).

[31] The term "dome" (a shield inselberg exposed above the tropical soil) must be distinguished from "desert dome," commonly mentioned in the British and American literature on deserts (especially on winter-cold deserts). Desert domes are much broader and flatter features of the basal weathering surface, covered by alluvial desert debris. This summit may rise to the surface, but is never exposed. Desert domes may carry domes or shield inselbergs perched upon them.

Since intensive weathering at the basal weathering surface works down along joints, which in shield areas can be anywhere from vertical to nearly horizontal, weathering often bypasses blocks of fresh rock, disintegrating the surrounding material until the block or *core stone (Grundblock)* floats free in the red loam or in the deeper zone of grus. These core stones remain *in situ* and show the original direction of bedding. Occasionally they may protrude above the upper soil surface, forming miniature shield inselbergs (see Figs. 40, 41, and 42c).

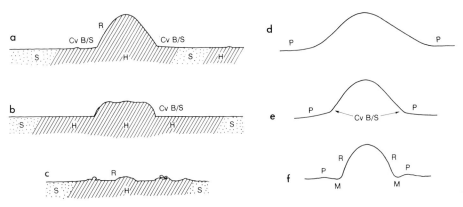

Fig. 42. Schematic profile of inselberg development. a-c: Relationships of inselbergs to outcropping harder (H) and softer (S) rocks. Inselbergs with steep rocky flanks (R) and sharp piedmont angles (Cv B/S) rarely coincide with lithic boundaries. (b drawn from Photo. 21.) d-f: Accentuation of the inselberg slopes through subcutaneous basal sapping. Rock pavements (P), sharp piedmont angles (Cv B/S), rock walls (R), and moats (M) appear as the rock slopes are gradually freed of soil cover, causing intense chemical weathering to cease. In frost-free climates the barren rock slopes can then weather back only very slowly. Particularly marked chemical decomposition produces moats.

Not every rainy season need be so heavy that rain water percolates all the way down to the basal weathering surface, joins the ground-water pockets, and ensures continued intensive weathering. Chemical weathering may be maintained even with discontinuous saturation in particularly heavy rainy seasons. Where the wet season is over six months long, basal saturation and continued latosol production are probably the rule. Where the season is shorter, saturation occurs only in particularly wet years, without substantially altering the overall process.

The basal weathering surface, with its knobby microrelief, forms an overall even plain, running more or less parallel (with occasionally more pronounced relief) to the wash divides and wash depressions (see Figs. 40 and 41) of the wash surface (to be described in Section 2.3.2.3). This is a natural consequence of the fact that it is this upper surface which relays all exogenous influences downward to create the basal surface of weathering. Only intensive chemical weathering and removal in solution take place at this basal surface; mechanical displacement of material in the humid tropics occurs only on the gentle slopes of the wash surface.

2.3.2.2 The Pedosphere in the Humid Tropical Lowlands

Erosional surfaces (that is, active etchplains in the humid and seasonal tropics) at low to medium-high altitude are covered by the thick soil sheets described above, usually consisting of red loam or latosol. (Other names, emphasizing particular characteristics include "ferrallite," "fersiallite," "oxisol," "luvisol," as well as "laterite" and "plinthite" for the indurated forms.) About 10-15% of the tropical lowlands are covered with a dark humus-rich "tirs" (soil with tirsification, following Kubiena, 1957); this is also called "regur" in the Indian Peninsula, while in Ethiopia it is called "grumusol" (A. Finck, 1963) or simply "black highland soil" (Büdel, 1954b). Semmel (1963) has found it next to red loam in low intramontane plains. Dark regur also occurs next to latosols in the etchplains of the Deccan, where it is by no means limited to basalts. According to Krebs (1939) it is associated with particular types of dry savanna, but in other places it is found together with wetter savanna types. It seems probable that this soil, rendered dark by its humus content, is formed in various quite different and as yet unexplained ways. A distribution map of the soil and its local names has been compiled by Finck (1963). We will refer collectively to "tirs soils."

Although the normal thick lowland latosols show far less profile development than our ectropic soils, this is not true of the thinner tirs soils. These also have more humus, montmorillonite, and illite, and less kaolinite, increasing their capacity for absorption and expansion. During the dry season they form gaping, often polygonal patterns of cracks (Büdel, 1954b), an extreme form of "gilgai," or tropical-subtropical patterned ground. According to Bremer (1959, 1965a, 1967a), Hallsworth et al. (1955), Mabbut (1963), and others, such patterned ground develops where there is a marked contrast between the rainy and dry seasons on flat ground having a high montmorillonite content, much humus, and clear horizons.[32]

The normal latosols are the most common soils on active tropical etchplains,

[32] All these are characteristic of tirs as well. Gilgais also occur frequently on alluvial soils, which, because of their spatially and chronologically discontinuous deposition, have horizons of a type. Gilgai structures such as the dry season polygons in tirs are not usually more than two m deep, though their diameter may range from 0.3 to 13 m. The protuberant centers of smaller polygons have been described as being 8-20 cm high. Gilgais show a very wide range of types, and along the edges of the tropics may grade into the desiccation cracks and salt patterns of the desert, as well as into the frost-patterned ground of the highlands. Where the protruding centers of the gilgai polygons are surrounded by both rings of cracks and of stones, they indicate that surface wash is not particularly effective (Bremer, 1965a). On the whole, gilgais are a very striking and interesting, but locally restricted type of tropical soil development.

and are thus in all known cases the actual agents for the mechanism of double planation which creates these etchplains. We therefore use the latosols here as a typical example of tropical relief-forming soils. Their special characteristics may best be recognized by contrasting them with ectropic soils, as an example of which we take gray-brown podzolic soils (see Fig. 6). This produces the following characterization of latosols.

[1] *Great thickness.* On active etchplains a norm of about six m can be assumed. Thicknesses of less than three or over 30 m are rare. Beneath this, the decomposition sphere may extend another 200 m down. Mature ectropic soils, on the other hand, are rarely over 1.5-2 m thick.

[2] *Very thin humus horizon.* Usually a few cm thick in latosols, this horizon quickly disappears with agricultural use. This is due to the rapid decomposition of organic substances: microbial activity speedily breaks them down into simpler compounds which combine readily with clay minerals. In this manner the humus derivatives become distributed evenly throughout the entire soil profile or solum. The ectropic relief-covering soils, on the other hand, possess a much thicker surficial humus horizon, whose lower boundary is very clearly marked (see Fig. 6). This horizon is 25-30 cm thick on the average, but may reach 50 cm or more in the chernozems.

[3] *Low pH values.* Ectropic relief-covering soils usually have high pH values, while tropical soils have low pH values and react acidly. This leads to the following features of intensive weathering:

[4] *High CO_2 content.* As we have already mentioned, the CO_2 content is much lower in the mid-latitudes and polar soils than in the tropical soils (see Section 2.3.2.1).

[5] *Carbonate leaching, calcretes.* At least trace amounts of calcium are nearly always present in the upper humus horizon of ectropic relief-covering soils. In the tropical red loams calcium is always completely leached out, sometimes even in the decomposition sphere. Bremer (1967a) names a case in Australia where Jurassic limestone under laterite has been fully decalcified to a depth of 30 m. This is one of the most striking features of intensive weathering. Limestone or loess beneath ectropic soils always maintains its original calcium content. Calcareous fragments may occur (e.g., as a result of cryoturbation) as a skeleton in the lower soil horizons, or even throughout the soil profile, and this is in fact the rule in rendzinas or rendzina-like soils. Where calcium occurs in the soils of the humid and seasonal tropics (a rare occurrence linked to pronounced dry seasons), it does so only in the form of calcareous concretions (calcretes).

Calcareous crusts develop in certain subtropical climates having prolonged dry seasons, e.g., in the herbaceous and desert steppes of the Atlas region on the N edge of the Sahara. Goudie (1973) collectively calls these crusts "calcretes," listing their distribution under thirty-six local names in various parts of the world. Most authors, however, take "calcrete" to refer to fragments of crusts or to individual calcareous concretions. The thickness of these crusts ranges between 0.5-1 m and 8-15 m, a maximum which I frequently found in S Morocco. Figure 39 shows how their thickness may be expanded by intercalated limnic layers. Goudie cites an extreme case of 60 m, which he found in the Molopo Valley on the S edge of the Kalahari Desert on the border between Botswana and S Africa. Rutte (1968) found that calcareous crusts are always thickest on the edge of sharp ravines in older calcareously encrusted deposits.

[6] *Silica leaching, desilicification, silcretes, gibber.* Yet another feature of intensive weathering in humid tropical red loams is the marked leaching of silica (SiO_2) from most minerals. This contrasts with the normally high silica content in ectropical soils. Only the pure quartz components of the parent rock (quartz granules from granite, related rocks, or quartz veins) remain preserved in the red loams, but even these are quickly comminuted, and form a secondary peak in the grain-size spectrum at 30-200 μm, next to the dominating clay fraction of under 2 μm (Bakker and Muller, 1957). Where the dry season is very marked, thick quartz concretions may form in the red loams. The most extreme, though uncommon, case of this consists of thick siliceous crusts or *silcretes*. Bremer (1967a) assumes that these develop under more humid (fully tropical) climates than do calcareous crusts, but under somewhat drier climates than laterite crusts (see below), and in areas of "very flat relief with little epeirogenic activity and constant climate" (p. 190). I myself have found silcrete fragments up to a cubic m in size floating in the thick red loams of the Ashanti Highlands of Ghana, where they are mined as highly desirable construction material. More frequent are the small cherty siliceous concretions (or other highly silicified rocks) forming *gibber*, or lag gravels, in areas where the red loam is slowly being removed. When such a soil matrix has been fully eliminated, lag gravels are important evidence of its former existence.

[7] *Illuviation of sesquioxides. Bauxite deposits.* The ectropic relief-covering soils generally contain only moderate amounts of sesquioxides (Al_2O_3 and other aluminum oxides, Fe_2O_3, FeOOH, and other iron oxides and hydroxides). In tropical red loams the greater eluviation of silicic acids is compensated for by enrichment of fairly insoluble sesquioxides (leading to the term "ferrallitic soil" for latosols). Iron is incorporated only to a limited extent into the clay minerals here. In amorphous form iron oxides display high surface activity, and therefore tend to be deposited in a film coating other mineral grains. This process, called *rubefication (Rubefizierung)* by Kubiena (1953, 1964) is the main cause of the red staining of many tropical soils (red

loams and latosols) and humid subtropical soils (terra rossa and red earth). Iron oxides produce more red staining, iron hydroxides more brown. Crystallized iron oxides form the minerals goethite and hematite. The aluminum oxides, largely gibbsite (hydrargillite), boehmite, and diaspor, are more commonly converted into clay minerals or incorporated into the interlayers of easily expanded clay minerals and organic compounds.

Extreme enrichment of Al_2O_3 and of other aluminum oxides produces *bauxites*, which, as raw material for aluminum, play an important economic role. Since aluminum deposits are nearly always found in Recent or relict etchplains, or in the fissures of the associated decomposition spheres, the study of old etchplains, hitherto regarded as purely academic, has recently become important for mineral prospecting, and the relevant field work has become an important branch of applied geomorphology. (See the Russian maps by Gerasimov and Sidorenko, 1971, and Sidorenko and Michailev, 1973, on etchplains of all ages and bauxite deposits in the Soviet Union; in Germany, also, important bauxite deposits have been found on old etchplains, see Fig. 63, also Bibus, 1973 and 1975.)

[8] *Further breakdown of clay minerals to 1:1 minerals. Intensive weathering.* In soils of warm-humid climates, many rock components weather to new clay minerals. Most of these have a layered lattice structure, that is, their atoms or ions are arrayed in parallel layers. In the ectropics this progresses only as far as the creation of 2:1 or mixed layer clay minerals, especially illite, vermiculite, and montmorillonite, which have spaces of 10-20 Å between the crystal layers. This relatively wide spacing can be further expanded by penetrating water (Scheffer and Schachtschabel, 1973). Moreover, the water can be adsorbed by adhering forcefully to the clay mineral surface, forming films which, in the case of montmorillonite, can be up to several hundred Å thick. The swelling capacity of 2:1 minerals is accordingly very large, as is their ability to form cracks in the soil upon drying out. This means that air can reach deep into the ground during dry periods, lowering the CO_2 content of the soil air and reducing the aggressiveness of the soil water both toward the soil components themselves and toward the bedrock below.

In the seasonal and humid tropics, constant high soil temperature, periodic saturation (an especially important factor), and more aggressively acid soil water combine to carry mineral transformation of the parent rock to a stage which we have named "intensive weathering." This is characterized by loss of silica or *desilicification* of the clay minerals, changing the 2:1 into 1:1 minerals, of which the most important are kaolinite and halloysite. These are much more compact than the 2:1 minerals, having a spacing of 7.2-10 Å, and are much more homogenous. Thanks to desilicification, kaolinite, the main component of all latosols, is 38-40% Al_2O_3 by weight. Water can barely penetrate such 1:1 minerals, and the film of adsorbed water clinging to the surface of kaolinite molecules is only around 50 Å thick. This means that kaolinite expands far less when wet, and does not crack as much or permit aeration when dry. Infiltration speed and soil fertility are consequently reduced, while the acidity and aggressiveness of the soil water are increased.

[9] *Relationship between the soil and the parent rock. Immobility of the red loam sheet on gentle slopes.* Ectropic relief-covering soils are characterized not only by climatically determined horizons (see Fig. 6), but also to a great degree by their parent rock. Thus we speak of sandy soils, calcareous soils (rendzinas, terra fuscas), loess soils, and crystalline soils, specifying the agronomically important component of parent material. This component is especially large in the mid-latitudes, since the chemically and biologically formed Holocene soils have often developed on a skeleton prepared by mechanical weathering during the Pleistocene cold stages, making the lowest portions of the profile particularly rich in fragments of parent rock.

Humid tropical soils are quite different. Their thickness isolates them from the parent rock, an effect heightened by the rapidity of intensive decomposition at the base of the solum. Weathering advances from there through fissures deeper into the bedrock, isolating the core stones. These core stones can be recognized by their substantial size (anywhere from around ten cm to several m in diameter), rounding (due to weathering), and especially "by the fact that their structural planes invariably agree with those of the basal knobs in the bedrock, since they are indeed merely chemically isolated extensions of these knobs" (Büdel, 1965, p. 38). In the seasonal tropics core stones are invariably packed firmly into the red loam sheet which is immobile on all slopes of up to 4° or more. Quartz veins may often extend as etched fragments straight out of the bedrock into the soil sheet, showing no offsets or other traces of movement. This is telling evidence for the immobility of the entire sheet of red loam on all gentle slopes. In the perhumid tropics this is less often the case: here deep-reaching soil movements have been observed on slopes of little over 4°, especially where man has destroyed the rainforest (Hagedorn, 1974).

[10] *Weaker profile development, lack of mature profile. Relief-forming soils.* The humid-tropical lowland soils, as we have seen, have no true humus horizon. Their substantial content of organisms and of humus derivatives is instead distributed throughout the entire profile. As Bremer (1973a and b) was the first to notice, any other type of profile development corresponding to the very regular horizons in mature ectropic soils is also much weaker here. At any rate, what many authors have called horizons (leaning too heavily on analogies with ectropic soils) actually derive from a number of quite different causes.

Most obvious is the fact that the soil coloring, which for most of the profile is a glowing yellow-red, cinnabar, or carmine red, regularly becomes paler toward its base, grading through a red or white flecked *mottled zone* into a pure white *pallid zone* rich in kaolin. Here iron oxides have been fully leached out by ground water (see Photo. 15 and Fig. 43). This basal lightening is common in relict latosol sheets, such as the basalt-capped sheet in the Hoggar Mountains of the Algerian Sahara (Büdel, 1955b). It is this feature which most resembles the horizons of the ectropic soils.

The other three, less similar forms of horizontal structuring also figure in A. Finck's division of latosols (1963), based largely on Kellogg (1950, 1960). The commonest type of structuring consists of more or less horizontal stringers of iron and manganese concretions, generally dark-colored, which usually wander sinuously and wedge in and out again. We will later come to know these as the precursors of laterite development. As last relicts, such "bean iron ores" or "pisolites" are important evidence of the former presence of red loam sheets. The previously mentioned pale quartz nodules (silcretes and gibber) and calcareous concretions (calcretes) are limited to regions with more marked dry seasons.

Harder to explain are the "stone lines," superficially similar stringers of coarser, usually slightly subangular clasts, which, however, are not concretions, but gravel or detrital fragments. These are not common, but when they appear, it is in the upper part of the profile, usually a few dm below the surface. This suggests that they originated through enrichment, especially where the composition of the soil changes just above the stone line. In the Somali Highlands the profile above such a stone line often changes radically to tirs-like, highly absorbent dark soil, particularly on slopes. Here stone lines may be interpreted as a consequence of soil movements. In other places their

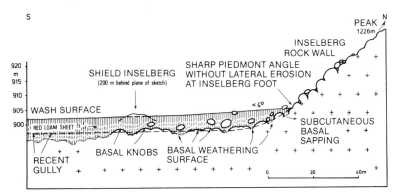

Fig. 43. Accentuation of the inselberg foot by subcutaneous basal sapping. (Near Kolar, E of Bangalore, in S India.) The inselberg slope has lost its soil cover through steepening, preventing further chemical weathering. Chemical weathering proceeds unimpeded on the basal weathering surface of the surrounding plain, and is even enhanced at the piedmont angle by increased rainwater supplied there from slope runoff. Since the slope no longer weathers back, a shallow moat develops (M in Fig. 42f). Drawn from Photo. 15.

15. Typical latosol profile (six m thick), red above, bleached below, at foot of an inselberg whose rock wall has been steepened by subcutaneous basal sapping. Near Kolar, 60 km E of Bangalore, S India. Photo used for Fig. 43. (Photo: Büdel, March 10, 1964.)

16. Active etchplain with rainy season wash film on the Niger above Niamey. Distant inselberg in the background. (Photo: Carol-Zürich, around 1955.)

environment suggests that they mark the lower boundary of the zone of plant roots and termite activities (Bremer, 1971). According to Lee and Wood (1971), this lower boundary lies at a depth of 60-120 cm, but the nests of certain termite species can be found in extreme cases as far down as 8.5 m. In any case, termites rework fine material and material rich in kaolin: they not only mechanically loosen the soil, but are also an integrated component of all humid tropical soils, promoting intensive weathering.

Thirdly, the latosols may appear "stratified," especially near the upper surface. This also has nothing to do with the horizons of ectropic soils, but results from wash effects (to be described in more detail in the next section). These dominate the upper surface during the rainy season, the wash rivulets constantly changing course and eroding in one place only to deposit in another. In the thick latosol profile, therefore, the distinction between "soil" and "sediment" is highly fluid, particularly near the larger rivers, whose suspended and traction loads, as we will see below, are of the same grain sizes as those prepared by intensive weathering in the soil.

On the whole these differences may be summarized as follows. The shallow ectropic relief-covering soils work toward a mature profile, expressed in clear horizons. This mature condition is quickly reached, for the ten thousand years of the Holocene have sufficed for this just as well as the several tens of thousands of years of the Eem and Holstein Interglacials, as shown by the paleosols (especially the B-horizons) of composite loess profiles. Once the mature profile is developed, the aggressiveness of weathering is much reduced and the soil's rate of downward growth greatly retarded. The fact that the soil's maturation coincides with the development of marked horizons may be due to the fact that water here flows nearly all year round from above downwards, following gravity, so that leaching of solutes and clay particles is always unidirectional.

In the humid-tropical lowland soils the high surface temperatures reach far into the ground, and the aggressiveness of the soil water increases with depth, thanks to generally high humus content and greater isolation from the free atmosphere. Proof of this is the great thickness of the soil sheet and the presence of these features throughout the entire profile (see Photo. 15).

The humid tropical soils grow quickly downwards, without achieving any similarly stable mature condition fixed by strict horizons. This may partly be due to the seasonally changing direction of soil-water flow here. Only during the rainy season does the water follow gravity (very slowly) downwards. In the dry season it rises against gravity upwards, partly by capillary action in the very fine-grained soil, partly in a gaseous state. The soil is further mixed by termites and other soil fauna. Thus in the absence of constant downward flushing, humus and other matter is distributed evenly throughout the soil profile.

The "relief-forming" trait of tropical soils is emphasized by two further important features. The first of these is the wide distribution of ground water at the basal surface of weathering. After a heavy rainy season it can be found everywhere, and after a weak one, at least in isolated pockets. The fact that it is areally effective over an average of many years even with discontinuous saturation (weak rainy seasons are especially common in the drier savanna types having an average of less than 4½ rainy months) is shown by the fact that bleaching (iron oxide eluviation) is almost universal in the deepest layers of these red loam sheets. When ground water can be tapped by wells at all on active tropical etchplains (and where it is not, as at the end of a very dry season, present only as adsorbed water), it can then be located with equal probability anywhere, at least in wash depressions. In the valley relief of the ectropics, ground water is found almost exclusively in the narrow valley floors. In the tropics on the other hand, widely distributed aggressive ground water attacks the substrate nearly everywhere.

The second feature supplements this situation at the soil's upper surface. Annual heavy surface wash causes repeated erosion (W. Penck, 1924), so that the tropical lowland soils, despite their great thickness, are far more actively integrated into the erosional process of double planation than our ectropic experience would lead us to think. Derivatives prepared at the basal weathering surface are of the very same grain size as those carried by even the tiniest rainy season rivulet on the wash surface. Thus the soil's entire thickness represents a transitional stage between preparation and erosion to a much greater degree than do mature ectropic relief-covering soils. We therefore call the tropical lowland soils not relief-covering soils (*Ortsböden*), but relief-forming soils (*Arbeitsböden*).[33] This is intended to emphasize their very different relief-forming functions. The shallow ectropic relief-covering soils, which effectively stagnate in their quickly reached maturity, have only a very disproportionate active function.

[11] *The age of the latosol soils and of their relief-forming mechanisms.* Active development of latosol soils in tropical lowlands is coupled to two climatic features. The first of these is constant high temperature. It is unimportant for soil development if, for instance, the monsoonal peritropic region on the E continental margins now and then suffers a brief winter cold spell, interrupting otherwise constant warmth with a few nights of frost—a factor which can have extremely selective effects on the vegetation. The second condition is more important: a rich supply of warm acid ground water which can maintain itself and remain active at the basal surface of weathering through the long months of the dry season. The soil can remain active even

---

[33] The literal translations of *Ortsböden* and *Arbeitsböden* would be ''*in situ* soils'' and ''working soils'' respectively. Since these terms, especially the first, would tend to be misleading, we have chosen to translate them with the more descriptive terms ''relief-covering'' and ''relief-forming'' soils, indicating the stagnant nature of the first and the active nature of the second.—Trans.

during long "winter" atmospheric dry seasons, as long as the year has at least 4½ humid months so that shifting pockets of ground water remain present on the basal surface for much of the year.

The thick soil blanket may nearly cease work when a climatic dry phase reduces the rainy season in the area concerned to, say, only two or three humid months, as in the drier parts of the Sahel Zone of N Africa today. As long as the soil sheet is not removed, it can, upon return to more humid savanna conditions, resume its work and continue as it was before the dry period hiatus. Active red loams forming today in the seasonal tropics (whose climate may fluctuate greatly from 4½ to 9 humid months) can maintain themselves in latent state through prolonged spells of greater aridity.

It seems likely that in the area of today's seasonal tropics the climate has remained the same (with at most brief fluctuations toward greater aridity or humidity) for most of the Tertiary; indeed, regions subject neither to marine transgressions nor to abrupt tectonic movements may in some instances have enjoyed the same climate since Late Cretaceous times. In any case, the climate has remained warm, for at no time during the geomorphic era are there any traces of an Ice Age hiatus here.

The latosol sheets of the seasonally tropical lowlands must accordingly be very old, certainly older than the ectropic relief-covering soils, which are entirely Holocene. But we cannot assume too great an age, for we have recognized that the latosol soils are active, that is, they are the transitory product of a continual adaptation to this climatic region, and eat steadily downwards while being removed from above. Towering over the etchplains on which the soil sheets lie are steep etchplain escarpments and inselbergs 100 or 200 m high. This means that enormous amounts of material have been planed away down to the level of today's surface, while the soil sheet which is the agent of this planation has been constantly renewed. It is thus rather the process of constant soil development that is so extremely old, and not the soil sheet itself. The soil sheet is more a transitory than an objective phenomenon.[34]

Nevertheless, parts of these soil sheets may be very old even in an objective sense, especially in areas of subsidence, where constant soil deposition has caused a transition from soil to soil sediment.

Very old tropical soil sheets look different from recent latosols. This is true, for example, of itabirites (Tienhaus, 1964a and b), which contain many of the world's iron ore deposits. Weathering sheets of such composition are not being formed today in any part of the world. Rubified sheets of buried

---

[34] A typical case of phenomenal existence is a *Föhn* (chinook) cloud, formed where the wind blows against a transverse mountain ridge. As long as the wind blows, the cloud as a phenomenon remains fixed in place. But the water droplets which make up the cloud build up constantly on the upwind side and evaporate to the cloud's lee. Thus the particles in the cloud have a lasting existence as phenomena, but only a brief existence as objects.

pre-Permian etchplains in the ectropics also differ from tropical latosols forming today. Latosol development may have become dominant only after the
angiosperms took over the continents. This great Late Cretaceous revolution
in the terrestrial biosphere coincided with other major changes which we will
discuss here parenthetically.

[12] *The global revolution between the Early Cretaceous and the Paleocene*
made it possible for flowering plants to spread as a dense, tall plant cover
throughout all suitable climatic zones of the earth. To what extent this was
accompanied in the humid tropicoid climatic zones by a change in soil development and in relief-forming mechanisms remains an open question. It is
also unclear whether relief-forming mechanisms like those found in the seasonal tropics today were responsible for the thick kaolinitic soils, decomposition spheres, and iron deposits of the E Franconian Alb, the oldest known
surface (an etchplain and inselberg relief) found in Central Europe, formed
during the Early Cretaceous and flooded during the Cenomanian.

The revolution coincided with other major events. The age of dinosaurs
closed with the Late Cretaceous (Schindewolf, 1954, 1960, 1963); as Schindewolf has often emphasized, this faunal change was by no means limited to
the extinction of the dinosaurs, for no fewer than eighteen major faunal
phylogenetic lineages died out "suddenly" (geologically speaking) at this
time. This was offset by the "explosive" appearance at the Cretaceous-Tertiary boundary (concurrent with the final spread of the angiosperms) of over
twenty major mammalian groups alone, possessing radically new anatomical
structures, and often displaying a very rapid increase in size of individuals.
One might say that the development of the mammals and birds was attendant
on the rise of the flowering plants. Though nearly half of these mammalian
groups died out in the Eocene, the remaining twelve still flourish today. No
rational explanation (such as Darwinism, selective pressures, adaptive radiation, etc.) can in my opinion suffice to explain these facts. Some as yet
unexplained force of nature must be at work here, revered in the past as a
creative spirit.

At the same time three major changes took place in the earth's structure.
In the Late Cretaceous the Old and New Worlds drifted apart from the mid-
Atlantic Ridge, forming the Atlantic Ocean. At the same time the first syn-
continental compression in the region of the alpinotype orogeny occurred,
taking the form of the decisive pre-Gosau orogeny in the Alps (which were
subsequently uplifted). Thirdly, the Late Cretaceous was accompanied by an
extraordinary worldwide transgression (the continents at that time were very
flat), and the earth went through one of its greatest phases of marine inundation. This must have helped to lessen climatic differences, an effect which
was reinforced throughout the Tertiary by the lack of polar ice caps and the
consequently reduced temperature differential between the higher and lower

latitudes. Relief fragments from before immediately pre-Cenomanian times are nowhere exposed. The earliest sizable relief fragments date from the Lower Tertiary. We therefore feel justified in calling the time from the Paleocene-Eocene to the present the "geomorphic era," and calling the Tertiary from the same point till well into the Pliocene the time of the "tropicoid paleo-Earth."

[13] *Aging and inactivation of latosols. Laterites, pisolites, and iron rinds.* As long as warm acid soil water is sufficiently available, latosol soils and their associated etchplains continue to be actively formed. During tropicoid paleo-earth times, minor climatic fluctuations and occasional arid periods must certainly have interrupted this process, but as already mentioned, once such a brief period had passed, the old mechanism of double planation is reactivated, and active development of the inherited surface continues.

At the Pliocene-Pleistocene transition such climatic fluctuations became more marked. The Ice Age hiatus reduced the region of the humid tropics to its present belt, bringing many broad surfaces with their ancient latosol sheets into arid climates. This rendered continued planation impossible here, and eventually led to their stripping.

Similar developments also occurred where surfaces in the humid tropics were subjected to rapid uplift. Dissection started to work inward from the outer rims, leading to edaphic desiccation of the adjacent latosol sheet, and interrupting any further active planation. Both these factors led to inactivation and finally to removal of the soil sheet.

Inactivation may intensify another process, which under certain conditions even occurs on active latosol sheets, increasing their life expectancy. Nontropical relief-covering soils display only distant echoes of this phenomenon, which may protect inactivated latosol sheets and their relict surfaces from erosion: we refer to the development of laterite, hard iron concretions, and iron crusts at or near the surface of the soft latosol blanket. Increasing laterization goes hand in hand with decreasing clay content.

This development may be followed through several stages. The first indications are found in nearly all latosol sheets as the stringers of manganese and iron concretions described above. These bean-iron ores or pisolites often form the last remnants of former red loam sheets, and are therefore important clues to its former existence. They are typically rounded, clearly concretionary iron nodules, not simply transport-rounded fragments of massive laterite crusts.

Greater enrichment of iron produces frameworks of hard iron crusts in the soft latosol. One may speak here of *in situ* development of a soil skeleton. Such crusts are also found in soil sediments and in clay and sand layers of the accompanying decomposition sphere. Remains of cellular frameworks form important clues to the former existence of red loam sheets or of the corresponding decomposition spheres.

The complete filling in of the iron crust forms a very hard, compact laterite, sometimes many m thick. This may develop slowly after climatic or edaphic desiccation, but may also form where excessive slash-and-burn cultivation has led to frequent burning of the savanna or its secondary vegetation. The resulting crusts, at first very slight but nonetheless irreversible, can cause serious soil damage within a few years. Even very young crusts may be so dense that fine material can be returned to the surface only by termites working through fresh weathering cracks (Büdel, 1952). Only along such cracks can savanna trees recolonize the area (see Troll, 1936).

Since the lateritic crusts, like the latosol sheets from which they develop, extend across old etchplains, they cut across the most varied of rock types, just like etchplains. The crust often forms a caprock which is uniformly hard, though not quite as continuous as the latosol sheet from which it derives (Bremer, 1967a). The upper surface of a laterite crust is partly marked by onion-like concentrically layered spheroids, and partly by slaggy weathering scars; it is often covered by a newly formed soil. Beneath the crust, latosol may often be found, still in its original soft condition. In Australia, iron crusts sometimes ten m thick overlie a mottled zone consisting of white clay with red and brown iron flecks, grading into a white kaolinitic pallid zone below (Walther, 1924).

These features make the massive laterite crusts one of the most striking phenomena of tropical soil development. Due to their irreversible hardness and longevity, they provide valuable evidence for the study of the relief history of an area, even when preserved fragmentarily in other climatic zones. The aging of relief-forming soils lends them a certain stability even when they have become inactivated.

A mass of literature is accordingly devoted to the subject of laterite, of which Maclaren (1906), Fermor (1911), Glinka (1914), J. Walther (1915, 1916, 1924), Harrassowitz (1926), Prescott and Pendleton (1952), Kubiena (1954, 1964), Finck (1963), Bremer (1967a), and Goudie (1973) represent only some of the main stages in the development of our knowledge. It is characteristic of this scattered literature that Goudie's book, presented as a summary, names only the first and sixth of the above pioneering works, besides suggesting a new name for laterite (''ferricrete''). Such semantic juggling does little to help us explain the phenomenon itself.

Early in this century (Maclaren, 1906) it was recognized that laterite sheets are not distributed evenly over remnant surfaces but rather follow old, shallow drainage lines, recognized by Bremer (1967a) as networks of shallow wash depressions. Here laterite development has been promoted not only by especially deep *in situ* weathering, but also by lateral supply of iron oxide. Continued planation wears down the unprotected areas next to the hard crust more rapidly, so that the remains of such hard crusts today crown low, remnant mesas with sharp erosional rims (breakaways). The surrounding Recent surface may lie 20-50 m lower. A laterite-caused *relief inversion* has therefore taken place in the course of gradual parallel surface lowering (see also footnote 39, p. 151, and Section 2.5.2.1).

New, secondary laterite development may then set in on the lower surface, for which the French researchers in particular assume an increased lateral supply of iron oxide (Tricart and Cailleux, 1961-1964, 1965-1972; Journaux, oral communication). They differentiate between the thicker, autochthonous "carapace ferrugineux" or "carapace lateritique" on the higher mesa, characterized by less quartz and silcrete, and the thinner, less solid, but more quartzitic "cuirasse ferrugineux" below, growing on the lower surfaces by lateral supply from, e.g., the mesa slopes. The word "carapace" is taken from the thick ingrown shell of a turtle, while the word "cuirrasse" refers to a lightly donned breastplate.

Of course the massive crusts on such mesas may, especially after climatic change, undergo renewed weathering and fissuring. Fragments of them may be preserved, due to their great hardness. Their angular shape distinguishes them from the rounded concretionary pisolites, which they otherwise resemble in color, composition, and weight. Such fragments are found as "lateritic rubble" on the slopes of mesas. On relict surfaces in arid regions, where they commonly occur together with pisolites, they indicate that the etchplain was formed in a more humid past.

A further clue to a past humid climate consists of chocolate-colored iron rinds, 1-3 cm, or at most 5 cm thick, covering the bedrock. These have been described by Bremer (1967a) in Australia and by Hagedorn (1971) in the Tibesti Mountains of the S Central Sahara. These rinds are found primarily on sandstones and quartzites, but also on granitic surfaces, and occasionally on the level tops of barren inselbergs in arid regions. I encountered them in the extremely arid Bani Depression S of the Moroccan Anti-Atlas. In the mechanically disintegrated debris covering the inselberg footslopes here, the rinds cover the oldest surfaces of fragments which were later shattered anew. As the above-named authors have stressed, iron rinds differ markedly in color, thickness, and composition from the desert varnish which covers most of the rock fragments there, and which is never more than 0.5-1 mm thick. Analysis of the thick rind shows that the most important constituent is FeOOH, just as in laterites. The fact that rinds are preserved on the footslopes of inselbergs, of all places, proves once again how tenaciously large relief elements from the humid past are preserved in today's arid climates.

Finally, traces of deep weathering along bedrock joints may be regarded as traces of old decomposition spheres whose weathering mantles proper have washed off. Bremer (1971) has included penitent stones (first described by Ackermann, 1962) in this category. The extraordinarily deep and closely set joints, following old lineaments in the Cambro-Ordovician sandstones, of the Tassili n'Adjer near Djanet in the Central Sahara must surely also be explained in this fashion. These widened joints are often many tens of m deep, with vertical, even overhanging walls. That they have been only minimally weathered in the desert climate of today is shown by the fact that the now famous rock paintings adorning their lower margins, dated at 6000-8000 B.P., are still almost completely fresh (Gardi and Neukomm-Tschudi, 1969).

2.3.2.3 The Wash Surface

The foregoing description of the basal weathering surface and latosol sheet as the actual agents of intensive weathering has acquainted us with the first part of the mechanism of double planation. We also listed the pedological residues which provide evidence, after the climate has turned drier or cooler, of the former domination both of the double planation mechanism and of the appropriate humid-tropical climate.

It must again be emphasized, however, that the prime evidence for the former presence of this mechanism is its relief product: the typical *etchplain and inselberg relief* that is still being formed in the humid tropical lowlands today. Thanks to the fact that its flat form is stamped into the hard bedrock, this relief has proved very durable even after climatic change; it may undergo traditional development but it remains recognizable even in altered, remnant form.

The etchplain and inselberg relief results from typical erosion on the second surface of the double planation mechanism, the *wash surface*. It is here that the final, flattening erosion occurs, which we have taken care to distinguish from the remaining kinds of areal denudation (i.e., slope erosion on steeper gradients and in other climates). The processes taking place at the basal weathering front in the decomposition sphere and pedosphere merely create the indispensable conditions for surface wash. How does transport occur upon this surface?

[1] *Precipitation, surface water film.* The rainy season closes in October in most northern hemisphere savannas (e.g., along the Niger, the Chari, or the Nile Rivers), and somewhat later along the E half of S India and on the Orinoco River. For months, a merciless heat then oppresses the land. The clayey soil is hard, especially where enriched by solutes which have been drawn up by evaporation of soil and ground water and precipitated near the surface. Due to the low absorption capacity of kaolinite and halloysite, desiccation cracks in the soil are very narrow. The pore volume of the soil, on the other hand, increases with greater fineness of particles, especially when these are platy (e.g., kaolinite) or rod-shaped (e.g., halloysite) instead of spherical (Scheffer and Schachtschabel, 1973, pp. 184-189). Such clayey soils, however, have fewer large pores which let water through, and more

middle-sized and fine pores. The middle-sized pores can be penetrated by bacteria (1-0.2 μm in diameter), but not by root hairs; they permit slow circulation of water, and their powerful molecular forces draw the soil water by capillary action far above the ground water table. The high number of bacteria in latosol soils probably increases the amount of middle-sized pores, making the soil more permeable despite the dense packing of its particles.

Larger soil fauna riddle the soil. At the end of the dry season the surface is covered with kaolin-rich fine matter brought up by these animals. This is easily attacked by raindrop impact and washed away. The vegetation (grass sod and herbaceous plants) is withered, patchy, and cropped, suffering especially under the inroads of natural and man-made fires. The fields of millet and other annual crops have been harvested, and the earth lies bare. Walking through the savanna, one rarely sees humus: the soil appears yellow or even red right at the very surface.

For one to two weeks, the clouds begin to pile up every day at noon into greater and greater masses, heralding the coming rainy season. This usually breaks suddenly one day with a heavy downpour: thunder and lightning blaze and crash like fireworks simultaneously on all sides. The state of mind of a European right before and during this tremendous discharge may be pictured, according to Hann (1910), "if one imagines the discomfort which one feels in Europe shortly before a summer storm breaks, increased tenfold" (p. 53).

But while a summer storm in Europe usually ceases after brief showers, the torrents of water in the tropics pour down for hours as though a host of closely crowded water faucets had been turned on up in the heavens. This continues nearly every afternoon for weeks and months. A half to one cm of water may fall per minute, or 5-10 l/m^2. In the ectropics, it is a rare extreme for even two liters to fall in a minute. In Central Europe, 140 l/m^2 in a day is nearly an absolute maximum, while in the tropical rainy season this is almost average. Were it not for percolation, evaporation, and runoff, this amount would yield a water column 14 cm high (see Photo. 16).

The first effect of precipitation on the dried-out ground is to produce a continuous water film a few mm to over one cm thick. The capacity of the soil to resist moistening (which I have in the past overestimated) actually lasts only for minutes, and is therefore unimportant to runoff. It can at most help to hasten the spread of the first water film, until the air has been driven out of the cracks into the soil.

[2] *Infiltration, soil water, and groundwater.* Heavy infiltration sets in with the first rains and with the creation of the water film. At first infiltration takes place through desiccation cracks, as long as these are open, then along root tubules and animal tunnels (made particularly by termites and worms), and through the coarse and medium-sized pores (possibly created largely by the abundant soil bacteria).

According to Herrmann (1968, 1970, 1971), the path of water percolating through the densely packed soil is determined largely by the pressure gradient. Many molecular forces acting on this water retard its downward progress, but once it is taken up as soil water, these same forces retain it tenaciously. This agrees with Bremer's statement (1971) that in these soils springs and ground water can hardly be found even at depths of 10-20 m, or even at 30 m. The poor yields obtained here also explain why "cities are not supplied with spring water, as is usual in Germany, but rather with water which in many cases must be piped for tens of km from reservoirs" (Bremer, 1971, p. 45). The ground water table is found in wells only at greath depth, if at all. This seems to me in itself a proof that the ground in and above the area of the basal weathering surface remains long saturated, permitting intensive decomposition.

According to Herrmann (1971), the water supplied by the shorter rainy seasons of the drier savannas (having only four to five humid months) is often insufficient to permeate completely the basal surface of weathering. Decomposition here does not proceed everywhere continuously, but in shifting pockets of ground water along the basal surface of weathering. Nevertheless, even the dry savannas are covered by thick latosol sheets, underneath which deep decomposition spheres are often present. This may be a result of past pluvials occurring within the framework of Tertiary and Pleistocene hygric oscillations typical of the peripheral seasonal tropics. The preservation and continued development (or at least the traditionally continued development) of the typical etchplain relief were not affected by such climatic fluctuations.

[3] *Swelling, puddling, and solution.* Infiltration is greatest in the first few days of the rainy season, while the open desiccation cracks ensure a constant supply of water from above. As soon as the ground is moistened through, the clay minerals expand, closing the cracks, while at the same time beginning surface runoff clogs the cracks from above. Infiltration is retarded until it keeps pace with the slowly streaming ground water far below. After this most of the copious precipitational water is removed by overland flow. A thick network of anastomosing water filaments is created anew with every downpour, uniting into a fairly continuous *water film* on the wash surface. Bremer (1971) stresses that this overland flow does not in itself automatically attack and transport the compact clays. A molecular process is necessary for this to occur, which Bremer labels *puddling (Aufschwemmung).* The uppermost layer of soil, of course, has the most water. Here the firmly packed soil particles become "increasingly mobile with respect to one another, as, due to the increased adsorption of water, the particles repel one another more and attract one another less" (Scheffer and Schachtschabel, 1973, p. 202). With increasing water uptake, the consistency of the clayey soil changes from a solid to a very thin liquid soup in which the individual particles become suspended in a sluggish or stagnant water film. They are taken up as suspended

17. Broad, fine sand bed of the Palar River, 90 km WSW of Madras. View looking W. In the background the etchplain escarpment of the inselberg range of Vellore. (Photo: Büdel, March 4, 1964.)

or traction load in the countless rivulets of the water film, and swept off toward the nearest stream or river. This proceeds even faster with the loose material brought to the surface by burrowing organisms: the loose material quickly enters suspension and is carried away.

The wash film carries not only suspended matter and fine sand, but also large quantities of solutes. The rivers expand many times in width during the rainy season, the amount of water carried increasing by an even greater factor, until the river swells to its farthest banks. It is self-evident, therefore, that these rivers must carry enormous amounts of dissolved matter, for it is in the tropics above all that many rocks are particularly susceptible to solution.[35]

[35] General figures have been compiled by Louis (1968a, pp. 92-93). Like Fahn's careful measurements (1975) on certain Alpine foreland rivers, they tend to show that dissolved loads of rivers in all climatic zones are considerably higher relative to total load than previously assumed. Particularly high values were shown in all zones by rivers flowing through deposits of salt, gypsum, and limestone. Other conditions being equal, the highest values are reached by tropical lowland rivers. Along the humid tropical to monsoonal coast of Australia (though this is largely mountainous), Douglas (1973) was able to show disproportionately high dissolved loads in 48 rivers, most of which were fairly small. According to Corbel (1959), 70% of the load transported by the lowland Congo River consisted of solutes. Following Louis, one may take this value as an average for humid tropical rivers flowing through moderate relief. This still forms an open field for geomorphology and hydrology.

The dissolved load derives less from the surface wash film than from the ground water, which becomes ion-enriched while circulating along the basal weathering surface. The surface wash film supplies the rivers with much of their suspended load.

During the dry season rivers on the etchplains of the seasonal tropics shrink to narrow rivulets or disappear entirely, leaving behind broad dry beds of fine sand (see Photo. 17).

[4] *Surface wash, rainy season rivulets, wash channels.* Intensive weathering provides such a fine-grained (partly soluble) substrate that the rainy season wash film is well able to transport it, and can perform much erosion. It is important to note that this areal overland flow and erosion remains constant for weeks and months. This process does not take place in an unbroken sheet of flood water. On these gentle gradients, movement occurs slowly and discontinuously; traction load is often carried only a few m and then dropped at a slight decrease in slope or at some small obstacle (a tuft of grass, a clump of leaves, an animal track, a termite hill, or a footpath). At the next cloudburst, water will pond here until it is forced into a slight detour, taking the material a short way further (Büdel, 1965). These *rainy season rivulets* thus form temporary connections between small puddles,[36] so that the water film is distributed unequally between the minute irregularities of the ground surface, and does not cover everything at once. But from one cloudburst to the next, and from one rainy season to the next, it washes without exception across the entire surface (see Photo. 16). Surface wash in the shifting network of tiny rainy season rivulets and wash channels is therefore the decisive process causing planation on the etchplains of the humid tropics. "The existence and the surface-forming effect [of planation] are linked to two important conditions: firstly, to the ubiquity of easily transported weathering products on the wash surface (which in turn depends on rapid intensive weathering and on a thick soil sheet which is fully decomposed right down to the basal weathering surface), and secondly, to the unique behavior of the larger rivers in this zone (which again depends on the same conditions)" (Büdel, 1965, p. 20). This will be considered in more detail below.

[5] *The micro-relief of active etchplains. Wash depressions and wash divides.* No terrestrial plain is as flat as a becalmed lake. Even reliefs which most resemble this ideal, e.g., lowland depositional plains and the flattest etchplains, are no exceptions. This is due very simply to the fact that even in these two cases, running water is the main agent of deposition and/or erosion. For water always flows downhill, and can therefore be morphodynamically effective only on a slope, however gentle it may be. This leads perforce to the creation of gently sloping surfaces.

Depositional surfaces slope unidirectionally. The outwash plains (*sandur*)

---

[36] On steeper slopes (up to 10°) and sandier ground, they form continuous wash channels, but these are not incised (Bremer 1967a, 1971).

of Iceland usually slope at 3‰, rising to over 10‰ near the moraines. The gravel fields of the Alpine forelands (e.g., the Münchener Schiefe Ebene and the Memminger Feld) also reach 12‰ (6°) near the terminal Würm moraines. In both cases the impression is fully one of a morphological plain.

This is even more true of tropical etchplains. The Tamilnad Plain of S India climbs from the Cormandel Coast by only 200 m over a 100 km stretch, having an overall gradient of only 2‰, or 0.11°. This is even less than that of the hyperflat outwash plains of Iceland, and is only one-sixth the gradient of the upper part of the Münchener Schiefe Ebene.

The outwash plains of Iceland also have a very gentle internal relief, consisting of shallow channels parallel to the slope, created by coalescing alluvial fans. During times of low meltwater supply, small streamlets may gather in these channels. An exact contour map would show how they all run in the same direction. But the streamlets neither created these channels nor the wash plain as a whole. Rather these were created by the fans deposited during the main periods of melting. The channels in the outwash plains can therefore only be classified in the very loose descriptive category of "a unidirectionally sloping concavity containing a river." A "valley," wherein a genetic relationship is present, is a special case according to our definition. This runs briefly: "a valley is an elongated, open-ended concavity created by the active linear erosion of a river and by the slope erosion controlled thereby" (for greater detail, see Büdel, 1970b).

This definition is based on the distinction between the two basic kinds of fluvial activity. The first is passive transportation of dissolved, suspended, and traction load, occurring automatically with waterflow. The second kind of work is active erosion, which by no means characterizes all flowing water. When the river's linear erosion (vertical, lateral, and sometimes headward erosion) outstrips areal denudation on the interfluves, the river creates a concavity, the slope and erosion of whose sides are controlled by fluvial erosion. Frequently not only the present, but also several past relief generations (in the form of valley terraces) may take part in the appearance of the whole. Only for such concavities, created by active fluvial erosion, do we use the term "valley."

Tropical etchplains, like the outwash plains of Iceland, undulate gently at right angles to the overall slope. The entire Tamilnad Plain of S India is composed of very gentle undulations, whose hollows I call *wash depressions* (*Spülmulden*; Büdel, 1957a, 1965). The very gentle convexities forming unobtrusive divides between these depressions we name *wash divides* (*Spülscheiden*; see Figs. 40 and 41). The flanks of these depressions and divides usually (about 95% of the time) slope at less than 20‰ (2% or 1.2°), and most are in fact less than 10‰ (1% or 0.6°). The crests of the divides are usually 200-500 m apart. These open-ended erosional concavities differ markedly from valleys not only by virtue of their hyperflat shape, but also in the following features:

They invariably have the same gently saucer-shaped cross-profile.

They are drained not by a perennial river, but by large rainy season rivulets. During the dry season the largest of these rivulets possess fine sand beds, which are not incised into the sauce-shaped profile. The grain size of this fine sand is the same as that of any rainy season rivulet.

At their bends the wash depressions show no alternation of undercut and slip-off slopes.

No working edges are present, and above all, no terraces.

From the floor of the wash depression to the low crest of the divide, the type and thickness of the soil remain essentially the same.

The wash processes remain the same throughout the cross-profile. This is not affected by the circumstance that the water may gather in the floor of the depression into somewhat larger, non-eroding rivulets.

Shield inselbergs may appear anywhere along the cross-profile.

The wash divides are usually not plateau remnants preserved from older relief generations, but current, active forms, fully integrated into the hyperflat relief. The sporadic remnants of older surfaces, e.g., inselbergs, are usually crowned and protected by laterite crusts, and display sharp breakaways.

As Bremer (1967a) showed in Australia, crusts were often formed by lateral supply of dissolved iron in the floors of former wash depressions (*cuirasses ferriguneuses*, to use the French nomenclature). Where crusts now cover the wash divides, they indicate a relief inversion during the course of parallel lowering of the surface. Displacement of wash depressions during erosional lowering is characteristic of this type of concavity, while displacement even of the largest rivers is far easier and more common on such surfaces than stream piracy in valleys. The bifurcation of the Casiquiare River in S Venezuela is a well-known present-day example. The gradual Upper Tertiary diversion of many of the S German Danubian tributaries into the Rhine system can only be explained by the feasibility of such relocations on the then existent etchplain relief (see Fig. 66); to assume countless cases of stream piracy for this purpose is unrealistic.

All these features show that wash depressions are not valleys in the genetic sense. They were not created by fluvial incision with subsequent slope lowering. Erosion on their flanks and floors proceeds independently and *in situ*, and is not controlled by any active fluvial erosion on the floor of the depressions. Fluvial erosion is in fact absent. Of course the rivulets which gather in the floor of the depression act passively as a base level for the slopes, but even in this role, the rivulets are fully integrated into the erosional mechanism of double planation.[37]

[37] When Louis (1964, 1968a, 1973) speaks of a "wash depression valley " (*Spülmuldental*), he is using the word "valley" in a far wider sense than our genetic sense defined above. He thereby runs the risk of confusing the highly distinct subaerial relief types of the zone of ectropic valley-cutting with the semi-humid to tropical zone of planation, which are governed by extremely

[6] *The valleyless rivers in the active etchplains.* These conclusions regarding wash depressions are, with a few limitations, also true for the larger rivers on the active etchplains (see Photos. 16 and 17).

In all the ectropics and in all mountains of the deserts and humid tropics, linear fluvial erosion clearly exceeds erosion on the interfluves. It therefore to a great extent determines the angle of the slopes and the type of erosion occurring there, creating the valleys and valley relief forms in these areas.

The active etchplains of the lowlands (and to some extent of the intermediate altitudes in the humid tropics) are a striking exception to this rule. The rivulets of the wash depressions collect into larger rivers, forming their local base level of erosion. Yet the rivers perform only the passive part of fluvial activity, that of transporting all material supplied to them from their flanks. Here we may note a remarkable phenomenon: the rivers carry much dissolved and suspended matter, but their traction load consists almost exclusively of fine sand of the very same grain size (30-200 μm) as that transported in all rainy season rivulets and wash channels. Moreover, the grain-size spectrum of the fluvially transported material corresponds almost exactly to that of the latosol soil itself (Bakker and Muller, 1957). This means that the river takes up and carries away only the material already prepared for it by intensive weathering in the latosol soils. The riverbeds themselves flow on top of these thick soil sheets. Thus fluvial work in the humid tropics "follows different rules from the non-tropics; it is characterized mainly by the washing out of weathered material, and differs from mechanical erosion in the ectropics" (Bremer, 1971, p. 155).

In the rainy season the rivers swell enormously, spilling laterally far out over the plain in which they flow. Many form natural levees, creating dammed-up lakes (river limans) in the lower reaches of their tributary wash depressions, where the water may remain ponded long after the end of the rainy season. Naturally the latosol soil becomes particularly saturated in these flooded areas. During the dry season, the river dwindles rapidly to a narrow rivulet running through the middle of its rainy-season bed (see Photo. 17). Often the river vanishes completely, leaving behind a broad dry bed of fine sand. Two features are worthy of note here.

Firstly, around 90-98% of all river courses on active etchplains have beds of sand only, without coarser gravel. Exceptions occur, for example, where old conglomerates or other easily crumbled rocks such as mylonite cross the river and produce gravel. In most cases, these cobbles are comminuted within a few hundred m. Even along gravel-carrying portions of the river, no fluvial down-cutting occurs.

Secondly, the sandy beds are not incised, and the surface slopes gently

different morphodynamic mechanisms. This, however, does not reduce the value of Louis's research.

right to the river. At most a bank 0.8-2 m high may be present, which changes with every rainy season.

Consequently these sandy riverbeds, like the rainy season rivulets, run for the majority of their courses over the latosol soils; they are completely separated by this soil blanket from the underlying bedrock, and rarely have the chance to cut into it (see Figs. 40 and 41). According to Herrmann (1970 and 1971), in regions with longer dry seasons even the permanent rivers lose all contact with the ground water circulating above the basal weathering surface, and are no longer fed by this (as is invariably the case in the humid ectropics), maintaining an independent regime in the upper levels of the soil.

The course of the riverbed may be interrupted by occasional basal knobs rising close to the ground surface. The river does not cut through these rocky barriers, but crosses them in waterfalls or rapids. That this is true even of larger rivers is proved by the cataracts on the Nile, the Congo, and the Amazon.[38] It was these cataracts which made it so difficult for the explorers of the sixteenth to nineteenth centuries to penetrate into the interior of S America and especially of Black Africa.

The stability of such rocky barriers in the longitudinal profiles of large rivers is shown by an observation made by Bakker (1965) in Surinam. The rocks of each set of rapids (locally called "sulas"), he found, are inhabited by a different species of snail. Biologists are aware of how much geologic time is needed for evolution to produce completely separate species: the rocky barriers must have existed for a long period during which the tropical river was unable to erode or cut through them. The river's longitudinal profile has remained suspended at the same rocky barrier for immense spans of time; between the steep breaks in slope of the rapids, it runs through very long flat stretches of latosol soil without ever coming into contact with bedrock. That it could not and did not erode the latosol soil is obvious; what is important was that it could not saw through the rocky barriers, because it was almost completely without erosional tools.

As Bremer (1971) has specially investigated, such rocky barriers across rivers are subject more to chemical than to mechanical erosion. Even so, rivers rarely pick up coarse cobbles and then only briefly. The rivers therefore do not create their own coarse erosional tools through mechanical vertical erosion, and due to their great dissolving power they quickly devour any coarser bedload produced at the rapids.

Like the wash channels, the rivers show no significant amounts of lateral erosion; their fine sand beds are bordered at most by very low banks, and in fact often have natural levees. They have no terraces, and their meanders show no differentiation between slip-off and undercut slopes.

---

[38] The Amazon itself flows through a depositional lowland, and is therefore navigable from its mouth to above Iquitos.

As a result, even the larger rivers on the active etchplains accomplish only the passive transportation inherent in all fluid motion. They essentially cannot perform any active vertical erosion, certainly not faster than intensive chemical erosion and its resultant planation can work down across the entire surface. Unlike all other relief regions of earth (including the tropical and subtropical mountains and the mid-latitudes up to the edge of the continental ice sheets), these plains are unique in that planation works downwards as fast as vertical erosion, creating plains instead of valleys. The rivers in their role of base levels of erosion and transportation vehicles are fully integrated into the entire network of planation.

The completed product shows that the creative forces involved must have been effective for extremely long periods of time within a considerable range of climates. Otherwise the etchplains which evenly bevel the most varied of rock types over thousands of km (including the etchplains of Australia, S India, the Sudan, and the other African savannas as well as those of Brazil and Guiana) could never have been produced.[39]

[7] *Effects on the basal weathering surface. Etchsurface depressions and etchsurface divides.* In most thick latosol soils, the wash surface relief of gentle divides and depressions is reflected on the basal surface of weathering as roughly parallel undulations. This is superimposed on a small-scale relief of basal knobs and shield inselbergs which measure a few m to over ten m in diameter. Hagedorn (1971) has suggested that these reflections of the external wash depressions and wash divides be given the names *etchsurface depressions (Rumpfmulden)* and *etchsurface divides (Rumpfschwellen)*. These terms become morphologically important when removal of the latosol sheet (through upheaval or through increased aridity) partially or wholly reveals the former basal weathering surface, which becomes subject at best to traditional development. In such stripped etchplains, of course, the highly variegated small-scale relief of the basal knobs is also exposed.

2.3.3 THE OVERALL APPEARANCE OF ACTIVE ETCHPLAINS

2.3.3.1 Inselbergs, Inselberg Ranges, Etchplain Passes, Wash Ramps, and Moats

The existence of isolated, steep-walled, and often sharp-footed[40] inselbergs and inselberg ranges towering over the active etchplains has long attracted

---

[39] As long as such a surface retains its homogeneous overall shape and continues to work under the same relief-forming mechanisms even after brief climatic changes, it seems to me unimportant to distinguish between monogenetic and polygenetic surfaces (Demangeot, 1975), even when the wash depressions and wash divides have shifted, or when laterite development has brought about a relief inversion.

[40] Inselbergs with gently concave footslopes are in the minority on active etchplains. Some

investigation, far more so in fact than the much more important, superordinate phenomenon of the surrounding plain. The plain is afterall the direct and primary product of the dominant erosional mechanism here. The inselbergs are simply untouched remnants left standing amidst the general planation, isolated remnants with very typical sharp edges (see Photos. 18-22). The literature which has accumulated since the time of Passarge (1895, 1923, 1928, 1929) and Bornhardt (1900), is legion. Yet all of these study inselbergs in isolation, divorced from the context of their surrounding plains. Bornhardt's

18. Small group of inselbergs on the latosol-covered Tamilnad Plain near Chingleput, 55 km SW of Madras, S India. (Photo: Büdel, March 6, 1964.)

early descriptions have given inselbergs the name "bornhardts" in English, a term still in use today. Further literature includes works by Brandt (1917), Thorbecke (1921), Waibel (1928), Mortensen (1929), Jessen (1936), Krebs (1942), King (1948), Kayser (1949), Pugh (1956), Büdel (1957a), Dresch (1957), Wilhelmy (1958a), Maull (1958), Ollier (1960), Tricart (1965a, 1965b), Thomas (1965, 1966, 1967), Bremer (1967a), Rust (1970), and Do Amaral (1973).

The connection between inselbergs and the tropical climatic zone (both humid and dry) as well as with the tropical lowlands was quickly apparent,

---

have a flat scree slope on one side and a sharp piedmont angle on the other. In such cases it is the sharp angle which is characteristic. While gentle footslopes occur in all climates, sharp piedmont angles lacking protective covering are found only in the humid tropics, so that the explanation of this unique phenomenon must be the primary goal of research. It will be found, we believe, that the explanation facilitates the understanding of gentle inselberg footslopes as well.

19. Sharp foot of the etchplain escarpment between the Tamilnad and Bangalore etchplains near Ambur on the Upper Palar (see Fig. 44). The low shield inselberg (partially exposed basal knob) in the left foreground is of the same granite as the escarpment in the background. The sharp piedmont angle of the escarpment does not coincide with a lithic boundary. (Photo: Büdel, March 7, 1964.)

20. Plateau inselberg near Krishnagiri, 80 km SW of Bangalore, in front of the etchplain escarpment between the Tamilnad and Bangalore Plains. The low shield inselberg in the foreground is of the same granite as the mountain in the background. Rainwater paths on the mountain's rocky sides show its great stability. Greater soil saturation at the mountain's foot is indicated by groves of trees there. (Photo: Büdel, March 8, 1964.)

21. The 350 m high plateau inselberg of Ayers Rock (nearly in the middle of Australia, in the Amadeus Basin, 340 km SE of Alice Springs). Extremely sharp piedmont angle between the mountain and the surrounding plain. The latter cuts smoothly across the same steeply dipping arkose of which the inselberg is composed. Photo used for sketch in Fig. 42b. (Aerial photo from Schwarzbach, 1970.)

22. Bell-shaped, bare-walled inselberg near Krishnagiri, near Photo. 20. Typical outlier inselberg. Steep blockhalde with sharp piedmont angle toward the etchplain, otherwise similar to Photo. 20. (Photo: Büdel, March 8, 1964.)

but only since about 1945 have inselbergs and etchplains begun to be viewed as expressions of a single relief type. Inselbergs do not form elongated ridges between wash depressions, but rather (and hence their name, literally meaning "island mountains"), these strikingly cylindrical relief forms are dotted across the entire surface. They may occur on wash divides, on wash slopes, and even near riverbeds, as has been reconfirmed by Jeje (1973). Where they form "rows" or "lines," these do not follow the wash divides, but run diagonally across the gently undulating relief of divides and depressions. This again suggests that both the surface and the inselbergs are formed independently by general erosion, and are not created by or composed of valleys.

The inselbergs themselves provide further proof of this. Their foot (often very sharp) is usually the only site of working piedmont angles, which though localized, are very marked. These show no signs of originating through lateral fluvial erosion: nowhere do we find river gravels, whirlpool potholes, or the like. The piedmont angles, moreover, lie at very different heights, according to whether the inselbergs sit higher or lower on the wash slopes or wash divides. Since they circumscribe only the individual inselbergs, they can nowhere be joined together in the manner of terraces. The only factor capable of forming these sharp working angles, therefore, is the general planation (see Photo. 18).

The way toward an explanation of the mechanisms involved remained impassable as long as inselbergs were regarded without considering their genetic connection to their surrounding etchplains. The early authors, especially, therefore often retreated to a geological viewpoint. It was found that inselbergs are preferentially composed of harder rock: dense granites, quartzitic metamorphics, thinly bedded sandstones and arcoses, quartzites, thick-bedded basalts or limestones, reef limestone, etc. This led to the simplistic notion that inselbergs were exceptionally resistant units (*Härtlinge*). Yet it should still have been obvious that the same pattern of lithic units is often exposed in the ectropics, without any accompanying active relief of inselberg-studded etchplains. Büdel (1957a, 1965) and Bremer (1965a, 1967a, 1973b) furthermore recognized that the hard lithic units making up such an inselberg often crop out all around this feature, underlying the surrounding surface (see Figs. 42a, b, and c, and Photos. 19 and 20). The surfaces beside the inselberg may cut across the same lithic units with no visible traces (Fig. 42a, right), or with only a slight pile of small shield inselbergs or tors (Fig. 42c). All these occurrences of inselbergs are indeed "lithically controlled," but each one differently. For geomorphology the appropriate question does not concern the fact *that* something occurs, but rather *how* it occurs.

Figure 42d shows an inselberg with gentle flanks and footslope, still largely covered by a red loam blanket. Here steepening of the flanks may continue, as shown in Fig. 42e. In Fig. 42f and a, the steep-walled, barren inselberg type has been reached, whose convex sides weather back with extreme slowness. Figure 42b corresponds well with the impressive Ayers Rock in Australia (Photo. 21), studied by Bremer (1965b). This plateau of solidly cemented, thin-bedded, nearly vertical arcose rises 350 m above the surrounding etchplain. Its walls are steep and nearly vertical along its entire eight km long circumference, and it has sharp piedmont angles lacking any traces of a gentle footslope or scree. Both the plateau and the surrounding, perfectly flat plain are composed of precisely the same rock. Using several generations of weathering forms such as tafonis and exfoliation, Bremer was able to show that the steep outer walls, though barren, are extremely stable, and have not weathered back over a very long time. This is also supported by the lack of any scree. Even a wash ramp is absent.

Larger complexes of inselbergs are especially common toward the interior edges of all etchplains. These complexes, called *inselberg ranges (Inselgebirge)*, are often marginally broken up by small fingers of the plain (triangular reentrants or surface apices, see Section 2.3.3.3). The ranges are frequently crowned by relict surface fragments, whose altitudes correspond to those of neighboring, higher etchplains. Even higher, isolated inselbergs often carry such remains. Such peaks (with or without relict surface fragments) have been named ''zonal inselbergs'' by Kayser (1949), while Büdel (1957a, 1965) speaks of *outlier inselbergs* (see Photo. 22).

All inselbergs (and inselberg ranges) are separated by etchplain extensions forming smooth corridors between them. I have named these *etchplain passes* (*Flächenpässe*; Büdel, 1965). Since etchplains surround all inselbergs, the plains themselves must largely consist of such passes. Had the passes been fluvially created, one would have to ascribe wildly meandering courses to the rivers, while fluvial sediments are lacking. Yet the surfaces reach from both sides into the passes, joining evenly and without a scarp. At most the slope on one side may be a few pro mille steeper than on the other. Yet a corridor between two inselbergs will not change in height even when it is many km wide. Within the passes are wash depressions which run off to both sides. As everywhere else on the etchplains, these wash depressions have no steep valley heads, but merge gently into the surface. This is in contrast to some etchplains of the perhumid tropics, as we will discuss in Section 2.4.

These features of the etchplain passes provide important evidence showing that such surfaces are produced by parallel lowering from above downwards. This can take place only through the relief-forming mechanism of double planation surfaces which we have described.

As will be shown below using Figs. 44, 45, and especially 49, this form of parallel lowering (corresponding to the theoretical *Primärrumpf* or primary peneplain, primary torso) is found wherever a semihumid-tropical climate and a tectonically stable setting permit double planation. This occurs right up to the abrupt lower edges of inselbergs, inselberg ranges, and etchplain escarpments, and even in small intramontane plains (see Figs. 47 and 48). This relief-forming mechanism therefore governs even the most recent surface extensions.

The rare exceptions to this rule are clearly related to variants of the above-named endogenic and climatic conditions. One of these, mentioned by Wirth-mann (oral communication) is located, characteristically enough, in the very rainy southern portion of the recently uplifted W Ghats. Here one climbs from the plain on the Malabar Coast up an escarpment over 1000 m high to the broad summit plain of the S Deccan (which includes the Bangalore Plain shown in the upper left of Fig. 44). This highly dissected scarp is a typical example of an "eroded slope," which according to Wirthmann is quickly eroded and weathered back. Instead of meeting the plain at a sharp piedmont angle, however, scarp and plain are separated by a "foothill zone" containing rapidly expanding valley embayments.

If erosion of these hills gradually created a continuous surface reaching right to the piedmont angle, then one could view this as an example of a peneplain (*Endrumpf*) in the sense of Davis (1912). It seems to me an exceptional example in which an escarpment undergoing dissection of the typical tropical mountain valley type is retreating faster than the surface with its relief-forming mechanisms can follow. In my opinion this is explained by the fact that the W Ghats are undergoing rapid uplift (the mechanism of double planation surfaces presupposes stability or at most weak endogenic uplift over long periods of time) and by the fact that the climate here closely resembles that of the perhumid tropics with their "partial planation."

The relict etchplains of the ectropics include many examples of surfaces which have been endogenously uplifted and peripherally dissected in the course of the Tertiary. With renewed downwarping, the valleys produced by dissection fill up with sediment. Should the climate remain seasonally tropic (especially where the relief-forming soil blanket has remained on the intact surface fragments), planation (or at least traditional planation) will resume across the entire block. The example of the Franconian Alb, to be described below (Section 3.3.1.5), will show how planation continued nearly throughout the Tertiary, reaching undisturbed across areas of episodic and localized dissection. This overall lowering occurred at a rate of 1-10 m per million years.

Naturally the creation of etchplains requires very long periods of time. But on the ancient shield areas of the continents, the requisite periods of relative tectonic quiet have certainly been present. The etchplains of Australia, for instance, have enjoyed an especially long period of undisturbed development, and have proved correspondingly durable.

The question remains, to what extent can such a surface expand laterally while undergoing parallel lowering. Often a surface will rise slightly (at a maximum gradient of barely 4°) toward the sharp piedmont angle of an inselberg. This *wash ramp* (*Spülsockel*; Büdel, 1965) is usually 100-300 m wide at the most. The red loam blanket becomes slightly thinner on the ramp (see Photo. 15 and Fig. 43), but reaches right up to the piedmont angle of the inselberg with thicknesses of 5-6 m. This shows that the sharp piedmont angle was created through *subcutaneous basal sapping* (*subkutane Rückwärtsdenudation*; Büdel, 1965) which, during parallel lowering of the surface, worked back laterally against the inselberg. The wash ramp is thus created by enforced footslope retreat.

Conditions for intensive weathering are especially good at piedmont angles, as shown by their supply not only of local rainwater, but also of water streaming off the mountain sides, especially where the latter have been steepened until devoid of soil. Many inselberg walls display rainwater paths marked in black by precipitated iron and manganese. The greater water content of the soil at the mountain foot is used to good advantage for plantations (see Photo. 20 for this and for the streaks of manganese and iron).

The foot of an inselberg is often notched (Dahlke, 1970). Intensified, this produces *moats*, or *marginal depressions* (*Bergfuss-Niederungen* or *Rand-senken*) mentioned by Passarge (1928-1929), Thorbecke (1921, 1927), Jessen (1936), and Pugh (1956), and studied more carefully by Twidale (1962), Hövermann (1966), and Hagedorn (1967b, 1971). Moats consist of greater thicknesses of soil toward the center of the wash ramp, directly at the mountain foot. When the climate has turned arid, the wind may blow away the soil, leaving the moat open. They are particularly common in the W Sahara.

Moats prove that subcutaneous basal sapping ceases once the wash ramp is created (Bremer, 1971; Hagedorn, oral communication); otherwise such a deep groove could never develop. According to Bremer, basal sapping ceases when the mountain slope above has become too steep to carry soil. Gently sloping inselberg flanks still have a soil blanket and can be weathered back. Once the slope has become oversteepened and barren, it becomes extraordinarily stable, as shown by Ayers Rock (Photo. 21) and Photos. 18-22.

This refutes two mutually antithetical doctrines maintained till now on the creation of etchplains: that of v. Wissmann (1951), which assumes continent-wide lateral erosion; and that of extreme resistance (the *Härtlingshypothese*), which assumes that the inselbergs were already delineated on a higher, original suface by the distribution of the harder rock units, and were then projected downwards onto the present surface.

2.3.3.2 The Rule of Diverging Denudation

The problems involved in the creation of sharp piedmont angles were largely explained by Bremer (1971, 1975), when she formulated the rule of divergent erosion. This is an important supplement to the theory of the double planation mechanism, and helps clarify the differences between tropical and ectropical relief development.[41]

---

[41] Originally Bremer referred to this as "differentiated weathering" (*differenzierte Verwitterung*), but later elaborated the term to "divergent weathering and erosion" (*divergierende Verwitterung und Abtragung*).

In the entire ectropics and in both the arid and humid tropical mountains, mass wasting follows the rule usually regarded as the terrestrial norm: the steeper the slope, the more effective the erosion. In the ectropics this is true from the highest pinnacles of the mountains down to the flatlands with their thick relief-covering soils. In the humid tropics, however, the situation is the exact opposite. The flatter the terrain, the more the relief-forming soils will be moistened right through to their base, the slower the water will flow through the dense soils, the more acid it will contain, and hence the greater the intensive weathering and attack upon the bedrock. This also means that the soil becomes saturated faster, leaving more water left over for surface wash. Thus in the humid tropical lowlands it is the flattest portions of the terrain which experience the greatest erosion, due to the planation mechanisms at work there. "This is the complete reverse of the erosional conditions in the ectropics, where mass wasting increases as the slope becomes steeper" (Bremer, 1971, p. 152). This further justifies the term "relief-forming soil," for this is here the agent of an erosion which increases in efficiency as the land becomes flatter. Only where the soil is present does erosion occur, continuing the development of flat surfaces. Where a soil is absent, as on the oversteepened flanks of inselbergs, erosion in this frost-free climate is greatly retarded. We therefore refer to this basic difference in mass-wasting as "divergent erosion."

The blanket of relief-forming soil covers all active etchplains. The overall gradient here rarely exceeds 2‰, the wash flanks being seldom more than 20‰ (2% or 1.2°); the wash ramps alone approach 4° (see Fig. 43 and Photo. 15). These surfaces of extreme erosion thus cover most of the humid tropical lowlands.

Where bedrock is exposed, as in inselbergs (Fig. 42a, b, and f), shield inselbergs, or tors (Fig. 42c), it enters a completely different climate environment. Rock surfaces here are, so to speak, edaphically arid. Physical weathering, consisting of small-scale exfoliation and grus weathering, works far more slowly, as Bremer has emphasized (Bremer, 1975).

The blanket of relief-forming soil may stretch over gentle inselberg footslopes and etchplain escarpments up to gradients of about 20°, while patches may be found even higher (see Figs. 42d and e). As long as slopes are covered by relief-forming soil, they can be weathered back. The surface may expand at their expense by means of subcutaneous lateral sapping, until a steep rock wall and moat are formed (Fig. 42f): this is how the "valleyless mountains" (*tallose Berge*) of Brandt (1917) are formed. Wilhelmy (1958a, Figs. 32, 33, 35, 52, and 53) shows some very good pictures of domes (*Glockenberge* or *Domberge*). In the captions to Figs. 55 and 56 Wilhelmy even gives a thumbnail sketch of their development when he says that "physical weathering works more slowly than the deep weathering and soil wash" (Wilhelmy, 1958a, p. 88).

At times blockhalden (boulder-strewn talus slopes) may occur on the lower slopes of barren domes (see Photo. 22), but many examples (see Photos. 20-22) show that this is rare. Now and then, of course, a block may fall from above, but it usually becomes incorporated into the relief-forming soil, where it is easily distinguished from core stones, for its structure is not aligned with that of the substrate, as is the case for core stones weathered out of the bedrock beneath (see Fig. 43). In this hothouse environment, blocks of both kinds are quickly decomposed and made ready for removal by surface wash.

Once the surface has formed a sharp angle with its steep backdrop of higher land, the overall aspect of the etchplain has reached maturity. Naturally this is only possible in areas of relative tectonic stability, where a seasonally tropicoid climate has dominated for a long period of time. Once formed, a mature etchplain is the *most durable continental relief type known*.

2.3.3.3 Triangular Reentrants (Plain Apices) and Intramontane Plains

Every etchplain, especially when mature, e.g., the Tamilnad Plain in S India (see Fig. 44), has *growth apices* both along its inner edge (in this case along the etchplain escarpment leading to the higher Bangalore Plain) and along inselbergs and inselberg ranges (here those of Vellore, and of the Jarvadi and Yelagiri Hills near the escarpment). Broad-based and tapering quickly, these growth apices ideally resemble equilateral triangles in plan, hence their other name, *triangular reentrants* (*Dreiecksbuchten*; Büdel, 1965). Similar forms are found along the edges of the etchplains of the Somali Highlands (Fig. 45), and they have been described by Bremer (1971, especially maps 7a, 9a, 10a, and 12a) from Nigeria. The literature also contains references to such forms along etchplain escarpments. We are therefore dealing with a general feature of etchplain escarpments.

At the base of the triangle, the reentrant is open to the plain, forming a growth apex or extension of the plain into the backdrop of higher land behind. Reentrants originate along zones of weakness in the bedrock. Their flanks, consisting of the etchplain escarpment (Fig. 44), are very steep and barren in places (indicating maturity and stability), but gentler flanks may still be soil-covered. Where the slopes are of intermediate steepness, the reentrant may grow: escarpment retreat with gradual steepening are possible here. This process begins at the base of the reentrant, so that it is widest there and narrows quickly. Following Büdel (1965, pp. 45-51), Bremer (1971, pp. 120-122, 1972) has described in detail the process of escarpment retreat and steepening, emphasizing that the backdrop is often steepest toward the reentrant's apex. Once this condition has been reached, the escarpment can no longer be substantially worn back.

In addition to this ideal equilateral type, narrower reentrants also occur, as well as stubbier, nearly crescent-shaped ones.

That triangular reentrants form growth apices of the surface is shown by

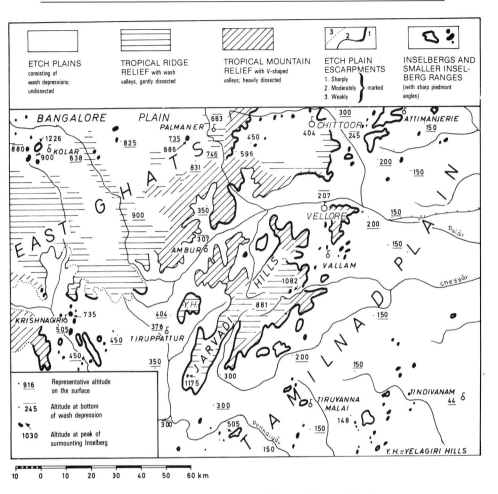

Fig. 44. Tropical relief types, on the E edge of the Highlands of Deccan (S India).

the fact that their overall slope does not fall from the sides of the triangle toward a central axis, but rather toward the plain. This slope is followed by the wash depressions: these do not fall toward a stream or river along a central axis, like the ribs of a beech leaf, but run parallel and independently toward the plain (compare valley drainage types in Fig. 50). This is shown in the cross-section in Fig. 46, whose position is indicated in Fig. 44. This section crosses three triangular reentrants in the Yelagiri Hills and the S edge of the Jarvadi Hills. The broadest reentrant is Irunapattu in the W. This reentrant, whose apex opens into an intramontane basin, is divided up by no fewer than four parallel wash depressions. The elongated valley-like reentrant of Kottanur is drained by only two, while the smaller, crescent-shaped Paramandal reen-

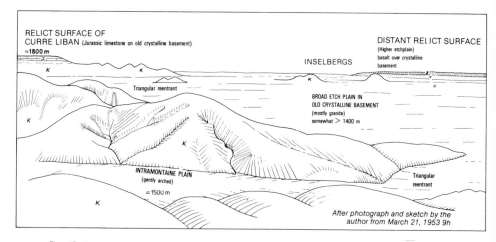

Fig. 45. Broad etchplain in the crystalline basement of the Somali Highlands. A small triangular reentrant (front right) reaches back into a small intramontane plain (front left). C: crystalline basement; J: Jurassic limestone; B: basalt cap.

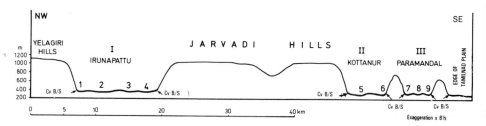

Fig. 46. Section across three triangular reentrants of the Tamilnad Plain, cutting from these into the inselberg ranges of the Yelagiri and Jarvadi Hills (see Fig. 44, below left of center). Embayments I, II, and III are occupied by several parallel wash depressions (1-9), proving that they are growth apices of planation, not products of lateral erosion by the rivers emerging from mountains at the triangular reentrant's tip.

trant again has three parallel wash depressions. Such reentrants are clearly not "valleys" cut and expanded by a single river. Their form can only be explained as growth apices of the plain onto which they open.

An analogous development is even clearer in the *intramontane basins (intramontanen Ebenen)*. These are flat-floored basins sunk abruptly into the higher tropical terrain (usually an uplifted relict surface). They literally represent islands of planation, in which the process may be observed in compressed and easily visible form.

The ectropics also have basins in mountainous country, usually associated with interbedded softer rocks or with areas of tectonic subsidence where a river was able to form a basin by eroding laterally. Examples of this are the Eferding Basin on the Danube above Ottensheim near Linz, or the Judenburg Basin in Styria. Thus it is not surprising that the basins in tropical mountain ranges were at first attributed to the same causes. Even Credner (1931, 1935, especially p. 42) arrived at this conclusion. Nevertheless, not only was he the first to call attention to this striking relief element, but he even made the prescient observation (as early as 1931), that "the intramontane basins are lower remnants inherited from a former, more widespread flat relief." We know more today about the mechanisms that result in such lowering than did Credner, who was forced to leave the question unanswered. Most later authors returned to the idea that such basins were created by fluvial lateral erosion.

Later investigations have shown that such an assumption may be invalid, even where conditions superficially appear suitable. This is the case, for instance, where a basin, drained by a single river entering and exiting through gorges, shows no signs of fluvial attack. As in the tropical lowlands, the rivers in such basins rarely carry gravels. The basin floor does not consist of gravel terraces, but is covered by a blanket of red loam, which, like that of the neighboring surfaces, is segmented by wash divides and punctured by larger or smaller inselbergs.

A characteristic example has been described by Bremer (1967a, 1975) from the Australian MacDonnells. Here intramontane plains are intercalated between inselberg ranges consisting largely of hogbacks. The larger rivers partly follow subsequent drainage lines between these, and partly break indiscriminately through the low hogback chains—a clear sign that they once lay on a higher surface which was unaffected by the bedding structures. Between these now weathered-out ridges lie a number of fluvially drained intramontane plains. According to Bremer (1957, p. 32), the flat floors of these plains lie "at the same altitude as the longitudinal valleys and surfaces of the forelands. The rivers of the MacDonnells also lack any major waterfalls or rapids in their gorges. Thus the intramontane basins cannot be explained by local base level of erosion [of the rivers running through them], that is, by assuming that fluvial down-cutting was retarded by a hard rock stratum, causing increasing lateral expansion upstream of the barrier (Credner, 1931, 1935; Semmel, 1963)."

But Bremer has found even more clinching evidence (1967a, 1975) that such basins were created not by lateral fluvial erosion, but by local planation from above downward following the mechanism of double planation. This is shown by several basins, but most clearly by Wild Eagle Plain, 50 km long and five km wide, in the MacDonnells (Fig. 47). This plain is drained by no fewer than four separate rivers, whose headwaters on the plain floor are joined by gentle etchplain passes. The rivers leave the plain through four separate gaps. *Had the intramontane plain been formed by the rivers, then four separate basins at different levels would have been formed, and not a unified surface.*

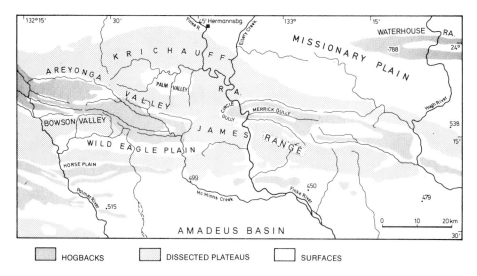

Fig. 47. Intramontane plain in the SW MacDonnells (Inner Australia). After Bremer (1975, Fig. 1). Like triangular reentrants, intramontane plains are isolated surfaces which work downward independently. Wild Eagle Plain is drained by several rivers whose drainages do not cut into the surface, but are united by even-floored etchplain passes. The rivers leave the plain through four separate water gaps.

The single plain surface can only be explained by independent and predominantly chemical lowering of the surface. The rivers played only the passive role of carrying away the prepared material, and can at most only briefly and secondarily have been involved in forming such a plain.

The drainage of a single intramontane plain through several gorges has since become the primary clue for recognizing an intramontane plain.

The Moroccan Anti-Atlas and the Bani Depression along its southern flank are pocked by basins (Büdel, 1975a) like those of the MacDonnells. In both areas these forms derive from humid-tropical Late Tertiary times, and in both cases the latosol sheets and tropical weathering traces have been almost entirely removed, but the major forms carved in the bedrock have been preserved with astonishing purity by the Pleistocene arid climate. It was Mensching (1955, p. 65) who first recognized that morphologically effective climatic differences exist today as in the Pleistocene between the High Atlas and the Anti-Atlas. In the High Atlas the Pleistocene lowering of the snow and solifluction limits affected large areas at high altitude (the peaks here are over 4000 m high); further S the snow and solifluction limits were sufficiently high that the Anti-Atlas (whose highest peaks are between 2000 and 2500 m) remained completely unaffected by snow, and almost unaltered by solifluction. This is therefore the northernmost African mountain chain in which the Ice Age left no significant morphological traces. Here the landform assemblage from the warm, humid pre-Pleistocene was preserved by the Pleistocene arid climate. The fact that a tropically warm, frost-free arid climate preserves relict forms particularly well will be discussed again in Section 2.5.

In the W Anti-Atlas, an especially abrupt basin is that of Tafrarout, whose very flat floor is almost exactly at 1000 m altitude. Both the basin and its entire mountain setting consist of the same granite. Photograph 23 gives a general view, while Photo. 24 shows detail from the abrupt basin edge. A series of neighboring basins are similarly sharp-edged. The basin of Taloua-line, nearly 50 km long, is completely adapted to soft Paleozoic beds. It is drained by two different rivers which exit through narrow gaps 10 km apart. The headwaters of these rivers are joined within the basin by a very gently arching etchplain pass region.

The E Anti-Atlas has even more of these well-defined intramontane plains. Many of these lie on the road from Aulouz (on the E end of the Sous Plain) across Tazenakht to the large depression of Ouarzazate, a distance of 220 km. After 60 km of traveling through small basins, one arrives at a magnificently flat, high basin near Tizi Aguerd. This is around 25 km^2 large, and at an elevation of 1400 m it is sunk only a few hundred m into the surrounding relict surface (at 1800-2000 m). East of this one crosses high passes into two further perfectly flat, high basins, each at around 1500 m altitude. On the basis of such high, isolated plains, one can generally agree with Louis (1964)

23. Intramontane basin of Tafrarout in the Anti-Atlas, 100 km SE of Agadir. View looking N. Most of basin margin very sharp, within a larger granite area. (Photo: Büdel, March 19, 1972.)

24. Sharp N margin of the intramontane basin of Tafrarout (see Photo. 23). The boulder-veneered slope and the sandy basin floor consist of the same granite. No boulders fall from the slope, no blockhalde forms at its foot. Only weak sanding occurs on the slope. (Photo: Büdel, March 19, 1972.)

that planation can occur independently of sea level even at great altitude. We must, however, stipulate that this can occur only in intramontane basins surrounded by higher mountainous terrain, and not on an open high plain. Dominating relict surfaces always originate at low altitude, and are later epeirogenetically uplifted, after which they are either preserved or undergo traditional development.

Summing up we may portray the development of an intramontane basin more or less according to the scheme in Fig. 48 (based partly on Fig. 5 in Bremer, 1973b).

I. On a latosol-covered relict surface (compare Credner, 1935) a shallow trough or swale develops, due to the presence of softer rock, to slight tectonic subsidence, to extreme shattering (mylonitization) of the rock, or to a combination of all of these. Water streams into the depression. Semmel (1963) mentions depressions in the Amhara Highland Surface (in Ethiopia) containing deeply weathered soil, whose stream cuts into unweathered bedrock below. Bremer (1975) says in this context that these forms may be regarded as incipient stages of intramontane basins. Busche (1974, p. 40) describes a case in the Tibesti Mountains (in the Sahara) where "basin lowering did not set in suddenly, but rather . . . gentle swales were first formed, cutting across hornfels and other resistant rocks; only subsequently did actual, more restricted

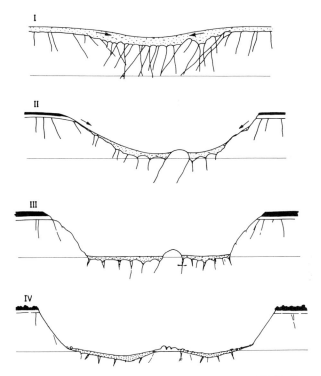

Fig. 48. Outline of intramontane plain evolution (largely after Bremer, 1965, Fig. 5).

I. Slight hollowing due to jointing and mylonitization, and/or tectonic subsidence. Water flows into the lowered area, thickening the pedosphere and decomposition sphere and increasing chemical weathering.

II. Continued downward working, especially on the floor which receives increased water supply from the sides. The upper slopes are bare; the summit plain is slowly becoming inactivated, and building of laterite crusts begins.

III. Mature form. Slopes largely free of soil, their feet accentuated. Continued downworking and evening of the deeply soil-covered floor. Continued growth of laterite crusts on the largely inactivated relict surface.

IV. Change to an arid-cold climate. Increased mechanical attack on slopes, which now become steepest above. Traditional pediments and glacis form at their foot (see Fig. 53). Weathering of the laterite crusts on the relict surface.

planation set in." Then "further lowering became limited to areas of more easily weathered rock."

Intensive weathering works down fastest along joints and fissures, so that the mechanism of double planation gradually, over very long periods of time, deepens shallow basins until reaching Stage II.

II. The basin is still trough-shaped and lacks steep walls, but the surrounding higher surface is at least peripherally inactivated, permitting laterite development. The basin floor and the lower slopes of its walls are still covered with latosol soil and are attacked according to the rule of divergent erosion; if gravel is available, some later fluvial erosion may also occur.

III. Should this process continue long enough, the basin matures, usually surrounded by steep, sharply footed escarpments. The basin's interior contains inselbergs of varying sizes, and etchplain passes unite the whole into a uniform basin floor. Several rivers may develop drainages here, leaving the basin through separate narrow gorges.

IV. We anticipate our later discussion here by describing the transformation of such an intramontane basin after the climate has turned arid and winter-cold. The surrounding walls are now mechanically weathered back somewhat, becoming steepest at the top. Traditional pediments or glacis radiate from their feet (where a gentle piedmont angle is now formed) toward the basin depths. Bremer (1973b) assumes that chemical decomposition may continue in the lowest parts of the basin, where water is most plentiful. Though inselbergs remain long-lived here, they are gradually bedecked with sediments, and slowly collapse into blocks. Near the basin walls the pediments carry fanglomerates (Stäblein, 1968); elsewhere they are covered by fine alluvium. During the course of time, limnic sediments may become deposited near the center of the basin. These often overlie old pipes or vents of red loam, the roots of the former pedosphere.

2.3.4 ETCHPLAIN ESCARPMENTS AND CUESTAS. ETCHPLAIN STAIRWAYS

The inner margin of an active etchplain leads by way of a riser to the next higher surface. The upper edge of this riser thus forms the outer margin of the higher surface. Such a riser or scarp between two etchplains is called an *etchplain escarpment (Rumpfstufe)*.

The schematic drawings of Figs. 43 and 48 show that active etchplain escarpments vary in the same way as the slopes of larger inselbergs or the walls of intramontane basins. As long as they are gently sloping and covered with relief-forming soil, they can be weathered back. An intermediate hilly area of *foothills (Fusshügel*; Credner, 1931, 1935) may exist at the foot of such an escarpment. These disappear when the scarp has become mature, leaving at most only a gentle rock pavement (Fig. 43) between the surface and the escarpment. Moats are far more common along such steep etchplain escarpments, especially along relict escarpments (e.g., on both sides of the Anti-Atlas), than has hitherto been recognized in the literature.

In bird's-eye view an etchplain escarpment is rarely straight, usually being extensively broken by triangular reentrants, according to its state of maturity. This contrasts markedly with fracture scarps (e.g., the E African Rift system), which are usually much straighter and higher.

Kayser (1949, 1957) quite early drew some important conclusions regarding the relationship between etchplain escarpments and cuestas. He showed that the old etchplains on the E flank of S Africa, reaching from the Transvaal to the gently arched fringe of the Drakensberg Mountains, cut evenly across strata of extremely different resistivities. When the same strata crop out on the steep E slope of the Great Escarpment they present a highly variegated appearance of treads and risers, in which each and every harder stratum is morphologically visible. Such morphologically visible petrovariance is evident only on the front slopes of escarpments, not on the relict surface which sets in above the highest erosional rim. This has been shown by Blume (1968, 1971, 1974) for a variety of climates. The same picture appears in the scarplands of SW Germany, as we will discuss in Section 3.3.1.4 using Späth's work (1973) in the Hassberge in Franconia as our basis. It seems quite clear to me that most of the long cuestas (*Schichtstufen*) of SW Germany were formed under a tropicoid paleoclimate in the same way as etchplain escarpments. On horizontal rock units of differing morphological hardnesses, escarpments will of course, as Blume has shown, tend to form erosional rims in any climate. With increasing steepness, the harder layers become much more marked than they are on the summit plain. In principle, however, the exogenic processes attack a cuesta landscape in the same way as they do heavily folded and thus (in hardness) more homogeneous rock series or rock units (Büdel, 1951, 1957d).

Similar results have been obtained by the work of Mortensen (1949), Hempel (1951), Hövermann (1953a and b), Mensching (1957a), Wirthmann (1961), Rohdenburg (1965), Spönemann (1966), Weise (1967), and Brosche (1969). As far as I know, Mortensen was the first to mention, and Hövermann was the first fully to recognize the general existence of etchplains in "structural scarp landscapes."

While the stratigraphy is revealed only in the treads of the scarp, being but hinted at on the broad intervening surface, Späth (1973) has pointed out that it was the periglacial processes of the Pleistocene cold stages which minutely weathered out each little ledge in the escarpment. In the Hassberge, however, the upper edge of the scarp cuts horizontally across the entire stratigraphic sequence, and is thus formed by beds of extremely different hardnesses.

This picture is supported by Bremer's investigations (1971, especially pp. 94-125) in Nigeria. Where sandy Cretaceous transgressive facies overlie crystalline basement, intramontane basins are contiguous to both rock types. Bremer writes that the "intramontane basins . . . are surrounded by slopes which must be described as etchplain escarpments (*Rumpfstufen*) in the crystalline rocks, and structural escarpments (*Schichtstufen*) in the Cretaceous. The one often grades into the other: their origin is the same" (Bremer, 1971, p. 120ff.).

This sheds new light on *etchplain stairways (Rumpftreppen)*. Active etch-plains in uplifted areas of the seasonal tropics and relict etchplains in the mountains of the mid-latitudes often lie in a series of steps one above the other. Just as all geomorphological theories were formulated in the mid-latitudes, so this phenomenon was also first investigated there. W. Penck's description (1924; he calls it a *piedmont staircase* or *Piedmont-Treppe*) gave rise to a very extensive literature on this subject. His view that such serially arranged surfaces can form without corresponding endogenic spurts of uplift through gradual and steady tilting of a block is theoretically possible, and may in a few cases actually be true, but the more natural case of jerky, interrupted tilting is probably more frequent. In this case we can identify the differing ages of the stairway treads by correlating the deposits found on them. Büdel (1935), working in the W Erzgebirge (right next to W. Penck's work area), distinguished an oldest central upland in the Fichtelberg peaks; below this a second, pre-Late Oligocene surface; a lower, third surface of Late Oligocene to Mid-Miocene age; and finally a broad and lowermost fourth etchplain from the Late Miocene to Late Pliocene (Fig. 58). More recent work by Gellert (1958, 1965) has largely substantiated this basic framework. One must remember, however, that such dating of etchplains refers only to the time of their first creation. Along with the older surfaces, they were then subjected, even under new and different climatic conditions, to traditional development.

Other features only discovered with the study of active etchplain stairways can also be identified on relict stairways. Körber's minute investigations in the Waldeck (in the German Mittelgebirge, 1956), identifying every possible surface, found 12 surfaces of varying ages, ranging between 300 and 800 m in altitude, or an average of a new surface every 40 m. Such minute differ-entiations between surfaces certainly occur in the tropics, but only as local forms which cannot be correlated over long distances. South of the Palar River, for instance (near Vellore in S India), one climbs from the inner edge of the Tamilnad Plain (Fig. 44) at 200 m up a single 600 m sweep to the next highest surface of Bangalore (at around 800 m altitude). North of the Palar, however, near Chittoor, a broad intermediate step is inserted at a height of 400-450 m. Local crustal displacements may raise lower surfaces and create local secondary escarpments, while river diversions open up new drainage areas for surfaces or even attack a step from behind; all serve to vary the picture, as shown by Bremer's map (1971) from Nigeria. Finally one must keep in mind that an active etchplain, though largely flat, is to a certain extent sculptured by wash depressions and wash divides, which may become locally accentuated into a "tropical ridge relief." The best discussions of relict etch-plain stairways in Central Europe, based on recent tropical studies, are in my opinion those of H. Fischer (1963-64), Quitzow (1969), Späth (1969, 1973), and Bibus (1971, 1973). We will discuss these again in Chapter 3.

2.3.5 AGE AND AGING OF ETCHPLAINS.
WASH VALLEYS

Aging and inactivation of etchplains, i.e., changes partially or wholly suspending the mechanism of double planation, may arise due to endogenic or exogenic causes. A tropical climate may turn arid but frost-free, a case which will be discussed in more detail in Section 2.5. This completely suspends planation across the entire surface, but preserves the relief carved into the bedrock by earlier humid-tropical conditions.

Endogenic factors such as epeirogenic subsidence or uplift may also lead to aging of the etchplain. Weak subsidence will cause superficial coating of the plain with sediments, e.g., soil sediment, through which the peaks of the higher inselbergs may continue to protrude (see Fig. 60). Examples of this may be found on the Tamilnad Plain.

Where uplift occurs along faults or steep monoclines, or where gentle arching of the surface occurs, etchplain stairways will form. As long as the new slope is covered with relief-forming soil, the lower surface will expand into the uplifted area by means of growth apices, until a steep escarpment has been formed (see Fig. 49). The lower surface now consists of an outer, older section (which is still undergoing active development), and a new, inner portion; these two often join as one surface. When dating an etchplain, it is wisest to investigate the age of the earliest part, while remembering that parallel lowering and inward expansion may have continued into far more recent times (right up to the present on currently active etchplains). Similar views have been expressed by von Gaertner (1968). Etchplains can continue

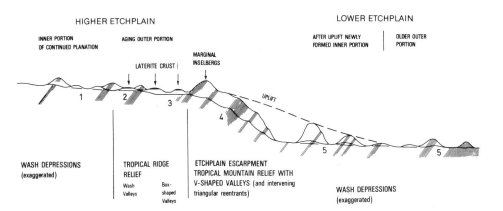

Fig. 49. Creation of an etchplain escarpment in an uplifted area. Continued planation on the lower surface; aging along the periphery of the higher etchplain. Rivers: (1) and (5) wash depressions. (2) Wash valley. (3) Shallow box-shaped valley. (4) Tropical mountain valley (usually stepped V-shaped valley with little gravel). Shaded areas: harder rock layers.

to form over long periods of geologic time, and can recommence activity after endogenic or climatic interruption. Once fully severed from the original, climatically controlled conditions necessary for its creation, an etchplain can continue to undergo traditional development (to be discussed in Section 2.3.9) and even continue to be weakly active.

In the case of Fig. 49, a lower surface has extended inward, transforming a formerly gently arched area into a steep etchplain escarpment, which then remains stable. Any harder beds will now clearly protrude on this slope (Kayser, 1949).

The higher surface at first has the same age of origin as the lower one, and also undergoes continued development, i.e., is lowered in parallel fashion. This mechanism works longer at the inner portion of the surface, where it can expand inward via triangular reentrants, for here the conditions are present for intensive weathering, namely, sufficient saturation of a thick relief-forming soil blanket. On the outer portion of the surfaces, this mechanism is quickly retarded. Working back from the scarp a small-scale *tropical mountain relief* quickly develops, with typical *tropical mountain valleys*. In longitudinal profile these are mostly stepped, having distinct nick points at every harder bed; here waterfalls develop during the rainy season (Fig. 50d). In cross-profile, tropical mountain valleys are narrow (though sometimes somewhat broader in their lower sections), V-shaped, and have no floor and little gravel. Usually they carry little sand or clay, having only a bare rock bed (Photo. 25, and river course 4 in Fig. 49).

The outer portion of the next higher etchplain surface also undergoes greater sculpturing. The rivers reach back from the scarp in low, box-shaped valleys with sandy stream beds (river course 3 in Fig. 49), merging into the surface as shallow swales *without* meandering sandy stream beds (course 2 in Fig. 49). A secondarily deposited, thicker latosol soil blanket covers the floors of these swales (Photo. 26). On the ribbons of *tropical ridge relief* between the swales, the latosol soil blanket is already discontinuous, and protruding basal knobs and shield inselbergs are more numerous than on the intact surface of the lower etchplain (Photo. 27). The mechanism of double planation becomes limited to narrow lines; yet even here it is chemical intensive weathering and not mechanical fluvial erosion which sets the pace for faster linear downcutting. The thick latosol covering the valley floor is removed only by surface wash and only during the rainy seasons. I have therefore named these valleys *wash valleys* (*Spültäler*; Büdel, 1965; also called *Kehltäler* by Louis, 1968b, p. 497). Photo. 26 shows such a wash valley on the periphery of the Bangalore Plain in S India; Photo. 27 shows the adjacent surface ribbon, with many slightly protruding rocks. Any parallel lowering of this outer portion of the higher surface is now impossible, for the relief-forming soil blanket is thin and discontinuous. Since water is removed through wash valleys, the soil is only very slowly renewed. Such stripped surfaces are very long-lived, but

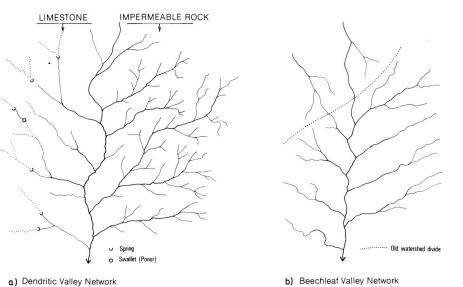

LIMESTONE   IMPERMEABLE ROCK

⌣ Spring
○ Swallet (Ponor)

.... Old watershed divide

a) Dendritic Valley Network

b) Beechleaf Valley Network

a) and **b)** :   Normal forms in unglaciated midlatitudes

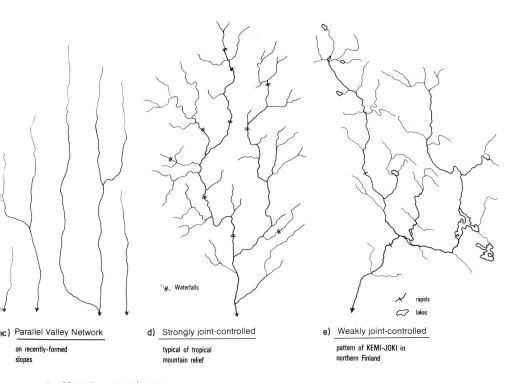

⌣ Waterfalls

⤨ rapids
⌒ lakes

c) Parallel Valley Network

on recently-formed
slopes

d) Strongly joint-controlled

typical of tropical
mountain relief

e) Weakly joint-controlled

pattern of KEMI-JOKI in
northern Finland

Fig. 50. Valley network types.

25. Typical steeply stepped V-shaped valley lacking both floor and gravel. Tropical mountain relief in the Entotto Mountains above Addis Ababa, around 2800 m altitude. Two large sets of waterfalls just downstream. Basalt, tirs soils, with young eucalyptus forest. (Photo: Büdel, February 22, 1953.)

(contrary to older views from Krebs, 1939, up to Rathjens, 1970), they can no longer be actively lowered, and can only be traditionally developed. Their preservation is in places aided by the growth of laterite crusts in the preserved latosol blanket.

In the inner portion of the higher surface where the latosol soil blanket is still well-preserved, the streams flow in wash depressions (Fig. 49, river course 1 and course 5 on the lower surface); planation continues, and insel-bergs can be hemmed-in and steepened. But this entire developmental complex weakens along the periphery of the higher surface; it becomes inactivated. Edaphic desiccation progresses inward from the periphery, not only above ground in the wash valleys, but also subcutaneously through weathering joints and flushed-out solution cracks of the underlying decomposition sphere. Even non-calcareous rocks are subject to such weathering and solution phenomena in the humid tropics. Weathering along joints and flushing out of cracks precede the creation of wash valleys and tropical mountain valleys.

26. Typical latosol-filled wash valley, free of sand and gravel, in tropical ridge relief near Hosur (120 km E of Bangalore). Located in a peripheral portion of the Bangalore Plain, undergoing slow dissection (see Fig. 49). (Photo: Büdel, March 12, 1964.)

27. Aging, partly inactivated outer portion of the Bangalore Plain (near Photo. 26), with thin, discontinuous latosol blanket. Basal knobs of the basal weathering surface (low shield inselbergs) exposed in many places. (Photo: Büdel, March 12, 1964.)

2.3.6 THE TROPICAL RIDGE RELIEF

The ectropic relief (especially after Pleistocene cold stage solifluidal reworking) is characterized by gradual transitions on convex-concave slopes, by ridges and slopes, and by mountains and valleys. Apart from a few high mountain ranges and occasional steep cliffs in the lower mountains, the majority of the relief is typified by slopes of between 2° and 15-20°.

Conditions are quite different in the tropics, particularly in the seasonal tropics. Here the double planation mechanism creates wide surfaces slanting at less than 2° in every direction. Thanks to divergent erosion, these abut sharply on steep or moderately steep slopes of over 20°, often up to 40° or more. The medium gradients to which we are so accustomed are the very ones which are rarest here. Gradual transitions are equally uncommon. Instead the eye is struck by sharp rims and angles which are invariably located at a distance from the stream courses.

An accordingly uncommon tropical relief type which approximates our conception of a hill country consists of the *tropical ridge relief*. We have already mentioned one variety of this landscape type, namely, that produced by wash valleys due to incipient aging along the periphery of a raised etchplain. This is clearly a *destructional landform*. A gently undulating landform assemblage can be found under corresponding conditions in many parts of the Sudan, e.g., in the Ashanti Highlands of Ghana.

A related relief type is not associated with the aging peripheries of uplifted etchplains, but covers parts of the lower, active etchplains. I first witnessed this relief type on low flights across the Upper Volta between Ouagadougou and Bamako (Büdel, 1952), and became better acquainted with it in the Somali Highlands (Büdel, 1954b). Photograph 28, taken in the Old African basement region in Borana Land, 120 km ESE of Ruspoli Lake, gives a general impression of the type. In native culture, fauna, and flora, this was the most undisturbed area of Africa available to me during my trips to Africa at the beginning of the 1950's.[42]

In the latter case particularly, it is questionable whether we are dealing with a destructional landscape, as described above, or with a *constructional landscape*, a type of "primary hill country" (*Primär-Hügelland*, based on the idea of a *Primär-Rumpf* or primary peneplain in the sense of W. Penck, 1924), i.e., a form by which the relief sphere of the semi-humid tropical

---

[42] Very large Nile crocodiles (some over six m long) lived in Rusopoli Lake. Many lions still lived there, finding abundant food. My diary has the following entry for March 20, 1953: "zebras in small troops, various kinds of antelopes, and long-necked gerenuks in herds of up to 20 heads, hyenas, jackals, and a great many ground squirrels." Birds included ostriches, a bustard, *Bucorax*, large vultures, bee eaters, and weaver birds. There were also many turtles. The brilliant red latosol soil was tunneled not only by ground squirrels, but also by termites, one species of which built rounded nests up to 0.6 m high, often surmounted by a needle-sharp spire up to three m high (see Photo. 28, lower left, with at least six more in the background).

28. Typical tropical ridge relief in Borana Land (S Ethiopia). Broad, trough-shaped wash valleys without sand or gravel beds, between gently arching ridges. Thick latosol blanket omnipresent. Extraordinary termitaria, over 3.5 m high. Near Neghelli, 200 km ESE of Ruspoli Lake. (Photo: Büdel, March 20, 1953.)

climate adapts to somewhat greater epeirogenic uplift. Such increased uplift could no longer be countered exogenically by the mere wash depressions in the etchplain surface, but only by a more active relief of wash valleys, that is, by incipient linear erosion. The ridges in Photo. 28 were still thickly covered with a latosol soil blanket, which was simply washed down more thickly onto the valley floors. For the most part, however, sandy dry beds were as rare here in the dry season as in Photo. 26. The problem of the tropical ridge relief deserves greater future investigation.

2.3.7 THE TROPICAL MOUNTAIN RELIEF

In discussing the relief of the tropical mountains, we will consider only the area below the forest limit, which in the seasonal tropics lies largely between 3500 and 4200 m, though in some cases it reaches somewhat higher (Troll, 1941, 1948, 1959). The bulk of our discussion will concern the warm altitudinal zone of the tierra caliente and tierra templada below around 2200-2500 m. Here the tropical mountain relief in our sense is fully developed. The region above the forest limit, though small in most tropical mountains, is more influenced by the Pleistocene cold stages.

Raised etchplains exist at all altitudes up to the altiplano of the tropical

Andes and altipiano of Ethiopia. Somewhat lower etchplains are also frequent in the remaining African highlands. In the N-S trending mountain chains of Indochina, Credner (1935) has mentioned etchplains up to 2000 m high. The highest etchplains are not covered with red loam, but with a tirs-like soil. On the broad highlands of Ethiopia, the settlements of the Christian farming and ruling class of the Amhara coincide well with the distribution of this highland soil, which is rich in humus and easily worked when moist. Underneath the tirs layer and its basal "stone lines," a red loam is frequently preserved. Along with this undoubtedly very old soil blanket, the etchplain assemblages (now largely inactivated) are often quite well preserved, even in non-calcareous regions. The relief here grades into a gentle ridge relief rather like that of the E Alpine "Rax landscape."

From these surface remains one crosses sharp rims to plunge down into narrow, wild, V-shaped valleys, much like those already discussed in the section on steepened etchplain escarpments (Section 2.3.5; see also river course 4 in Fig. 49) but on a much vaster scale. These are found in many tropical high mountains. Photograph 29 and Fig. 10 show a typical example from the Highland of Semién in Ethiopia.

Tropical mountain valleys differ in many respects from ectropical perigla-

29. Dever Canyon in the Highlands of Semién (N part of the Amhara Highlands, 470 km N of Addis Ababa). Typical many-stepped tropical mountain valley, sunk approximately 1300 m into the surrounding gently rolling, segmented surface of the paleorelief. Highlands thickly settled; heavily forested gorges lack permanent settlements. Photo used for Fig. 10. (Photo: Büdel, April 10, 1953.)

cial valleys (Photo. 10, Fig. 14), whose broad gravel floors reach smoothly deep into the mountain cores. Tropical valleys have steep longitudinal profiles with many steps. Each resistant rock bed causes another nick point. No floor is present: the valley is only as wide as the river during the rainy season, and gravel is largely absent (Photo. 25). The torrents of the rainy season have little bed load, i.e., little in the way of coarse tools for mechanical erosion, such bed load as they bear being mostly sand. This means that even here linear valley-cutting occurs largely chemically, through intensive weathering, flushing, and solution. The lines along which decomposition occurs preferentially follow zones of weakness in the rock, so that the valley network in tropical mountains often reflects the jointing of the bedrock (Fig. 50d). From the edge of the mountain massif these valleys climb steeply, reaching but a short way horizontally into the mountain block. Given equal conditions of petrovariance, epeirovariance, height and area, a tropical mountain will peripherally be very steeply dissected, yet its core will remain relatively untouched compared to an unglaciated ectropical mountain range (Photo. 29). We agree with Louis (1957) and K. Fischer (1963) that the highest and lowest planes of the relief are far closer together in the tropical than in the ectropical example. Yet the deep and rugged valleys of the tropical highlands provide the most impressive valley scenery on earth. Their microclimate is also distinct from that of their surroundings (Troll, 1952). In the Highlands of Semién (Photo. 29, Fig. 10), they reach from the cool, well-settled plateau down into the hot, luxuriantly forested depths of the canyons. The Amhara peasants avoid these canyons so much that they have become a refuge for social outcasts (Büdel, 1954b, 1955a).

Tropical mountain relief exercises an important influence on human geography, shown in Europe by the small side valleys in the S Mediterranean mountains. Here settlements are preferentially located on slopes, peaks, and ridges, avoiding the nearly untraversable, often flooded, and formerly malarial torrential gorges. These valleys are far from providing access ways into the mountains: rather the roads climb in heady switchbacks across the slopes and over the ridges. Not for nothing are the Italians known as the masters of mountain road and railway construction. In Ethiopia the entire Amhara Highland forms a giant natural fortress surrounded on all sides by precipices. Here the Christian peasants defended themselves successfully for 1200 years against the Islamic nomads. Canyons reaching back into the mountain block divide up the summit plain into isolated highland isles. At the extreme altitude of nearly 4000 m, barley is cultivated, and cattle and horses are bred. While in the European Alps one gazes from the settled lands of the broad valleys upward to the lofty pinnacles of the mountains, in Semién (the highest part of the Amhara Highlands), one peers from the settled highland isles down into craggy wild gorges: the Matterhorn in reverse (see Photo. 30).

2.3.8 Modification of the Tropical Ridge Relief
After Climatic Change

Valley reliefs throughout the world are easily modified. On flat surfaces, erosion attacks everywhere with approximately equal intensity, while in a valley, erosion is largely limited to lines. Even when new relief-forming mechanisms arise, their main agents, be these different kinds of rivers or glaciers, are still closely bound to the previously established lines, which are therefore the first to be changed. This means that once a new relief generation has set in after a change in climate, every valley relief is quickly altered.

The mid-latitude high mountain valleys were already in existence by the end of the Tertiary. In accordance with the prevailing climate, these took the form of tropical mountain valleys. In the Pleistocene cold stages, they were then altered in two ways. In the periglacial regions, rapid down-cutting took place, evening out all steps in the longitudinal profile and forming broad valley floors and terraces. Where stepped valleys from formerly tropicoid mountains became glaciated, their cross-profiles were broadened into the well-known U-shaped glacial troughs. Their longitudinal profiles were not only not smoothed, but the steps were even scooped out. The mechanical manner in which this came about has been very convincingly explained by Louis (1952). The principle of inheritance from pre-glacial tropicoid valley steps was recognized independently by Bakker (1965) and Büdel (1965). This explanation has later been substantiated by Späth (1969) in the Glockner region. More recent investigations have shown that the pre-glacial relief shows through even more in the Alpine valleys.

2.3.9 Preservation and Traditionally
Continued Development of Etchplains.
Genetic, Active, and Dynamic Geomorphology

Unlike valley landscapes, which quickly adapt to a new relief style, etch-plains are relief elements of astounding durability. This is due to their very flat gradients, which are less than 2°, and frequently less than 1°. In the drier and cooler morphoclimatic zones, it is these very gradients which experience the least erosion; even the most intense erosional processes of the ectropics have little effect on them. Yet it is on these hyperflat slopes, where the erosional processes of all other climatic zones are minimal, that the specific denudational mechanisms of the seasonal tropics are strongest. Once such a surface is subjected (without much epeirogenic disturbance) to a change of climate, the new processes unfolding upon it will be unable to alter it much, even over very long periods of time. For this reason extensive plains from the tropicoid paleo-earth have remained with minimal alteration as a record of the pre-Pleistocene landscape in all climatic zones from the equator to the pole. The preservation of these etchplains is one of the most important rules of geomorphology.

It has often been noted that etchplains are particularly durable despite very radical climatic changes. But this general observation must be augmented by the more specialized concept of the *traditional development* of etchplains, a concept which, founded by Bremer (1971), was formulated by Büdel (1971), based on similar experiences.

To elucidate this concept we will choose a simple example common in the ectropic mountains, where not only climatic change but also epeirogenic uplift has occurred (see Fig. 51). Between a higher slope I (a mountain slope or older etchplain escarpment) and a lower slope III (a valley side or more recent etchplain escarpment) stretches the raised etchplain (relict plain) II. This lies between piedmont angle A (Cv B/S A) and working rim B (Cx B/S B). The climate now changes, becoming ectropic or periglacial. Both slope erosion IV on the higher slope, and stream Va (lightly incised) meet at their local base of erosion, piedmont angle A, where they may even be assumed to aggrade slightly (dotted area VI). In any case, slope erosion (IVb) on the nearly flat surface II is reduced practically to nil (see above). Even stream Vb runs along the previously formed surface II for a short way with very little incision, for its erosive powers are greatly reduced on this very gentle slope. Only in the steeper areas below working rim B can it (now Vc) once more cut down and back. Thus in a new climate favoring valley-cutting, clear attack on relict surface II occurs only along its lower margin. Even the new slope erosion (IVc) will only commence on the lower steep scarp III.

Even with a new climate in which planation does not occur, then, the overall form of etchplain II remains fairly unchanged, being firmly stamped

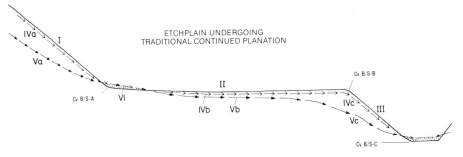

Fig. 51. Raised etchplain undergoing traditional continued development after change of climate. (I) Higher slope (e.g., old etchplain escarpment). (II) Raised etchplain (relict surface). (III) Lower slope undergoing current dissection. (IV) Slope erosion, a on I, b on II, c on III. (V) Fluvial erosion, a on I, b on II, c on III. Cv B/S A, Cx B/S B, Cv B/S C: concave and convex breaks in slope between I, II, and III. IV and V adapt themselves to the inherited relief assemblage. For further explanation see text, Section 2.3.9.

onto the bedrock. The relief-forming soil alone (from the time of the etch-plain's creation) is to a great extent removed. Slight aggradation takes place at piedmont angle A, while slight incision occurs at working rim B. Apart from these relatively minor changes, it is clear that we cannot speak here merely of the preservation of overall form II. Rather the processes of slope erosion (IVb) and fluvial erosion (Va, b, c), governed by the new climate, are forced by the pre-existing surface to act in a fashion which is in effect planar. The critical feature is that these processes, otherwise adapted to the cutting of valleys, could never work in a planar fashion were this surface not already present. They could never create the surface on their own. They can, however, continue working on an already present surface in a traditional manner. This explains the term "traditional."

We can present many further examples. Consider Fig. 51 without the lower slope III. Relict surface II is no longer bounded by a sharp working rim B, but stretches at equal gradient to a further base level of erosion, such as a large interior basin or marine beach. Dissection from below would be lacking even in a climate of valley-cutting: the processes controlled by the new climate would adapt to the pre-existing flat form across its entirety, and traditionally continue its development.

We have till now assumed that the new climate is one of valley-cutting. Let us now consider the case in which an etchplain is not uplifted, but becomes subject to a frost-free arid climate. Without considerable uplift, valley-cutting will not take place. The latosol will be very gradually stripped or converted to a desert soil, but no dissection will take place and no significant attack will be made on the bedrock base of the relief. Inselbergs and even very small features on the basal surface created in the former semi-humid tropical climate will long remain undisturbed. These new processes disturb the pre-existing surface even less, and accordingly traditionally preserve and continue its development all the more. It was in such areas that Bremer (1965b, 1967a) and Büdel (1955b, 1975a) developed these conclusions. We will discuss them in more detail later.

Let it simply be mentioned here that in research conducted by various authors in arid regions, the processes which traditionally continue the development of ancient, inselberg-studded etchplains have been regarded as the ones which actively created the etchplains and their inselbergs.

The danger of false logic can be illustrated by the following example. A geomorphologist of a generation which no longer read A. Penck attempted to solve the problem of Alpine glacial trough shoulders by "modern" methods. He thoroughly investigated the types of soil-forming processes, grain-size distributions, clay and heavy mineral content, orientation of boulders in the soil (which he evaluated as signs of recent movement), annual freeze-and-thaw cycles, vegetation, precipitation, and wash processes, as these occur today on the typical flat shoulders of the U-shaped glacial valleys. His conclusion was that these processes created the trough shoulders of the Alps. His evidence for the certitude of these results was the indubitable precision of the analysis. Yet he had investigated only those processes which today weakly continue the traditional development of the trough shoulders.

This leads us to a very basic conclusion. Current processes investigated by the dynamic geomorphologist take place on most of the earth (as in the ectropics and the arid tropics) on a pre-existing stage. The parent relief was not created by current processes (many of which have established themselves only in the brief post-glacial period), but is inherited from older, much more powerful processes, which in many cases were active for very long periods of time. This is obvious in the case of the glacial trough shoulder, which received its last major remodeling subglacially; but it is also true for much greater portions of the earth where the inherited relief setting was created by a quite different subaerial climate. Here much more detailed investigation is required to establish what type of relief preceded the current processes, and what relief forms were actually created.

Such an investigation must always distinguish three things:

I. The pre-existing stage set by the parent relief.
II. The scope of the alterations produced by Recent processes (usually in the brief post-glacial period).
III. The means (dynamics) by which this alteration has occured.

Here I is the task of genetic geomorphology, and II that of dynamic geomorphology. Point III, which combines the other two, has hitherto not been granted an independent position in the framework of geomorphology. We have therefore named this "active geomorphology" (Büdel, 1971).

2.3.10 THE TRANSFORMATION OF ETCHPLAINS

One of the most important tasks of climatic and climato-genetic geomorphology is to distinguish as clearly as possible between the processes and effects of fully active planation, and those whose activities barely suffice for the traditional expansion of the pre-existing plain.

The relationship between the inherited etchplain and the processes affecting it today differs from one climatic zone to another. In the zone of excessive planation, the active current processes are a hardly distinguishable continuation of the past processes. Here the Pleistocene hiatus is lacking, or was sufficiently modified as to have little effect on relief-development.

The situation in the ectropics is quite different. Here the radical climatic change of the Pleistocene had lasting effects on planation, ranging from traditional development to complete destruction. Between these two extremes are many cases where the bedrock framework of the old surface was so disturbed or altered that one may no longer speak of traditionally continued planation, yet where the former surface stamped onto the bedrock still glimmers through the overall landscape. We speak here of *transformation* of the etchplain. Two examples will be provided.

In winter-cold deserts, frost mechanically attacks the barren rock slopes of inselbergs, etchplain escarpments, and steep mountain rims, creating debris fans and glacis in the forelands, and pediments (to be discussed in Section 2.6) along the mountain edges. These may considerably alter or veil the old etchplain relief. Nevertheless, the basic structure of the former etchplain in many ways still determines the total relief.

A further example may be found in the humid ectropics. The Plio-Pleistocene climatic change forced the rivers into an erosional cycle which dissected the old surfaces and reduced them to relict plains crowning many different kinds of mountains. The surfaces then to some extent underwent traditional development as depicted schematically in Fig. 51. Where dissection continued, the old surfaces were converted into approximately parallel ribbons and ridges, leaving effectively nothing of the old bedrock basis of the former surface. Yet indirect evidence of the old surface is still clearly present, in that the inherited ridges all have approximately the same height. This is a typical case of transformation. These more or less equally high ribbons furthermore show little morphologic expression of petrovariance. Isolated spots may preserve more direct evidence of the relict surface, e.g., penitent rocks, deep sink holes, pipes (possibly with fossiliferous red loam), scattered iron rinds, remains of laterite crusts, bean iron ore, pisolites, quartz concretions (silcretes), and secondarily deposited lag gravels such as the E Alpine *Augensteine* (well-rounded quartz pebbles). Indirect evidence may consist of the corresponding deposits of weathered material (according to mineral composition, grain-size spectrum, and fossil content) in neighboring areas of subsidence.

Similar examples are provided by the glacially reworked relief assemblages of the Alps. Firn field levels (*Firnfeldniveaus*), cirque floor terraces (*Karterrassen*), and trough shoulders have largely turned out to be the transformed remains of former etchplain stairways. Here the pre-glacial relief assemblage has been heavily reworked, yet it is still sufficiently visible that Späth (1973) and others have been able to reconstruct fairly continuous surface systems.

2.3.11 Fossil Etchplains as Evidence of Past Humid Tropical Climates

For many relief types the climatically controlled mechanisms responsible have been so well explained by now, that a well-preserved relict relief may often be used as unambiguous evidence of the corresponding mechanism and climate. Furthermore, enormously long spans of stable conditions are required to create a mature relief. This makes them a far surer index for long-term climatic periods (of particularly effective climates at that) than, e.g., paleo-organic remains, which up till now have been the primary evidence for past climates.

Thoughts have been voiced on this subject by Schwarzbach (1961), but they may be carried still further in my opinion. Paleo-organic remains are often sporadic, and need not be typical of the overall climate of a large area. A modern-day example of this is provided by the Hauash River. Descending from the Amhara Highlands, this runs 400 km out into the Dankali Desert before terminating in a lake near Assaitta. Along this entire course it is accompanied by a very dense but narrow gallery forest, in places only two km broad, which contrasts sharply with the surrounding total desert. This forest, like the tropical rainforests far away, is a refuge for a purely humid tropical fauna and flora, with crocodiles, as well as apes, hyenas, parrots, and rainforest birds. Desert animals also come to water at the river, of course, but the only fossils preserved here would be of those animals buried in the river muds. These would primarily be such as would indicate a humid rain-forest or savanna climate, and would give a completely false picture of the concurrent landscape and relief of the Dankali Desert.

Furthermore, many plant and animal genera lose value as climatic indicators as we go back into Tertiary time. Many lineages have adapted to marked climatic changes during their evolution. While some species are closely tied to climate, close relatives may be very adaptable. In contrast to this, the long-term dominance of the relief-forming mechanism needed to stamp a mature relief type and its attributes into the lithosphere offers sure and comprehensive evidence that for very long periods the climatic conditions did not transgress the limits needed to keep the mechanisms in operation.

For a long time, particularly since the work of A. Penck and Brückner (1909), relict forms created by glaciers have been interpreted climatically. In the last three decades, periglacial forms, and lately even the valleys created by the ice rind effect, have been increasingly accepted as clues to past relief-forming mechanisms and climates. We have attempted in this section to show that relict etchplains, thanks to their unique relief-forming mechanisms and their exceptional preservability, are particularly striking evidence of long-term past climatic conditions.

2.4 The Inner Tropical Zone of Partial Planation

2.4.1 COMPARISON BETWEEN THE RELIEF
OF THE SEASONAL AND OF THE PERHUMID TROPICS

The rainforest belt of the perhumid tropics, with its nine to twelve humid months, has little in the way of a dry season. This is the chief difference between it and the seasonal tropics, which have sharply alternating dry and rainy seasons and varying degrees of savanna vegetation. It is this climatic alternation which in stable areas gave rise to planation.

The perhumid tropics lack this continuous relief-building. They have not been nearly as well studied geomorphologically as the savanna countries, due to the hot humid climate, the difficulty (thanks to the forest covering) of seeing the relief, either on the ground or from the air, and (for the same reason) the lack of precise geological and topographic maps. Thus we have so far only managed to collect isolated observations, which cannot as yet be united into any cohesive picture. Nevertheless, the perhumid tropics resemble the seasonal tropics so much that they may be considered a variation thereof. The area covered by the savannas is in any case much larger.

Despite many gaps in our geomorphologic knowledge of the perhumid tropics, three main features may be distinguished.

On the old, fairly stable shields, the overall relief is the same as that of the seasonal tropics. In Guinea, in the Congo Basin, on Ceylon and N Borneo, in Guiana, Roraima, and the Brazilian uplands, we find broad etchplains surmounted by inselbergs and even steep domes (*Domberge* or *Glockenberge*). The question is whether this relief type is still being generally (not just partially) formed today, or whether much of it is inherited from a past climate with longer dry seasons. Wilhelmy, on the basis of his considerable experience in tropical S America, concludes that "in the Late Tertiary erosional areas at low altitude in today's perhumid tropics were dominated by a seasonally wet climate with surface wash" (Wilhelmy, 1974, p. 344). Wilhelmy also explains the barren rock inselbergs, found with their feet buried in weathering material on the crystalline margins of the Amazon Basin, as inactivated relicts.

A change from perhumid to semi-humid tropical conditions may have been brought about during the Tertiary by continental drift, which particularly affected the southern continents (in which the majority of the rainforest belt is located today). Moreover, our narrow equatorial W wind belt may have arisen due to major changes in world atmospheric circulation at the Plio-Pleistocene boundary, connected with the spread of the polar caps and the confining of the entire tropical belt to the Intertropical Convergence Zone (ITCZ). The inner-tropical W wind belt may previously have been less well defined, extending in places much further N and S, and may have permitted the existence of much longer dry seasons. Due to the widespread etchplain-like relict reliefs in the above-named regions, I have chosen to call this the zone of *partial planation*, whereby the word "partial" is intended not only geographically but chronologically. That areas of recent deposition such as the Amazon lowlands were not erosional surfaces is of course understood.

A second region of the perhumid tropics, the mountains, resembles the equivalent savanna region far less. In the perhumid tropics, all gradients up to over 50° are still covered by thick soil sheets with decomposition spheres. Wirthmann (1973, p. 226) stresses that even on ancient basalts of the fully humid upwind sides of the Hawaiian Islands, "unvegetated, unweathered rock is hardly to be found (except for fresh soil slip scars and stream channels) even on 80° gradients."

This intensified chemical weathering greatly accelerates erosion on all these slopes. Even the dense root networks of the forests are torn through. Landslides and soil slips are extremely common, especially where the jungle has been cut for agricultural purposes. This has been noted particularly by Verstappen (1974) and Hagedorn (1974) in Sumatra. According to Wilhelmy (1974, p. 333) "*derrumbes* (landslides) occur in great numbers every year on the slopes of the Colombian Andes, and leave deep scars and trails as red wounds in the green forests covering the steep mountain sides." In the fully tropical mountain chains of New Guinea, Behrmann (1924, 1927) described large-scale landslides which quickly steepen the slopes and convert the ridges into "scar crests" or "niched ridges" (*Nischengrate*). At the same time he found that gentler slopes have a widespread system of "subsylvan soil flow" (*subsilvines Bodenfliessen*). The soil sheet, saturated to suspension, flows slowly downhill beneath the superficial layer of densely matted tree roots.

Linear erosion cuts deeply into the easily transported and constantly saturated kaolinitic soils. One scar often borders directly on the next. Yet the slope scars usually cut only through the thick soil sheet: lacking any coarser gravel, they cannot cut into the bedrock. As in the mountains of the seasonal tropics, the larger valleys have narrow cross-profiles, are well-stepped, and climb steeply up into the mountain heights. According to Wirthmann (1973), however, it is primarily the drier downwind sides of the Hawaiian Islands that display this form so typical of the semi-humid tropical mountains. The perhumid upwind sides facing the trade winds have broader valleys with tremendous valley heads. This evidence of vigorous erosion and dissection has caused some authors to view this as a further zone of preferential valley-cutting, and to accord it a place beside the polar zone of excessive valley-cutting.

Another reason for this view is that the wide, inactivated lowland etchplains, though showing a preponderance of flat relief forms, bear traces of dissection and other features deviating from the picture of current relief-forming mechanisms on active, seasonally tropic etchplains. We will discuss these later in context. First, however, we must examine the question of soil development and the relationship of the soil sheet to the thick vegetation cover of the tropical rainforest.

2.4.2 SOIL DEVELOPMENT AND THE RAINFOREST IN THE PERHUMID TROPICS

The soil sheet in the humid tropic erosional lowlands intensifies the latosol features we know from the active etchplains of the semi-humid tropics. Here also we have intensive weathering, with even more eluviation of calcium and

silicic acid, and here again we find formation of 1:1 clay minerals and ses-
quioxide enrichment. A careful look at the profiles described by Kubiena
(1953, 1957) and Finck (1963) for tropical red and brown loams shows that
what are described there as horizons are far less marked than those of ectropic
relief-covering soil. The only exceptions (described below) consist of greater
podzolization and intercalated soil sediment. This soil sediment helps make
these latosols (which Finck also calls oxisols) much thicker in the perhumid
regions than in the seasonal tropics.

Here again, then, we have a continuous sheet of red to "red-brown loams
without visible horizons" (Walter, 1970, p. 49). These soils are extremely
acid and nutrient-poor, with pH values as low as 4.0. This seems discordant
with their very luxuriant vegetation, but the tropical rainforest draws very few
nutrients from the subsoil, and some species, especially the epiphytes, live
directly from the nitrogen in the air. According to biologists, the tropical
rainforest leads a largely independent existence above the red loam sheet of
the subsoil. Walter even goes so far as to say that "the entire nutritional
supply required by the forest is contained in the phytomass above ground"
(Walter, 1970, p. 50). While we may take his "entire" here with a grain of
salt, it is certain that the rainforest lives off self-sufficient biologic recycling
between living biomass and litter, to a far greater extent than, say, an ectropic
forest. In the uniform climate, a constant shower of leaves and dead branches
falls to the ground, followed in time by the huge jungle giants, rotten with
age. The amount of falling leaves alone may amount in one day to 5-10 g/m²
dry weight. Yet the amount of total litter falling in a day is much greater. In
the constant humidity, heat, and darkness (only 0.1-1% of the sunshine
strikes the earth in the gloom of the rainforest), the litter decomposes more
rapidly than in any other climate. "The wood is destroyed by termites, whose
presence in the jungle is not immediately apparent, as their nests are underground. The construction of a test site in the Congo was hampered by the
fact that up to 25% of the area involved consisted of termite nests" (Walter,
1970, p. 50).

Litter here is therefore mineralized and made available to the plant roots
faster than anywhere else. For this reason the feeder roots remain in the
humus-rich upper layer, which, with average thicknesses of 10-20 cm, is
hardly thicker than that of the savanna. In the rainforest of Manaus on the
Amazon, Went and Stark (1968) found that the feeder roots of trees on sandy,
nutrient-poor soils have a mycorrhiza at depths of only 2-15 cm. This mycorrhiza, a dense network of fungal hyphae in the litter zone itself, helps the
roots take up nutrients and hinders eluviation of nutrients by rains. Walter
(1970) mentions examples on sandy substrates in the Amazon Basin, in Thai-
land, and in Indonesia, where through extreme production of humic and
carbonic acid, the jungle has brought about a self-induced degradation, a sort
of podzolization of the subsoil. Above this are thick horizons of undegraded

humus and even peat bogs. But such cases, approaching the podzolic profile development of ectropic relief-covering soils, are exceptions. "On the whole, humus enrichment is an exception in the tropics, and organic substances in the ground are highly susceptible to mineralization" (Finck, 1963, p. 98). This occurs largely through fungi (at 85% atmospheric humidity and 20-25°C) and bacteria (at a maximum of 98% atmospheric humidity and 30-50°C). Thus, despite rapid disintegration, the soil completely retains the character of a relief-forming soil. This is also indicated by the thickness of the underlying decomposition sphere, which in the mining pits on Bangka Island in Indonesia reaches 150 m deep, far below sea level (Cissarz, oral communication).

The taproots sent down by many jungle trees into the red loam subsoil serve more for mechanical anchoring than for uptake of nourishment. Other jungle trees have evolved buttress systems for support.

Beneath its thick canopy, the rainforest has created its own microclimate, which is even more uniform than the atmospheric climate above. The entire system of biologic cycles is intimately linked to the existence of the rainforest, and is highly sensitive to disturbance. Where the forest has been razed and burned, the constant rains fall directly onto the ashes, leaching out the topsoil nutrients which were mineralized in the fire (Walter, 1970). Only a small fraction of the nutrients is absorbed by the clay minerals, making them available for the domesticated plants of shifting cultivation. But this limited supply of nutrients is exhausted after a maximum of three years. Left on its own, the burned area will be invaded by secondary forest. Even when this is allowed to stand 12-15 years (as was formerly usual), it will achieve nothing remotely approaching the verdant luxuriance of the primary forest. If the secondary forest is burned off at more frequent intervals (as is unfortunately common today), the nutrients will be increasingly leached out. On some soils the vegetation may become so depleted that only ferns and certain coarse grasses can survive, such as the infamous alang-alang grass of Indonesia. On the whole, no climatic zone on earth has a soil so susceptible to human interference as that of the tropical rainforest.

Biologists have asked how a rainforest could have evolved such a dependence on self-created circulation of materials. It seems probable that this occurred when the rock was not as deeply weathered as today, and the tree roots still contacted the parent rock. This could have been the case during a drier climatic phase. It is worth repeating here that during the long period of the pre-Pleistocene tropicoid paleo-earth, the entire atmospheric circulation was far weaker than today. The tropical climatic belt was probably far less well defined, meaning that the equatorial W wind belt was more broken up and its climate marked by drier seasons. Hence there must have been a less marked contrast with the humid savanna. This would mean that much of today's rainforest belt had less intensive soil development, and possibly that the tree roots had easier contact with the rocky substrate. In addition (see above), secondary oscillations in humidity in the rainforests and the humid savanna could have occurred for several reasons both in the Tertiary and in the Quaternary.

This theory may also explain the astounding, though rarely questioned fact that despite their far deeper soils, the perhumid tropics have basically the same relief assemblage (inselberg-studded etchplains) as the seasonal tropics. Weathering in the latter is far shallower, but it leads to complete clay mineralization; wash effects are far greater, for at the end of the dry season the vegetation is drastically depleted, so that the heavy downpours of the beginning rainy season attack an almost unprotected soil. The effectiveness of wash in the rainforest is hampered by the canopy and the absorptiveness of the soil litter. The major relief assemblage of the perhumid tropics could be explained by marked dry seasons during a long pre-Pleistocene period, but the question remains open.

2.4.3 Fluvial Activity, Subrosion, and Partial Planation Along with Soil Karst and Soil Dissection in the Perhumid Tropics

We do not know to what extent these conditions of soil development apply to the present relief-forming mechanisms of the perhumid erosional lowlands. We are only certain that partial planation here (partial in that it is less dominant and less widespread) is on the whole carried on by relief-forming mechanisms similar to those of the excessive planation zone in the seasonal tropics. Planation is particularly active near the savannas, while in the interior of the rainforest belt, it may consist only of continued development, perhaps with some alteration, of the pre-existing surface.

The white-water rivers[43] of the lowlands run completely at the level of the surface, even forming natural levees. Characteristically, however, the most extreme form of levee-building in S America does not occur in the rainforest area, but in savanna having 8-9 humid months, the Pantanal of the Matto Grosso, found by Wilhelmy (1957, 1958a). This enormous pro-Andean area of subsidence has many inland delta features, yet lacks an actual swamp. Gradient-poor rivers flow from the E scarp westwards across the subsiding zone of the Paraguay River, supplying clays and sands for the rather thin alluvial sheets. These factors provide the necessary conditions for the creation of levees and (often cut-off) meander loops, the two features which distinguish the Pantanal. Circular meander loops form levees round the loop interior;

---

[43] White-water rivers, rich in suspended and sandy material, usually originate in the higher mountain country (the Andes, the Roraima, or the Brazilian Mountains) having the clayey soil profile normal to this climate. The clearwater and particularly the blackwater rivers arise in rainforests having deep podzolization or bogs overlying podzols. These carry some fine sand, but very little clayey suspended matter.

water then ponds in these circumvallated areas, forming *levee-rimmed lakes (Umlaufseen)*. Much of the Pantanal is sprinkled with these round lakes (called "Tiger's Eyes" by the natives), making the landscape one of the most extraordinary on earth. Although the surface is largely covered by sediments, the recently subsided and partially broken basement is often but a short distance beneath. "No other interpretation is possible for the old blocks which thrust out of the broad lowlands like inselbergs" (Wilhelmy, 1957, p. 48). The eastern edge of the Pantanal consists of a scarp which, according to Wilhelmy's map, is frequently broken up by deep embayments. One may therefore surmise that the Pantanal substrate consists of a partially buried etchplain.

Beyond this lies a broad, segmented etchplain landscape on the N slopes of Guayana between the Orinoco Bend to the W and French Guiana to the E. These etchplains were first described by Passarge (1903, 1933b). Their N edge is formed in ancient crystalline rocks, to the S of which rises an extensive sheet of sandstone, peaking in the Roraima at 2630 m. Surrounding these areas we find wide etchplains with high inselbergs in the surrounding crystalline rocks.

More recent investigations have been undertaken in this area by Bakker (1957a, 1958, 1960) and his colleagues on the very broad etchplains (on granite) studded with very rugged, barren domes in N Surinam. Their studies dealt largely with specific questions of granite weathering and soil development, as well as with the origin of rapids and the cause of their extreme stability. That the results fit well into the framework of double planation is evident in many places in Bakker's text and in the pictures. In a humid tropical rainforest climate lacking marked dry seasons (Hann, 1910), therefore, we have a planation process whose basic features all resemble those described for the seasonal tropics. Yet we must reckon with the possibility that in the recent geologic past brief periods of savanna climates with excessive planation (in the sense of Zonneveld, 1968, 1972a) have occurred along the N margins of the major rainforest belt of S America. The etchplains of Surinam, then, particularly those along the very humid coasts, may have already been formed. To what extent and precisely where a savanna climate may have been present in the Plio-Pleistocene remains an open question. Bakker's investigations have established that the humid rainforest climate of today has long been present there, and that it has continued (at least traditionally) to develop the inherited etchplain without changing to any kind of valley-cutting.

The S slopes of the great Guiana shield (called the Roraima after its highest divide in Brazil, the region of the white-water Rio Branco) have a rainforest climate, apart from a few drier spots with fewer than nine humid months (Lauer, 1952). Here Bremer (1973a) found a new erosional system in the region of the Rio Branco and Rio Jari (flowing toward the base of the Amazon Delta) as well as on the Amazonian slopes of the Brazilian mountain country in the province of Rondônia (an area of nine humid months). This system may be regarded as a modified version of the double planation mechanism. Here again, we do not know to what extent older flat surfaces preceded the existing system. The currently existing mechanism is explained by Bremer as follows.

The lower surfaces, formed largely of Tertiary sediments, lie on both sides of the Amazon River at a relative height of up to 30 m, and have an independent erosional regime. Above them are the higher surfaces of the true etchplains, studded with shield and dome inselbergs on the broad crystalline shield to both sides of the river. The microrelief of these surfaces shows two significant deviations from the familiar picture of wash depressions and wash divides, deviations which according to Bremer are interrelated.

The first of these is that the surfaces are dissected by a coarse network of narrow but usually shallow gullies, with steep walls of 20° or more (Photo. 31). These gullies are only 8-15 m deep and may expose bedrock without cutting into it. Photograph 31 from Bremer (reproduced here) shows the situation particularly clearly. The surface segments generally do not slope toward these gullies, and gently convex transitions between the surface and the gullies are rare. The gullies sometimes arise out of gentle swales, but usually they start at steep, almost vertical heads nearly 12 m deep. "The gulley-cutting and the incised surface are largely independent of each other" (Bremer, 1973a, p. 204).

31. Typical partial planation in the perhumid tropics (partially destroyed rainforest in Roraima, N Brazil). Soil karst: shallow closed depressions in a network of narrow gullies cutting through the soil sheet. This soil sheet maintains planation. (Photo: Bremer, October 1970.)

As shown in Photo. 31, this is accompanied by a second striking feature. Both the lower surfaces on the Tertiary sands and clayey sands of the Rio Jari in the Lower Amazon area, and the higher surfaces reaching up to 800 m altitude on the crystalline shield of Rondônia and Roraima are sprinkled with variously sized closed depressions. These show that surficial wash does not take place across the entire surface, but is directed toward the closed depressions in the middle of the individual surface segments. Occasionally a depression will be drained by a channel running above ground to the nearest gully (see Photo. 31), but for the most part these depressions are drained subterraneously, and water and material (usually in dissolved form) are drained underground to the gullies. This subterranean removal, or *subrosion*, follows root tubules, animal burrows, and soil horizons (e.g., the plane formed by the root zone, or the plane above the mottled zone). Bremer (1973a) was able to find drainage channels of subrosion in wells and tin mining pits. These channels can reach 50 cm in diameter and may cause settling; the resulting depressions in the highly decomposed soil are largely filled in by sand. The appearance of the etchplains as a whole remains unchanged. Bremer describes the Roraima in particular as having "very flat surfaces stretching evenly over broad expanses. . . . These are surmounted by inselbergs sometimes several hundred m high in granite and basalt, with crystalline ridges and a sandstone plateau" (Bremer, 1973a, p. 213; see our Photo. 32). The etchplain extends into rainforest to the S, and cuts right across a savanna island in the N. This agrees well with a parallel example from the dry savanna of the Somali Highlands (Photo. 33). In overall appearance, then, there are no major differences between the savanna and the forest, though the microrelief of gully-defined surface segments containing water-filled depressions is more common in the rainforest. Also (though this is also found in some parts of the savanna), larger rivers and their floodplains are incised 15-20 m into the surface. In general planation produces the same relief assemblage of inselberg-studded etchplains in the savannas as in the rainforests (though here with the addition of gullies and depressions). From this Bremer concludes that the shallow depressions, subrosion, and gullying do not interrupt planation in the rainforest. "Planation and valley-cutting proceed simultaneously" (Bremer, 1973a, p. 219). Bremer considers that the major difference is groundwater activity. "In the savanna, heavy infiltration occurs at the beginning of the rainy season, until soil expansion has closed all soil cracks. Surface wash then becomes relatively intense, while soil water and groundwater movements are apparently weak, for wells are rare and produce little. In the tropical rainforest, subsurface water circulation seems to be much more extensive, for wells are commoner" (Bremer, 1973a, p. 218). There are two reasons for this. First, the dense rainforest vegetation generally retards the runoff which is so extraordinarily effective in the savannas at the start of every rainy season. Furthermore, a high percentage of precipitation water flows down the tree trunks into the soil and is guided by roots along set paths. These remain open

32. Etchplain in the humid Roraima savanna (N Brazil). Shallow, water-filled, closed depressions occur here also (midground, in front of the inselberg group). (Photo: Bremer, October 1970.)

33. Inselberg group in the dry savanna of the Somali Highlands (S Ethiopia, 210 km ESE of Ruspoli Lake). Despite drier climate, the major forms closely resemble those of Photo. 32, indicating the range of climates capable of forming etchplains by means of double planation surfaces. (Photo: Büdel, March 20, 1953.)

through short dry seasons, and are gradually enlarged, presumably creating the permanent tubules of the soil horizons; in the savanna, on the other hand, soil cracking during the dry season repeatedly destroys any such tubules. The rainforest has no lack of tree roots (including dead roots consumed by termites), and water, supplied all year long, can move constantly through the same passages. According to Bremer, underground water currents may even help create the gullies.

These conclusions have meanwhile been supplemented. Fittkau (1970) mentions undrained, partly saline closed depressions near the Xingu divide (around 1000 km S of Santarém on the Amazon). Such lakes, according to him, are typical of the entire Brazilian Highlands along the SE edge of the Hylea. The "dambos" found by Ackermann (1936; networks of narrow, thickly grown, shallow valleys in the broad savanna etchplains of Zambia) closely resemble our above description, though lacking the closed depressions; this region in any case lies in savanna having only 5-6 humid months (Lauer, 1952). Löffler (1974) has found a complete analogue to Bremer's description in the lowlands of New Guinea. Here he noted "sinkholes and blind valleys leading via *subterranean piping* or *subrosion channels* into gullies." He describes these, under the name of *pseudokarst*, as follows: "The principal condition for its development appears to be a steep hydraulic gradient, the presence of unconsolidated or weakly consolidated sediments of predominantly fine grain (clay and silt) that disperse readily, and an appreciable permeability of the material. As a geomorphic process, piping is closely associated with gullying, where it is largely responsible for the initiation and headward extensions of the gullies." (Löffler, 1974, p. 18). This fully corroborates Bremer's view, which Löffler does not mention.

The morphoclimatic description of the inner tropics as a zone of partial planation is in my opinion supported by these results, particularly if one takes the word "partial" to mean not only partial in space, but also in time, i.e., discontinuous.

Subrosion, or piping, is geomorphically important. It occasionally arises in other humid climatic zones, but never to the extent described above. Mears (in Fairbridge, 1968, p. 849) and Rathjens (1973, largely in Afghanistan and other arid regions) have described cases of piping. The narrow valleys, usually cutting through the soil layer only, which frequently develop out of such piping are also called "gullies."

This entire complex of shallow sinkholes, subrosion pipes, and widely spaced feeder gullies is in my opinion poorly described by the term "pseudokarst" employed by some authors, and is far better described as *soil karst* (*Bodenkarst*). The closed depressions occurring here could be called *soil dolines* (*Bodendolinen*), while the associated system of narrow gullies could be covered by the term *soil dissection* (*Bodenzertalung*), in contradistinction to the term "soil erosion" (*Bodenerosion*), which applies uniformly over large areas.

Even in the inner tropics, soil dissection may intensify locally until it uniformly covers broad areas, e.g., the lower terraces (usually of Plio-Pleistocene sediments) of the Amazon system, below the relative height of 30 m (Klammer, 1971). The shape of the higher of these terraces makes it difficult to tell whether it is a fluvial terrace or a surface undergoing present erosion. Journaux particularly (1975) has shown that the deep soil blanket on the thickly wooded terra firma to both sides of the recent alluvial deposits of the

Amazon is closely dissected in a manner approximating that of soil erosion. Elsewhere the soil blanket contains "orange halves," approximate spheres, which, under a brilliant orange covering of loam, often show several layers of coarser material. According to Klammer (1971), considerable Pleistocene glacio-eustatic changes in sea level affected this area. As the constantly saturated soil here has but a loose foundation, soil movements, tearing, and slumping presumably figure in the as yet poorly investigated total picture.

2.4.4 DEEP-REACHING KARST SOLUTION
IN THE PERHUMID TROPICS

We have already spoken of the peculiarities of tropical limestone karst, and we have already mentioned that while *tower* or *cone karst* (*Turm-* or *Kegelkarst*), described especially by H. Lehmann (1936, 1953, 1955; see Fig. 52), is both striking and common, it is by no means the only form of karstification in the seasonal and perhumid tropics (Gerstenhauer, 1967). To this we must add that these karst towers or *mogotes* (always found in closely packed

34. Tropical cone karst in Kwangsi, S China (around 300 km W of Canton, below the Tropic of Cancer). Cone karst has been described almost exclusively from the perhumid tropics (see Fig. 52). This picture proves that fully corresponding relief features may develop in the summer-wet monsoonal subtropics. Precipitation here 1000 mm, April to September; mean January temperature circa +16°, mean July temperature +27°. (Photo: Wulf Diether, Graf zu Castell, 1935.)

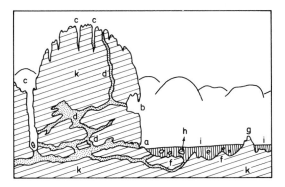

Fig. 52. Tropical cone karst as a special case of inselberg development (after a drawing by H. Lehmann, 1953). (a) Piedmont angle of karst cone (*foot cave* or *Fusshöhle* formed by solution sapping at the level of the karst surface i). (b) Older piedmont angle of former, higher karst surface (*partial cave* or *Halbhöhle*). (c) Open karst pipes and jamas. (d) Partially open joints, pipes, and caves, partly clogged by red loam. (e) Red loam sheet on the karst surface i. (f) Karren or lapies-shaped basal weathering surface. (g) Karren rock surface (corresponding to a shield inselberg) protruding through the red loam sheet. (h) Core stones. (i) Karst surface (equals wash surface). (k) Stratified limestone bedrock.

swarms) may, in conjunction with the surrounding red-loam-covered surface, be viewed as a form of our familiar etchplain, dotted with densely crowded inselbergs or domes (see Photo. 34 and Fig. 52).

On the one hand this extreme karstification testifies to the intensity of karst solution (as of all other chemical weathering) in the perhumid tropics. But at the same time, the similarity to the inselberg-studded etchplains dominating all rock types in the lowlands supports the theory that petrovariance is fairly insignificant for the relief forms of the seasonal and perhumid tropics.

The thick latosol soil sheets, wherein soil karstification occurs, are effectively independent of lithology. Karstification extends below the soil sheet into the bedrock, even when this is not limestone. Other rocks with appropriate composition and heavy jointing also fall prey to vigorous decomposition at depth, leading at times to the creation of depressions in the ground surface.

Such karst-like phenomena develop in basalt only when the latter are fresh. In this condition subterranean drainage is favored by jointing, interbedded loose tuff or pumice, and above all by the tubes commonly formed by the last streams of glowing lava, hot solutions, or gases flowing through the cooling and solidifying mass. Water circulates freely through the hardened passageways, enlarging them by solution. Even in the cool Icelandic climate these factors have led to karstification, favored in the recently volcanic areas of the island by rapid heating of the groundwater. I found many surficial basalt dolines there, whose underground extensions are formed by lava tubes. The largest of these passages, according to Schwarzbach *et al.* (1971), occur in the lava field of Hallundarhraun, and are at least 23 km long, having a cross-section of up to 10 × 10 m. Their interiors show the flow marks of the gradually sinking lava, and their walls and ceilings were remelted by hot gases in a final phase. Iceland also has places where brooks disappear into doline-like swallets of tubule-riddled lava sheets, an example quoted by Schwarzbach (1967, p. 395) being Brekkua near Hredavatn. Even more numerous are the waterfalls issuing from steep basalt cliffs, as at Hraunfossar near Stori As (74 km NE of Reykjavik), where a whole chain of small waterfalls, issuing from a basalt cliff barely 15 m high, plunges into the Hvita River (Photo. 35). Larger examples are found at Storidalur (near the S coast, opposite the Westmänner Islands). Even the natural basalt bridge near Ófaerufossar, through which tumbles a waterfall, was in my opinion (though there are also other interpretations) preceded by a subterranean cavern just like the many natural bridges in limestone karst areas (e.g., in the Ardèche on the French Plateau Centrale). Iceland has more waterfalls than any other ectropic country, due to the valley steps formed by a combination of glacial and volcanic processes (Photo. 36).

But it is above all in the tropics that heavy surficial weathering and flushing of degradational products gradually induce subterranean solution of basalt. This agrees with the conclusion reached by Wirthmann (1970, p. 193) that "even when all drainage starts off subterraneously (as in many fresh basalts),

35. Waterfalls issuing from a 15 m high undercut slope of basalt at Hraunfossar, flowing into the Hvita-Vatn River above Stori As, 75 km NE of Reykjavik (Iceland), as an example of the permeability of fresh basalt sheets. (Photo: Büdel, July 23, 1971.)

36. Gullfoss waterfall on Iceland, 90 km ENE of Reykjavik. The three sets of falls total nearly 50 m. Such breaks in profile are common in the tropical mountains. In the ec-tropics, they occur only where favorable features coincide: here, hard, thick, horizontal sheets of fresh basalt, emphasis of steps by Pleistocene glacial scouring, lack of smoothing out by Holocene permafrost or ice rind, and primary joints and caves in fresh basalt. Upon caving in, these caves and joints form steep gorges, such as that below the falls depicted here. Iceland has more waterfalls than any other ectropic country. (Photo: Büdel, July 21, 1971.)

soil development soon interferes and forces the creation of a valley landscape.''

Wirthmann found large-scale, true karstification in the peridotites of New Caledonia (1965, 1970). This rock unit covers one-third of the island, which in size is slightly smaller than New Jersey. Peridotite is extremely unstable under chemical attack, with the result that an area of solid peridotite on the SE end of the island (which has almost no dry season) has several dozen true sinkholes, up to 40 m deep and up to 250 m wide. ''The fact that many major joints run vertically may also favor their development.'' Yet karst drainage on peridotite ''is arrested far earlier than in carbonate karst, for despite peridotite's solubility, it leaves many weathering and solution residues which quickly plug up karst tubules and sinkholes. Most dolines on peridotite in New Caledonia contain water for several days after a rain, and some are even permanently water-filled, for once the decomposed material and laterite are thoroughly waterlogged, they become fairly impermeable'' (Wirthmann, 1970, p. 193). Yet karst development is much more far-reaching in peridotite than in basalt.

Incipient subterranean drainage and uvula-like karst basins are also found on other rocks in the tropics. Bremer (1971, pp. 67-68) has described such forms in the Tertiary sedimentary sheets of Nigeria as shallow depressions having diameters of 1-1.5 km and depths of up to 13 m. Similar forms, reworked (not created) by aeolian activity, have been found by Hagedorn (1971, pp. 54-56) on Paleozoic sedimentary surfaces in the S of the Tibesti Forelands, and Busche (1974, pp. 44) has also found them on the higher basalt sheets of the mountains and on the sandstones of the Plateau du Mangueni NW of the Tibesti Mountains, where they are associated with very deep traces of chemical weathering. These basins may also be due to a Tertiary humid climate, to which Klaer (1970) and Kaiser (1972) also attribute the very deep weathering of granites found in the Tibesti Mountains.

Observations and reports on more superficial, small-scale forms of karst, such as crystalline and especially granite karren, tafonis, and the like, have been collected by Klaer (1956) and summarized by Wilhelmy (1958a, 1974). These are found ranging from the etesian subtropics into the inner tropics, where granite karren may be over 50 cm deep. The relationship between solution and decomposition, as in the above examples of basic rocks, is not fully clear. Jungfer's observations (1974) in the high mountains around the basin of the Deh-Bala oasis in S Iran (SW of Yazd) are notable. Here the basin floor lies at 2100 to 2600 m. Up to around 3200 m its steep rock walls are composed of granite, above which lie Cretaceous limestones, dolomites, and marls up to a maximum of 4500 m high. The Cretaceous slopes are nearly barren, while the granites are covered by a relatively dense vegetation. The higher regions carry permanent snow patches which supply copious amounts of water, particularly during the summer. Much of this travels through the Cretaceous sediments, which must be highly porous, into the granite, which stores it. Not only the water for the 6000 inhabitants of the oasis, but also for every qanat in the area is drawn from this granite. Such a function has hitherto been unknown for granites, and further research would be highly desirable here.

The same is true for a report by Thorp (1967) from N Nigeria. Here Tertiary etchplains cut across the African basement and stocks of Jurassic granite contained therein. The granite contains areas of heavy jointing as well as areas which are readily decomposed. Where these two coincide, shallow open or closed basins up to several km in diameter are particularly numerous. The depths of these basins relative to the lowest point of the surroundings is not recorded, but the contour lines given suggest that in at least one case they are drained through narrow water gaps sloping unidirectionally. If these are not polje-like karst depressions, they are at least pronounced intramontane basins. One of the basins depicted in Thorp's map (1967, Fig. 7) even shows the drainage typical of such basins, having two separate water gaps. We will discuss the genetic relationship between such basins and the true poljes of the Dinaric karst in Section 3.4.3 (see also Büdel, 1973).

2.5 The Warm Arid Zone of Plains Preservation and Traditionally Continued Planation

We now leave the humid tropics, and turn to the great arid belts of the earth. This region is found on the W coasts and continental interiors of the trade-wind belt along the Tropics of Cancer and Capricorn. We have already mentioned the distinction between the low, frost-free arid zones (thermally still part of the tropics) where surface preservation occurs, and the winter-cold, frost-prone arid zones (found in the continental interiors and stretching into the subtropics and ectropics) where more pronounced surface overprinting takes place. The distribution of both zones is shown in Fig. 13 (endpaper). The tropical arid zones may be termed the *trade-wind deserts*, while the subtropics and ectropic arid zones belong to the *inland deserts*.

The warm arid zone of the trade-wind deserts has two main areas of distribution. The Saharo-Arabic region covers over 13 million km² (Europe covers 10 million km²), while the second major landsurface of this zone, Inner Australia, is barely 5 million km². Together with the smaller trade-wind deserts of S Africa and the American W coasts, this tropical arid region occupies nearly 14% of the terrestrial surface, while the tropics as a whole cover over 32%.

The most salient geomorphological feature of this zone is the major relief assemblage (e.g., that of the Sahara) inherited from pre-Pleistocene humid periods, and consisting of broad etchplains studded with inselbergs and inselberg ranges, etchplain passes, triangular reentrants, and intramontane plains. About 75% of this relief has survived the dry periods which dominated most of the Quaternary. Only the remaining 25% or so has undergone traditional planation or even transformation. In Inner Australia this percentage is even less than in the Saharo-Arabic region. We will begin by discussing the latter.

2.5.1 THE SAHARO-ARABIC ARID REGION

Following the first modern Sahara crossings by H. Barth (in 1850 and after), G. Rohlfs (in 1865 and after), and G. Nachtigal (in 1869 and after), the Sahara, the largest desert of the world, has been the subject of considerable research. This was facilitated by its geographic proximity to Europe, and by the fact that colonial rule (largely French, but also English in the Nile area and Italian in Lybia and Danakil) was extensive and long-term. Many Frenchmen have dedicated themselves to research here, including Monod (1947),

Capot-Rey (1953), Cailleux (1951), Cailleux and Tricart (1959), Dresch (1957, 1959) and Rognon (1967a and b), to name a few of the most important.

Physio-geographical research focused largely on the desert nature of the Sahara, and particularly on the Late Pleistocene-Holocene alternation of humid and dry phases. The most recent work on this is by, among others, Butzer (1957, 1958, 1961) and Geyh and Jäkel (1974). The latter's summary provides a fairly well-substantiated picture, which we will reproduce in brief.

The transition from the Late Pleistocene to the Holocene was marked by greater humidity, becoming drier, however, between 9700 B.C. and 8500 B.C. After this followed a phase of above-average humidity, lasting from around 8500 B.C. to a transitional phase at around 5100 B.C. At this time an arid period set in, lasting until around 4000 B.C. This was succeeded by a further humid phase between 4000 and 2700 B.C., coinciding with the rise of the Old Kingdom in Egypt. Between 2700 and 1700 B.C. the climate turned drier, after which alternated periods of aridity and humidity of several hundred years each. The major periods of settlement always lagged shortly behind the onset of the favorable humid periods. Thus the Neolithic revolution in the Sahara took place around 6000 B.C., more or less in the middle of the favorable period between 8500 and 5100 B.C.

Investigating the dry and humid phases (pluvials and interpluvials) of the Pleistocene is more difficult. It seems certain, however, that these climatic fluctuations had about the same amplitude as those of the Holocene, which proves that these pluvials cannot be correlated with the ectropic glacial stages. Pluvials can be divided into those with more etesian winter rain and those with more tropical summer rain.

The investigation of the Quaternary climatic history of the Sahara has hitherto rested largely on paleobiological, archeological, and sedimentological evidence, paying little heed to pedological and geomorphological evidence. These disciplines however, have gained considerable importance with the study of the pre-Pleistocene relief genesis of the Sahara, for it soon became clear that the great climatic change from the Tertiary to the Quaternary was well documented in the relief history of this region. The study of the Saharan relief generations has since become quite important (Büdel, 1955b), and has produced the following hypothesis regarding climatic history.

The decisive change in climate and relief which affected the Sahara took place toward the end of the Late Tertiary, largely due to intensified atmospheric circulation at the Plio-Pleistocene transition, which in turn was caused by the development of the polar cold-air caps and the continental ice sheets. Previously even the high latitudes had enjoyed a tropicoid, moderate trade-wind climate with very gradual temperature reductions toward the poles. Maritime influences further ameliorated the terrestrial climate, for the Tertiary marine regression was very gradual. Meanwhile uplift in the region of the previous alpidic orogeny did not begin until after the Mid-Miocene. The terrestrial climate thus enjoyed higher and more uniform temperatures throughout the Tertiary. The mean temperatures of the equatorial and polar regions (around 70° N. lat.) in the Early Tertiary were probably $+25°$ and $+15°C$ respectively, in the Late Tertiary $+25°$ and $+10°C$, and today (in a Quaternary interstadial) $+25°$ and $-10°C$. From the Late Tertiary to the Pleistocene interstadials, the difference in these annual means grew from about 15° to 35°, and in the winter of the glacial maxima, probably to 45°, or many times the difference which existed during the Late Tertiary.

We have already learned to recognize the typical tropical relief assemblage as the surest evidence for long-term, supraordinate climatic periods (see Section 2.3.11). It follows that we find etchplains which were undergoing development or at least traditional development in the Early Tertiary as far N as N Scandinavia and Spitzbergen, and etchplains from the Late Pliocene in the tropics and into the mid-latitudes. Where these have been but moderately uplifted, they have not been dissected, and latosol-like soil relicts may be preserved in protected areas.

The abrupt temperature drop over the polar caps (which formed in the Earliest Pleistocene before the Günz or Nebraskan glaciation) must have enormously accelerated atmospheric circulation. As the polar caps spread into today's mid-latitudes, influxes of polar air must have spurred the cyclonic regime which today dominates the W wind drift in the latitudes between 35° and 75°. Parts of these latitudes were iced over during the Pleistocene, but above 45° latitude they were mostly covered by thinly forested to unforested periglacial regions, whose features we have already described. In this climate, the old etchplains became severely dissected, terminating the pre-Pleistocene planation phase throughout the ectropics.

An important question regards the changes which this climatic revolution wrought upon the relief in the modern trade-wind desert belt. It is generally agreed that an etchplain relief existed here in the Late Cretaceous and Tertiary, resembling those preserved in the mid-latitudes and active in the seasonal tropics. Opinion is also more or less unanimous that a frost-free desert climate is particularly suited to preserving inactivated relief features, even when these are very steep. And finally, it has to a certain extent been accepted that a desert climate (which, with the above-mentioned pluvial phases, dominated the entire Quaternary) changed the old etchplain relief, replacing its soil sheets, particularly in low areas, with desert-type rubble sheets. In this manner these low-lying areas were altered through an "arid morphodynamic system" (Mensching, 1968).

Here, however, opinions differ. Many researchers, especially in France, Britain, and America, tend to set the period of humid tropical development as far back in the Tertiary as possible, assuming a correspondingly early advent of full desert conditions, and an accordingly thorough transformation of the original relief. Some advocates of this view go so far as to regard the entire Saharan etchplain and inselberg relief as a product, not of a former humid tropical climate, but of a long-term and still active desert climate.

We will elaborate on this below; here we shall list only those points suggesting a different interpretation of the Saharan etchplain relief.

(1) We know from Sections 2.3.3 to 2.3.5 what extraordinarily long periods of relatively constant, seasonally tropical climate are necessary to create such continent-wide etchplain reliefs by chemical weathering. The brief Quaternary period with its markedly reduced chemical weathering is incapable of such an accomplishment.

(2) We have already expounded on the complicated mechanisms of double planation surfaces and divergent erosion, which alone can produce this unique relief style. It seems impossible that a totally different ''arid morphodynamic system'' could produce the same relief, and the dynamics which would be responsible for this remain as yet unexplained.

(3) It seems unlikely that while amazingly uniform etchplains were beveled from the equator into the polar regions during the long Tertiary period of the tropicoid paleo-earth, the region of today's trade-wind deserts should have been an exception where this did not occur, and where similar etchplains were subsequently produced by a quite different mechanism in a much shorter time period.

(4) The most important post-Cretaceous change in atmospheric circulation took place at the onset of the Pleistocene. Throughout the Tertiary there is neither pedological nor paleontological evidence of any similar major change. It was the start of the Ice Age which brought about another drastic faunal break (Schindewolf, 1954). The expansion of the polar air caps and the shifting of the powerful W wind drift toward the lower mid-latitudes must have hemmed in and intensified the tropical wind system. Only then could wide and continuous trade-wind deserts arise. Previously the Sahara may have been no more than a few ill-defined islands of aridity.

(5) The Quaternary desert climate in much of the Sahara is characterized by low fluvio-eolian sandplains. These will be amply described below. In places they are segmented by terraces a few m or tens of m high. Even on the equatorial margin of the Sahel, they are at most 50 m high (Mensching, 1970b, p. 41). The terraces rarely cut into bedrock, for fluvio-eolian sandplains usually overlie clayey substrates. Removal of the older soils and the younger fluvio-eolian sands often reveals a well-preserved basal knob relief from the former basal weathering surface (see Photos. 44 and 45).

We therefore conclude that fully desert conditions can only have been widespread in the Sahara since the Plio-Pleistocene transition. These desert conditions have been alleviated by slight humid phases, but *the dominant developmental processes have only caused traditionally continued planation of the old etchplain-inselberg relief*, the basis of the Saharan relief assemblage today. The flat expanses of the low fluvio-eolian sandplains could never have formed had not *excessive planation stamped this flat form into the bedrock* during the long humid-tropical climate of the pre-Pleistocene past.

2.5.1.1 The Anti-Atlas and the Hoggar Mountains.
Marl Sand Facies and Fluvio-Eolian Sandplains in General

The above conclusions pertain to the flatlands which cover more than 70% of the Sahara. Over these loom the Saharan mountains, characterized by different and more dissected arid forms. The mountains include: the Anti-Atlas (up to 2531 m high); the Hoggar Mountains (3003 m high) with outposts in the Adrar des Iforas (up to around 1500 m high); the Air Mountains (up to around 2000 m high); and the Highlands of Tibesti (up to 3415 m high) along with the uplifted areas of Ennedi (up to 1450 m) and Darfur (up to 3088 m) adjoining the Tibesti to the SE.

The NW Sahara between the Anti-Atlas and the Hoggar mountains (1500 km apart) is especially characteristic of the Saharan flatlands. The S edge of the Anti-Atlas reaches down to 28° lat. N, while most of the Hoggar lies S of the Tropic of Cancer (23.5° lat. N). The main axis of the desert runs perpendicularly between these, at about 27° lat. N. It is along this axis that precipitation is lowest, totaling three to five rainfalls per decade.

It is here that we can best answer the question of the extent to which the desert climate causes not only traditional planation, but actual transformation of the parent relief. As already mentioned, some researchers believe that "the deserts" (including the frost-free trade-wind deserts!) are controlled by their own erosional system, equal in result to that of the seasonally wet tropics (which has long been at work) and that of the periglacial regions, though comparatively brief in force. These researchers seem blissfully undisturbed by the fact that the product of their desert processes strongly resembles the product of the seasonally tropic relief-forming mechanism. Their assumption is based on the existence of pediments and glacis[44] along the edges of mountain ranges in the winter-cold subtropic and ectropic deserts. The researchers who regard this as a form of arid etchplain creation speak here of "pedimentation" or even "pediplanation." Recent work by Mensching (1973) rightly stresses glacis development. But the differences between this system and the major terrestrial relief templates (as we have described them) strike me as obvious for various reasons (see Fig. 53, to be explained below).

For one thing, pediments, even when present, are not independent of the epeirovariance and petrovariance in the desert area concerned. They are as-

---

[44] Büdel and other German authors use the term *Pediment* to refer to a narrow notch carved into the bedrock at the foot of the mountain, while *Glacis* is the fan of loose sediments that spreads out from there into the foreland (see Section 2.5.1.4). They use the term *Fussfläche* (literally "foot surface," a mistranslation of "Pediment") to refer to the entire pediment/glacis complex and to avoid genetic implications. We will employ the neutral "piedmont surface" to translate *Fussfläche*, designating the pediment/glacis complex. —Trans.

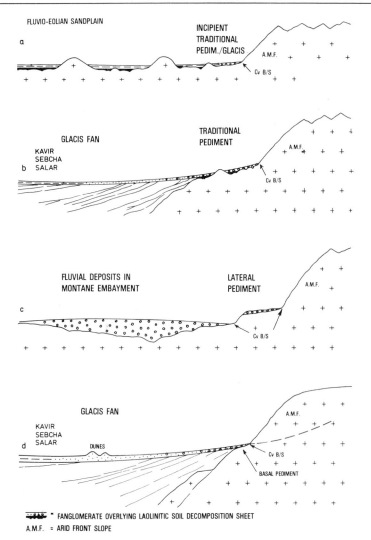

Fig. 53. Types of piedmont surfaces (glacis surfaces and pediments). (a) Old etchplain, now in a warm arid climate. Former latosol sheet largely replaced by a fluvio-eolian sandplain. Slight traditionally continued development of basal surface. Right: incipient *traditional pedimentation* or *glacis* at foot of arid mountain front. (b) Basin and range structure in winter-cold arid climate. Right: exposed etchplain margin at foot of arid mountain front, now buried under fanglomerates and converted into a *traditional pediment*. From here wide *glacis fans* spread across the unconsolidated basin sediments, passing from discordant to concordant stratification. (c) Recent *lateral pediment* forming along a river aggrading in a montane embayment (view toward mountains). Older pediment with corresponding gravel sheets preserved like a terrace above. (d) Glacis fan similar to that of b, formed without previous etchplain (present in a and b) and without valley mouth (as in c), through backweathering of arid mountain front. Weathering works slightly into the mountain, forming a narrow *basal pediment*. Intermediates between cases b and d are especially numerous.

sociated with specific, usually narrow, epeirogenetically outlined areas, and radiate from the fringes of isolated mountain chains into the surrounding basins. Pediments are most striking in arid regions with basin and range structure, where they have been best described. Secondly, even in these fringe areas, they are governed not by an internal mechanism working vertically from above downward, but from a distance, by the debris spreading out from the neighboring mountains. The actual mechanism producing these narrow ledges in the bedrock remains unexplained, despite a veritable flood of literature on the subject. Too little attention has been concentrated on whether and how much the pediment surfaces of the winter-cold deserts were predefined by older etchplains. While traditional planation continued in the fluvio-eolian sandplains of the warm deserts, the surfaces underwent greater transformation in winter-cold deserts. Inherited forms would still have been involved. The important point is that pediments are weak or lacking in many arid regions, even when epeirogenetically suitable fringes are present, especially in frost-free trade-wind deserts.

Hövermann (1963, 1967, 1972) was the first to point out that sizeable pediments are found only along the mountain fringes of winter-cold arid regions, where frost can continually loosen masses of debris which are then intermittently swept away across the unvegetated footslopes by occasional cloudbursts. The process is unique and intermittent, and to become effective, the climatic conditions responsible must remain stable for long periods of time.

This agrees with the geographic distribution of the phenomenon. Broad, active pediments and glacis are found in the relatively old continental desert of Tibet, in especially well developed form in the dasht plains of Central Iran (Bobeck, 1934, 1961; Weise, 1974, and others), and in the driest parts of the SW U.S. Inactive, single, or composite pediments have been described by Mensching (1958b, 1964a, 1973) from the Iberian Peninsula and along the N of the Maghreb countries; by Seuffert (1970) along the margins of the Sardinian graben; and by Stäblein (1968, 1973) in the Palatinate along the Rhine graben and in Central Spain. The piedmont surfaces of the latter have recently been reinterpreted in a very notable publication by K. Fischer (1974). A similar reinterpretation has also been suggested by Fink (1961b, 1973) for the E edge of the Alps, and by Pécsi (1968), Székely (1970), and others along the internal arc of the Carpathians. I myself have seen the piedmont surfaces of the internal and southern edges of the S Carpathians and of the United States. Most of these inactivated features have been shown to have a highest pediment or glacis (*Dachpediment, Dachglacis*) dating from the Earliest Pleistocene, i.e., they belong to an interval between the tropicoid paleo-earth period of etchplain development and the Ice Age Pleistocene. Whereas valley-cutting generally prevailed during the Pleistocene, it appears that terrace-like surfaces were cut into the highest pediments along some mountain edges. The later works of Székely and K. Fischer have shown that a hot humid phase during the Late Pliocene may have been instrumental in creating these stepped pediments.

The question now remains whether pediments, comparable, e.g., to the dasht plains of Iran, exist at all in the frost-free (or nearly frost-free) deserts of the trade-wind belt, and if so, to what extent. While some workers assume the answer to be self-evident, it is disputed by others.

When, twenty-six years ago, I first visited the Central Sahara (Büdel, 1952, 1955b), this question had never been investigated there. Since then studies have been undertaken by, among others, Rognon (1967b) in the Hoggar Mountains, and Mensching (1964a and b, 1968, 1969, 1970b), and Mensching, Giessner, and Stuckmann (1970) in the SW Sahara. Working from the Bardai Station in the Tibesti, very thorough investigations in the E Sahara have been pursued by Hövermann (1963, 1967), Hagedorn (1971), Ergenzinger (1971-1972), Busche (1974), Grunert (1972a and b, 1975), Pachur (1975), and others. These studies have provided much new knowledge on the relief development there.

Yet in order to decide whether or not pediments occur in permanently hot deserts, I felt it necessary to take a new look at the W Sahara.

In the N Maghreb, e.g., along the Moulouya Valley fringing the Middle Atlas in Morocco, Mensching (1958a and b, 1960b) and Mensching and Raynal (1954) describe well-developed pediment series. I therefore felt it incumbent upon me to investigate the extent to which this phenomenon extended into the W Sahara (compare Büdel, 1975a, for the following discussion).

Along the High Atlas the piedmont surfaces become inconspicuous, unless one considers the gravel fans radiating into the recently subsided basin of Ouarzazate as glacis surfaces.

Along the Anti-Atlas I found no pediments at all. On the Tiznit surface, the largest embayment running from the Sous Plain into these mountains (which are covered largely by desert and desert steppe), I found many traces of a true relict etchplain cutting smoothly across steeply dipping Silurian and Devonian sandstones. On it lies a red loam sheet several m thick, grading downward into a red decomposition sphere another eight m deep. Along some rivers (e.g., the Oued Massa) lenses of allochthonous cobbles appear in this loam blanket, and it contains thick calcareous crusts with heavily weathered upper surfaces: a sign of the blanket's great age. Near the mountain edge (e.g., E of Tiznit), low inselbergs protrude through the loam sheet.

The most striking feature is the junction of this surface with the Anti-Atlas. The surface does not rise toward the mountains, as would a pediment, but for the most part falls toward them, forming a pronounced, though shallow moat along their foot. From here the mountains climb steeply via a narrow, concave footslope (found in nearly all climates) to a broad summit plain, forming a gently domed etchplain stairway whose highest point lies between 800 and 2000 m. This is studded with various central upland groups which climb another few hundred m, reaching a maximum of 2531 m altitude.

A characteristic feature of the entire Anti-Atlas is that this domed surface is pocked with intramontane plains. Some of these are very broad and shallow, while others (some covering areas of over 100 km²) are sunk deep into the plain. Their floors are even and sharp-edged (Photos. 23 and 24), and are often dotted with inselbergs. The valleys draining these basins are usually narrow and gently sloping, while their V-shaped tributary gorges (like those of all fairly warm countries) nearly always have many steps and almost no gravel, closely resembling the tropical example in Photo. 25. They show no trace of the floodplains so characteristic of the periglacial valleys running evenly into the heart of the Mittelgebirge.

We have already demonstrated (Section 2.3.3.3) that intramontane basins were not formed by lateral fluvial erosion, but by areal downworking following chemical decomposition where this was accelerated by easily decomposed rock, extreme jointing, mylonitization, or even to some extent by local tectonic subsidence (see Fig. 48). Such an origin is indicated when a basin contains several rivers draining centrifugally through separate gaps. Naturally, these rivers drained the basin throughout its development, for the material now lacking in the basin was removed through their water gaps.

The basin of Tafrarout in the Anti-Atlas is particularly deep and sharply set within a stock of granite (Photo. 23). The basin walls are strewn with blocks (Photo. 24) which, though often precariously balanced, are quite stable. Only sand washes off these barren block slopes into the basin. There is no trace of a blockhalde, of scree, or even of a pediment. A very ancient form is preserved here, as in all the other intramontane basins of the Anti-Atlas. Such basins, drained through several narrow gorges, are particularly common along the S of this mountain range.

Here the junction of the Anti-Atlas with the Bani Depression for the most part forms a sharp angle. The Bani Depression, 10-20 km wide, contains many hogbacks and hogback inselbergs, a particularly marked example of the former being the Djebel Bani. The dominant relief elements, however, are longitudinal troughs between the hogbacks.

Some of the larger rivers running S out of the mountains into the Bani Depression open onto wide gravel floors (activated a few days a year in the annual winter floods); these floors are accompanied by low marl sand terraces resembling those found in all other mountains of the Sahara and the Sinai Peninsula and even in the larger intervening inselberg ranges (see Photo. 37), a distribution which shows that the terraces must be of climatic (regional), not tectonic (local) origin. Figure 54 shows the terrace distribution in the SW of the Hoggar Mountains.

The regional climatic interpretation is supported by the uniform structure

37. Granite inselbergs with broad, eight m high marl sand terrace above a fluvio-eolian sandplain (S of Meniet, Central Sahara, 280 km NNW of Tamanrasset). (Photo: Büdel, December 30, 1972.)

and environment of all these terraces. In mountain valleys they are usually 6-12 m high; only at narrows, e.g., in gorges leaving the intramontane plains, do they climb higher, sometimes reaching over 20 m. In the Hoggar, where they run smoothly across valley steps, they can even tower 40 m above the present valley floor (Büdel, 1955b, p. 107). They consist exclusively of thinly bedded yellow-red to reddish-brown marly sand, and almost never incorporate coarser material from adjacent block-strewn slopes. The terraces are without exception depositional; they have never attacked the rocky slopes next to them and at no time continue as erosional terraces onto the slopes. They constitute foreign bodies in the valleys.[45] The large valleys running S out of the Anti-Atlas show this foreign character especially well. At the narrow canyon mouths where the rivers issue from the mountains, the marl sand terraces are often only remnant "collars," perched 20 m or more up on the rocky slopes (see Photo. 38). They bear no rubble whatever from the slopes above. At such a height, they must certainly predate the Würm Ice Age, which again corroborates the stability of the rocky slopes above.

Chemical weathering, the sole possible producer of such fine matter, is insignificant here today. Yet the masses of fine matter in today's terraces indicate that such decomposition must once have been widespread in the

[45] A connection between the marl sand terraces and the relict red loam sheets could be established only in the interior of the Hoggar (Büdel, 1955b; Kubiena, 1955; Rognon, 1967b; Mensching, 1970b).

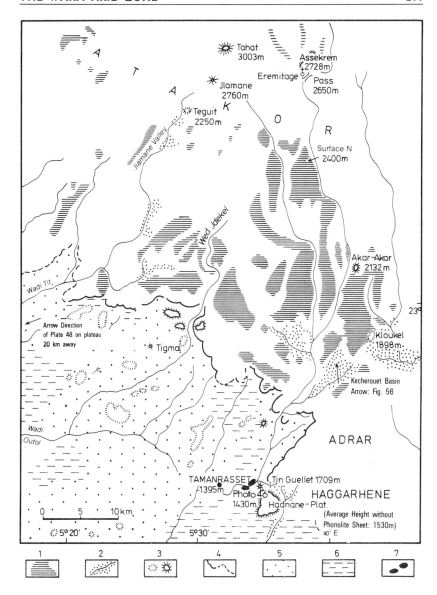

Fig. 54. Relief generations on the SW flank of the Hoggar Mountains (Central Sahara). (1) Old etchplains, usually with thick red loam blanket preserved under basalt flows. (2) Marl sand terraces and parabolic slopes in wider valleys and basins of mountains and mountain fringes. (3) Mountains, plateaus, and large inselbergs. (4) Foot of Hoggar Mountains at fluvio-eolian sandplain (broken line where course unclear or not investigated). (5) Fluvio-eolian sandplains. (6) Low basal knobs of relict basal surface of weathering, protruding above sandplains. (7) Red loam remnants E of Tamanrasset.

30. W precipice of the Highlands of Semién, between the gentle paleorelief, upper right, at around 4000 m, and the Sudanese lowlands. Massive series of thick basalt flows interlayered with tuff. (Photo: Büdel, April 9, 1953.)

38. Remnant collar of a 20 m high marl sand terrace in the valley narrows of Foum el Hassane, where it runs through the Djebel Bani in S Morocco. One of three separate exits from an intra-montane plain, 175 km SSE of Agadir. (Photo: Büdel, March 1, 1972.)

Sahara. In the Hoggar Mountains, some of this marly sand derives from red loam remnants, now covered with basalts of probably Late Tertiary age. Up to 50% kaolin in clay fraction and reaching 20 m in thickness, the loam remnants overlie an etchplain surface. Both the surface and the loam sheet testify to a pre-Pleistocene seasonally tropic climate.

Etchplains exist at various heights in the mountains of the Sahara, but rarely are paleosols preserved on them as fully as under the Hoggar basalt sheets. Here the marl sand of the terraces demonstrably derives at least in part from pre-Pleistocene soil sheets. The uniformity of the terraces throughout the Saharan mountains suggests that the terraces outside the Hoggar Mountains also derive at least in part from loam blankets which covered a tropical plateau relief resembling that still preserved in the Anti-Atlas. Even while this plateau relief was being riddled by intramontane plains and peripherally dissected, planation could probably continue at its center, as long as soil sheets were still present.

Periods of weakened chemical loam development reoccurred during the Pleistocene pluvials. The older, easily transported fine matter was reworked in each pluvial, and new matter washed down and added to it. Thus only one terrace is the general rule,[46] rather than one terrace per pluvial. This contrasts with periglacial valleys of Central Europe, where one clear-cut terrace represents each ice age. Multiple terraces developed only at higher altitudes, if at all, in the Sahara. The ice rind effect, with its rhythmic lowering and widening of the valley floors, was lacking, and a purely depositional terrace maintained a single level during the various periods of fine material production and wash. The easily transported matter was continually redeposited, but neither its base nor its surface was significantly lowered. In the forelands, later alluvium may even overlie the older.

In the broad basins and lowland ribbons of the Bani Depression, slashed by inselberg-like hogbacks, the marl sands lie relatively intact across broad areas; these are regions of transitional deposition. East of Goulimine is a flat lowland strip 25 km long and 8 km wide. Its gently arched floor is flanked on both sides by shallow moats, not by pediments. The loam deposits here, over 10 m thick, contain a number of calcareous crusts several m thick and varying in age, resting on plant-bearing limnic limestones (see Fig. 39). These may correspond in age to the limnic limestones of the Villefranche beneath some calcareous crusts in the Sous Plain. Here again the very solid crusts are weathered on top, showing that both the crust and the loam filling are by no means recent. The upper weathering layer contains scattered, well-rounded, allochthonous cobbles showing signs of later redeposition. Some of the fine

---

[46] Terraces are found in the Hoggar on rare occasions: these are then double, the two steps being a few m apart, and located at extreme heights in isolated valleys. Such terraces are more common in the Tibesti (Hagedorn, 1971; Busche, 1974; Grunert, 1972a, 1975).

matter must be very old, while the overall arched form of this wide, elongated basin floor must have been renewed relatively recently.

One must accustom oneself to the idea that toward the equator, cold stage influences will have formed terrace sequences as we know them only at high altitudes. At lower elevations, the fine-grained terraces show little difference in height, where separated at all. This is especially true in basins and along mountain edges, as in the Bani Depression and the surrounding foreland.

On the whole, these marl sand deposits are less a terrace in the usual sense, than a facies, created each time a pronounced pluvial phase with savanna vegetation and chemical soil development occurred in the Sahara. In the higher mountains, an intercalated terrace of this facies may be correlated with a particular chronological stage in the relief history. In the depositional areas of the mountain fringes and forelands, the highly mobile facies bodies unite into a single level, whose outward shape is recent, but whose content may derive from various pluvials.

That the deposition of the Bani Depression sediments started long ago is shown by another feature in the same area, E of Goulimine, where the elongated basin joins the gentle footslopes of the Djebel Taijert hogback to the S. Quartzite fragments which have tumbled onto the footslopes from above include some with chocolate-colored iron coatings up to three cm thick, resembling ancient rinds found by Hagedorn (1971) in the Tibesti. Pisolites are not uncommon in the red loam matrix of the Djebel Taijert rubble blanket, together with allochthonous quartz cobbles displaying deep iron-stained scars of ancient warm wet weathering. The rinds contain over 25% iron hydroxide, a compound formed only in warm humid tropical conditions. This agrees with the almost complete leaching of calcite, and with the fact that what little clay there is contains hydrargillite. The lower slopes therefore also preserve an ancient form, and must be extremely stable. The marl sand was thus deposited in an ancient, perhaps pre-Pleistocene basin, whose sides have since been neither undercut by lateral erosion nor veiled by glacis development.

South of the Bani Depression and of the Oued Draa draining its southern margin, the tectonic structure flattens out, and the Cretaceous beds overlying the Paleozoics smooth out into gently undulating sheets. Shallow synclines, such as that underlying the Plateau du Tademeït (450 × 200 km large), are rimmed by sharp escarpments. Yet the nearly flat surface is only adapted to the bedding structure along a narrow peripheral zone, and otherwise arches gently across the shallow syncline. This surface is covered by thick, red loam-like fine material, topped with a solid crust of black-coated gravels which give the whole a typical serir appearance. There can be no doubt that this is an ancient etchplain undergoing traditional development. The valleys cutting back from the stepped escarpment are typical tropical mountain valleys, having narrow V-shaped cross-profiles and stepped longitudinal profiles.

The escarpment itself is broken up by typical triangular reentrants. In front of these, and in front of the smaller, lower steps are outlier inselbergs, some still with hard Cretaceous caps. Such inselbergs stud the broad fluvio-eolian sandplains in the basin of In Salah as well, and have the same sharp feet as the granite inselbergs dotting the fluvio-eolian sandplain further S (see Photos. 39 and 40). I could find no basic difference between the inselbergs with hard sedimentary caps, and those weathered out of the plutonic rocks of the African shield: even the color contrast between the darkly crusted, stable inselberg walls and the palely glittering sandplain is the same in both cases.

Toward the central dome of the Hoggar, Cambro-Ordovician sandstones, representing the oldest sediments overlying the African granite basement, crop out beneath the Cretaceous strata. These sandstones form the well-known Tassili escarpments encircling the central dome. Deeply washed-out joints forming bizarre, closely set rock walls and turrets have made the E Tassili n'Adjer (near Djanet) into a tourist attraction. These forms are not so much steep mountains as enormous penitent rocks, while the narrow passages between their vertical walls are not valleys, but flushed-out joints.

The joints run vertically through the horizontally bedded sandstone. That no subaerial deepening or widening occurs here today is proved by the rock paintings, dated back as far as 9000 B.P. Many of these are well-preserved, even in rock shelters, proving the extreme durability of even the steepest rock faces in the frost-free desert.

The granite dome of the central Hoggar bulges up out of the surrounding Cambro-Silurian sandstones. The overall dome is the size of the Baltic Shield, but the actual basalt-covered mountain range at the center is much smaller. The broad, gently ascending periphery of the shield is covered by the omnipresent fluvio-eolian sandplain, punctuated by groups of barren granite inselbergs of all sizes. These include everything from steep low shield inselbergs (see Photos. 39 and 40) to bell-shaped domes (Photo. 41) and pyramids (Photo. 42) rising mirage-like over the sands. The pictures reproduced here clearly show that no pediments radiate from these inselbergs, which instead are often surrounded by moats (Photo. 43).

Where the sandplain has been removed by wash or deflation, the knobs of the former basal weathering surface may be partially exhumed (see Photos. 39 and 41, foreground, and the aerial view of Photo. 44), exposing the joints etched by former decomposition at the basal surface. The old red loam blanket, of course, was stripped much faster than the bedrock beneath, and its material worked into the sandplain. The high sand content of the latter derives from the fact that protruding inselbergs and basal knobs in this climate are subject only to block disintegration (whose products remain lying on the slopes) and to sanding. Exposed knobs are gradually reduced to block heaps (*Blockgipfel*) through boulder cleavage (heat and salt wedging). The blocks remain *in situ*, gradually crumbling into sand, which is rapidly washed or blown away (see Photo. 45). The resemblance to the wind corridors in Photo. 47 is unmistakable.

The top layer of the sandplain is always of nearly pure sand. Yet dunes

39. Shield inselbergs on Central Saharan fluvio-eolian sandplain. Foreground: several exposed basal knobs of the relict etchsurface. High conical mountain in background, right. Granitic forelands of the Hoggar Mountains, NW of Tesnou, 250 km N of Tamanrasset. (Photo: Büdel, December 30, 1972.)

40. Smooth granite dome with sharp foot: no pediment-like transition to the fluvio-eolian sandplain. Lower left and center right: low basal knobs and shield inselbergs of the same granite as the higher mountain. Same site as Photo. 39. (Photo: Büdel, December 30, 1972.)

41. Steep granite inselbergs, up to 200 m high, in the Central Saharan fluvio-eolian sandplain. Sharp piedmont angles without gentle concave footslopes. Basal knobs and shield inselbergs protrude though the sandplain. Lower left, dunes. (Photo: Büdel, March 1, 1973.)

42. Fluvio-eolian sandplain with far-off jagged granite inselbergs. Central Sahara, same region as Photo. 41. (Photo: Büdel, December 30, 1972.)

43. Moat of tall, barren, granite inselberg near the oasis of Tesnou, 240 km N of Tamanrasset. The partially block-veneered convex slopes produce only sand. The sandplain drops markedly toward the sharp piedmont angle. Below the sand is a loam sheet, used by the inhabitants for making dried bricks. (Photo: Büdel, December 30, 1972.)

46. Thick sheet of relict red loam on higher area in fluvio-eolian sandplain near Tamanrasset. Photograph used for Fig. 55. For position of site, see Fig. 54. (Photo: Büdel, January 3, 1973.)

44. Stripped terrain of basal knobs, deriving from the basal surface of a former etchplain, with patches of fluvio-eolian sandplain. Near In Amguel, 145 km NNW of Tamanrasset. Upper right, higher inselberg ridge with sharp foot. (Photo: Büdel, January 4, 1973.)

45. Same as Photo. 44, closer up. Lineaments of the African basement clearly visible on the basal surface. Signs of wind erosion. Compare Photo. 47. (Photo: Büdel, January 4, 1973.)

(usually star-shaped, due to the alternation between trade and etesian winds) are rare, probably because of the sheet of yellow loam which commonly lies but a few handbreadths below the surface. This loam represents the reworked and sand-mixed remnants of the former red loam blanket. In many places it is excavated by the native inhabitants to make sun-dried bricks (see Photo. 43).

Where the sandplain ascends slightly at the former wash divides and etch-plain divides, it has remained relatively unworked by arid wash processes (i.e., by the wadis), and is affected only by deflation today. Here we find relict sheets of remnant red loam, described as such by Kubiena (1955) on the highest central part of the Tamanrasset surface E of the Tamanrasset Oasis, just before the rise to the Hadriane Plateau (see Fig. 55). This paleosol is 70% illite and only 5% kaolinite, yet in provenance and structure it closely resembles the soils of the seasonal tropics. It is brilliant red, completely chemical in origin, and has no connection with the desert climate of today (Photo. 46). Over it lies a serir scattering of coarse vein quartz, and veins

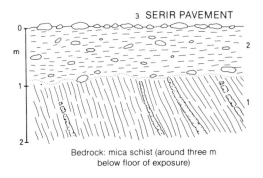

Bedrock: mica schist (around three m
below floor of exposure)

Fig. 55. Relict red loam E of Tamanrasset (for location, see Fig. 54, number 7). Found on flat low hill, 14 m above the broad bed of Wadi Tamanrasset and 10 m above the surrounding fluvio-eolian sandplain. (1) Relict red loam, nearly four m thick, deriving from full decomposition of underlying mica schist. Isolated quartz veins still show steeply dipping bedding. (2) Redeposited red soil sediment, containing individual well-rounded cobbles, mostly of vein quartz. (3) Serir pavement of nearly pure vein quartz, with occasional well-rounded quartzite and basaltic clasts (see also Photo. 46).

of the same quartz run undisturbed up through the soil, almost to the surface (see Photo. 46 and Fig. 55). Mixing is found only in the top 60-80 cm, which contain some well-rounded cobbles of allochthonous basalt and quartzite: this represents the transition from deeper, *in situ* soil to soil sediment. The underlying bedrock consists of mica schist with frequent quartz veins.

The sheet of relict red loam is fully integrated into the etchplain and inselberg relief dominating the entire Hoggar area, and lies at about 1430 m on the central part of the Tamanrasset surface near the phonolite-capped Hadriane Plateau. Barely visible in the background of Photo. 46, this plateau runs from S to N (right to left in the photograph) at an altitude of 1560-1700 m, or an average of around 1630 m, and therefore stands around 200 m above the loam-covered surface in Photo. 46. The phonolite, nearly 100 m thick, covers an old etchplain in the crystalline rock, which therefore has an altitude of around 1530 m, and lies only 100 m above the surface in Photo. 46.

Above the lower W portion of the Tamanrasset surface (NW of the Tamanrasset Oasis), again at a relative height of 100 m, lies a dissected relict surface in crystalline rock which is not covered by volcanics (Photo. 48). The Tamanrasset surface between this relict plain in the W and the basalt plain of the Hadriane Plateau in the E has therefore been lowered by about 100 m. The relict red loam sheet on it proves that this occurred areally, not in the desert climate of today, but in a previous long humid period. The Pleistocene pluvials were neither long enough, nor climato-mechanically effective enough for this. We may therefore assume that the lowering occurred during a pre-Pleistocene humid period.

The relief picture shows that the entire etchplain and inselberg relief of the NW Sahara must have been created during a very long Tertiary humid period. The later Quaternary arid period has only altered and thinned much of the former soil sheet, but has not affected the major relict structures, and on the whole has only traditionally continued their development. The present wadi ribbons in the fluvio-eolian sandplain have cut at most 5-15 m into the Tamanrasset surface, and have hardly affected the bedrock at all.

The extensive preservation of the paleo-relief in the nearly frost-free desert climate of the Sahara is best shown by the fact that the mountains and inselbergs mentioned, e.g., the Anti-Atlas, the Tassili scarps, and the inselbergs and inselberg ranges of Photos. 39-44 all lack any trace of a pediment or of a gently sloping glacis or debris footslope. In many cases, especially in Photo. 43, the sandplain drops into a shallow moat next to the granite inselberg range; this is the case along most of the Anti-Atlas. On granite slopes the blocks, once weathered free, lie unmoved (Photos. 24, 40, and 43). Material sanded off is removed by the fluvio-eolian plain. Had these mountain slopes been subject to more erosion and backweathering during the Quaternary, a pediment, a glacis, or at least a broad debris footslope would be present at their foot. Yet there is not the slightest trace of any such thing in the entire region between the Anti-Atlas and the Hoggar.

2.5.1.2 A Closer Look at the Fluvio-Eolian Sandplain
and Other Present-Day Surface Features
as Further Indications of Traditional Development

The fluvio-eolian sandplain, as we have stressed, has its own microrelief, mainly of larger wadis with gently saucer-shaped cross-profiles cutting a few m (at least a few dm) into the sandplain (see Photo. 41). Mensching (1970b) labels the deeper wadi floors "kori" surfaces. Other authors have called the main sandplain surface above them "Lower Terraces," which, however, have nothing to do with the Würmian Lower Terraces (*Niederterrassen*) of the Holocene valley floors in the mid-latitude periglacial regions. That both levels of the sandplain belong together is best shown by the fact that the upper courses of the wadi branches hardly cut into the sandplain at all.

That these intermittent streams (*Trockenflüsse*) cut no deeper is explained by the fact that they carry only sand and no coarser erosional tools. Moreover, they flow but rarely, and then only within the local area of rainfall, so that their sand load is quickly redeposited.

The main source of sand is the erosional process dominating all protrusions in the frost-free deserts, namely, the process of sanding. As Meckelein (1959) has stressed, the smaller basal knobs are most affected by this process, disintegrating largely through cleavage (*Kernsprünge*). Unlike the smooth-walled inselbergs, these basal knobs underwent chemical joint widening beneath their former loamy or sandy soil cover, and therefore present more surface area for sanding. Boulders may even disintegrate from within (see Photo. 39 lower left, and Photos. 44 and 45). Variations on this have been described by Wilhelmy (1958a, 1974, 1975), as "hollow blocks" (*Hohlblöcke*) and "turtle shells" (*Schildkrötenschalen*). Smaller protrusions are thus the first to succumb. They are gradually absorbed by the sandplain and are slowly worn off along with it. The sandplain does not create the basal knobs, as does the latosol blanket of the seasonal tropics, but rather *vice versa, the disintegrating basal knobs feed the sandplain*.

On open sandplains not protected by mountains, the westerly and northerly etesian winds and the more typical easterly trade winds blow the sand into great dune fields or *ergs*, as the Arabs call them. Although these constitute the popular picture of the Sahara, they cover only about 20% of its surface. The dunes may be longitudinal, running parallel with the wind, or slightly sinuous transverse dunes at right angles to the wind. The latter dunes are often broken up into crescent-shaped barchans with gently sloping convex upwind sides and steep concave lees. Barchans often occur in irregular swarms and are typical of the dunefields blowing across the sand or salt-clay flats (*sebkhas*, called *playas* in the American deserts). Isolated dunes, sometimes over 50 m high, may also occur in such flats (compare Meckelein, 1959, Fig. 33), and may be star-shaped, especially with frequently shifting winds. Dune development was reduced considerably during the pluvials; hence different generations of dunes may, through degree of weathering and through intervening soil layers, be laterally or vertically distinguishable.

Ripple-marked sandfields and sheets of drifting sand, such as those found by Hagedorn (1971) in the Tibesti Highland area, provide a transition from fluvio-eolian sandplains to ergs.

No Arabic terms for fluvio-eolian sandplains or for the steeply dissected mountain deserts have entered the scientific literature. Even "hamada," often used for mountain desert in the older literature, is now usually used differently, as described below. Probably both native inhabitants and visitors regard the sandplains and the mountain deserts as so typical as to render any special term for them unnecessary; at any rate, my repeated inquiries in this regard met with no success.

"Gravel desert" or "pebble desert" on the other hand, is covered by no less than three Arabic terms. These are not scientifically as well defined as "erg" (dune-covered desert) and "sebkha" (salt-clay flat). Thus "reg," "serir," and "hamada" tend to overlap, both in nature and in the literature. We will attempt to define them as broadly as possible here.

Mensching (1970b) uses *reg* to mean a clayey desert surface, usually at low elevation and of recent origin, covered by a gravel sheet of pebbles sorted out of the substrate by washing or deflation of fine material. No xenoliths are present. More common are the *serir* surfaces, similar in origin, but usually older and somewhat higher. These contain coarser particles sorted out from an otherwise fine-grained substrate, along with well-rounded allochthonous pebbles. We have already described serirs from the Bani Depression (Fig. 39) and on the relict red loams near Tamanrasset (Photo. 46). Most marl sand terraces bear a loose serir sheet. Meckelein (1959) defines serir as a fine gravel desert of angular to rounded clasts of partially allochthonous origin. These surfaces may be very extensive. They are "plains running flatly or with gentle undulations, sometimes for distances of over 100-300 km, forming landscapes of almost unrivaled featurelessness. . . . Oppressively uniform, usually devoid of plant life, these [serir] gravel deserts are sometimes located in large basins" (Meckelein, 1959, p. 52). Hagedorn (1971), referring mainly to the lack of plant cover (which is even sparser here than in any other desert), justifiably speaks of the "desert in the desert."

Hamada surfaces usually consist of coarse gravel pavement. These are common on older plateaus, such as the Cretaceous Plateau du Tademeït. With no adjacent higher land, hamada surfaces usually lack allochthonous components. Some authors, such as Mensching (1970a) use the term not only for the surficial gravel on a high surface, but for the plateau itself. A hamada would therefore be a plateau desert.

These differing interpretations of the Arabic terminology are insignificant in view of the basic features shared by these pavements. They all indicate that an ancient, very loamy, chemically formed weathering sheet (possibly with pre-Pleistocene and intra-Pleistocene components) is being surficially removed, but that in many places these sheets are still present in considerable thicknesses. The sorted pavement created by the combined erosional mechanisms simultaneously retards these same mechanisms with increasing effectiveness as the pavement gets thicker. Surface wash (running water) is certainly important, but on the desert axis especially, as on the Plateau du Tademeït, heavy rains and wash processes occur only at intervals of years. Powerful winds are far more frequent, and play an important role in forming pavements.

The *wind-corraded surfaces* recently investigated by Hagedorn (1968, 1971) and Mainguet (1968), where the wind has stripped away the soil cover and now attacks the substrate itself, are of particular interest. Hagedorn studied this principally on the rise of Borku (S of the Tibesti Highlands, at about 10° lat. N), where a very broad eolian relief has formed on what must have been a gently rising, low altitude, rocky mass in the surrounding gravels and sandplains. This was attacked by the powerful trade winds typical of this latitude and of the passage between the Tibesti and the Ennedi Highlands, blowing with extreme regularity from NE to SW. Attacking the easily transported, recent sediments, these winds have carved the easily mobilized sediments into streamlined convexities or *yardangs* protruding only 0.5 to 4 m above the intervening *wind corridors (Windgassen)* (Hagedorn, 1971, p. 114).

It is morphoclimatically more significant, however, that wind corrasion, according to Hagedorn, also molds the bedrock itself of the Borku rise. The old etchplain stretches here for hundreds of km in all directions. Hagedorn's illustrations (especially 1968, Plate 1, and 1971, Figs. 75-84), one of which is reproduced here in Photo. 47, show a classic wind-corraded, basal knob relief. As on all exposed basal weathering surfaces, the lineaments of the former decomposition sphere are clearly revealed. "Joints running parallel to the wind are sculptured into wind corridors, while joints running in other directions remain as small notches in the aerodynamic forms" (Hagedorn, 1971, p. 115). This is of course especially true of the bedrock *wind ridges*, which run with striking uniformity parallel to the wind between the wind corridors. Eolian corrasion has also overprinted an old river network here which was likewise joint-controlled, and which in my opinion may derive from the basal surface of an old sandplain which has since been blown away. Parts of this former river system were adapted to the pre-existing plain, and did not cut into it: further proof that even this fluvially formed predecessor of today's wind-corraded relief bore the character of an etchplain.

The wind-entrained particles which blast the rocks of Borku today derive from the surrounding sand and gravel desert. According to Pachur (1967), the main grain sizes involved are between 0.15 and 0.35 mm (optimum 0.25 mm). As the wind carries these particles close to the ground, it is most effective in the already established wind corridors. The striking wind ridges left between may be but a few m wide, reach lengths of 500-1000 m, and generally stand only a few m above the sand floor of the corridors. In some places higher elevations tower up to 50 m or even 100 m above the surroundings, forming broader "heads" behind which trail the elongated "normal" wind ridges (see Photo. 47, bottom right). The highest of these heads or knolls was described by Hagedorn (1971, in his interpretational sketch of Fig.

47. Wind sculpted features: eolian knobs and wind corridors in the Borku Mountains, 700 km NE of Lake Chad. Wind-formed relict etchplain base, after Hagedorn, 1971, Fig. 79. He writes: "Joints and faults crossing the NE-SW trending wind-shaped forms are clearly visible. In the upper right quarter, wind-oriented forms could only partly develop against the grain of structures following other directions" (p. 199, translation given on p. 236). The structure of the African shield apparently allowed a group of higher shield inselbergs to form on the previous etchplain surface. (Photo: Institut Géographique National, Paris.)

79, reproduced here as Photo. 47) as a "non-eolian mountain peak." We may view this as an old shield inselberg, rising above the zone of eolian activity near the ground (see Photo. 47, somewhat above center, especially in the right half of the picture).

On the whole, our conclusion is that these old etchplains "are subject to little further development today. [Current] development is limited largely to the preservation of their flat character and the evening out of such microrelief as is still present" (Hagedorn, 1971, p. 121). This then, is a classic example of *traditional development of an old surface*.

2.5.1.3 The Relief Generations of the Hoggar Mountains

A number of works cover the Hoggar relief generations, including my own treatments (Büdel, 1952, 1955b, 1975a). These have been refined upon and added to by an extremely detailed account of Rognon's (1967b), Mensching's reaction to that (1969), and Mensching's own research (1970a). On the whole these studies have tended to support my views.

The 300-600 km wide foreland plain which surrounds the Hoggar Mountains proper mounts from around 600 m at the edge of the Tassili to around 1400 m toward the interior. This plain, studded with inselbergs and veiled with fluvio-eolian sands, cuts smoothly across a variety of crystalline units. Its summit is crowned by the Hoggar mountain range itself, a steep-walled oval around 100 × 150 km in size. This is a sort of giant inselberg range, on whose margins surfaces at various heights are capped by young volcanic sheets. The corresponding volcanic necks are grouped toward the center. The edges of the high block may locally coincide with its crystalline base, but in general they correspond as little to the petrovariance as do the smaller neighboring inselbergs (Photos. 39-44) and plateau fragments (Photo. 48).

The differing lithologies of the granitic-migmatic base and of the recent volcanics and the resulting structures were heavily stressed by Rognon (1967b). But the petrovariance is revealed largely in the valleys, not on the surface.

From the sandplain of Tamanrasset one climbs steeply and more or less directly to the broad relict surface surrounding the volcanic necks of the central upland of Atakor (its highest peak, Tahat, is at 3003 m). The series of surfaces climbs gently from the lowest, outermost one at 1500-1600 cm, to the highest one, around Atakor, at 2400 m. The ascent is step-like but does not involve a genuine etchplain stairway. True, the peripheral surface ribbons may to some extent be younger, and the highest inner surfaces older, yet these different levels are hardly separated by clear etchplain escarpments, but rather by zones of highly dissected relief. Clearly a once uniform surface created during prolonged tectonic stability was uplifted and the various segments displaced with respect to one another. The outer lower portions may then

48. Dissected etchplain above fluvio-eolian sandplain, 35 km W of Tamanrasset. (Photo: Büdel, January 4, 1973.)

have continued to be active, or at least traditionally active for longer than the higher ones.

Otherwise these relict surface ribbons are fairly uniform, despite differences in height. Their broad flat surfaces "have largely been preserved intact. They form typical examples of true etchplains cutting with almost perfect smoothness across a heavily folded crystalline basement of greatly varying lithologic hardness. . . . The steeply dipping quartz beds are particularly striking: though beveled smoothly by the old surface, they form knife-like ridges across the present valley slopes" (Büdel, 1955b, p. 102ff.). The surfaces even bear low inselbergs with rather steep domes and gently concave footslopes, e.g., the Kloukel (28 km NE of Tamanrasset, Fig. 54). One thousand eight hundred ninety-eight meters high, this towers 150 m above a well-preserved relict surface at 1760 m. Both consist of the same mica schist, though quartz veins are concentrated around the Kloukel.

That these relict surfaces are products of a past seasonally tropic climate is shown not only by their well-preserved relief assemblage, but also by their nearly continuous deep latosol cover, which in some places is over 20 m thick. This makes them unquestionably true etchplains. According to analyses by Correns (oral communication), Kubiena (1955), and Rognon (1967a), the latosol sheets contain over 50% kaolin. The upper, brilliant red horizon is frequently preserved, along with the mottled zone and the pale to bluish-white basal layers below.

This is of great importance. At the time when a flat etchplain with relief-developing soil was being created, a savanna climate with marked dry seaons and a rainy season of at least 4½ to 5 months must have prevailed in the region of today's Hoggar Mountains. These were not mountains at the time; otherwise the highly fluid basaltic flows would not have collected there. The wet climate therefore cannot have been associated merely with a humid mountain "island," but must have prevailed in a wide area. This again supports the idea of an old etchplain relief in the NW Sahara.

In the Hoggar Mountains, these extensive etchplain fragments with their latosol soils are preserved only where protected by old basalt sheets, which extend across thousands of square km (see Figs. 54 and 56). Only 20-30 m thick, the basalt sheets nowhere descend into the dissected fringes of the relict plain fragments today. Thus they could not have found a valley network already carved into the old surface, but poured as highly fluid lavas across a broad, quite uniform, inselberg-studded etchplain. The source of these oldest lavas can no longer be found.

Even the 20 m of relief-creating soil preserved today presupposes a tropically humid climate lasting for hundreds of thousands of years. But tens of millions of years must surely have been necessary to reduce the old African shield by the hundreds of m required to produce such a featureless etchplain landscape. The beginning of this period, then, may be set in the Early Tertiary,

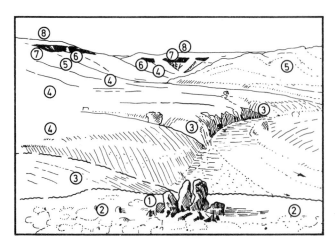

Fig. 56. Wide area in valley of the Hoggar Mountains, Kecheruet Basin, 22 km NE of Tamanrasset. View looking N from a location 1650 m high; background 1800 m high. (Drawing by Büdel, March 7, 1951.) (1) Frost-shattered basalt block. (2) Traces of needle-ice patterned ground. (3) Seven m high wadi bank, with marl sand terrace above. (4) Broad, gently concave, parabolic slope cutting across (5) tightly folded ancient crystalline rock, up to (6) rim of dominating Tertiary etchplain. This is covered by (7) an old latosol sheet over 12 m thick, brilliant red above, with pale white kaolinitic loam below. Above this lies (8) an old basaltic sheet up to 20 m thick.

as I concluded in 1955. There is no reason for extending the period of etchplain activity back into the Mesozoic. The basalt flows, probably of Miocene age, brought the Hoggar's first relief generation of etchplain development to a close, and at the same time heralded the new period of further volcanic outbreaks and general upheaval.

The next phase of volcanic activity produced mainly phonolite and other acidic material. Being tougher, this has been better preserved in plugs and intrusive domes forming the highest peaks here today. Lava flows from this phase are rare, and, as on the Hadriane Plateau, cover a higher energy relief lacking kaolinitic soil sheets. This phase of eruption must therefore have been preceded by some uplift and linear erosion. Like Büdel (1955b), Rognon (1967a) assigns this phase, terminating the second Hoggar relief generation, to the Pliocene. He stresses that the Villafranche, forming to some extent the final phase of this period, was still extremely humid, and he supports this with much paleobotanic evidence.

Rognon (1967a) also agrees with me that not until subsequent Pleistocene phases did the final uplift take place, while today's drainage concurrently developed in a pronounced arid climate. The valleys were carved out nearly to their present depth by the Early Pleistocene, and were filled up again by two further, lesser phases of basaltic eruption. In the mountain interior they were then dug out again to their present depth, while in the peripheral areas the former depths were not always regained.

These valleys have a highly characteristic shape. The small tributaries have markedly stepped, very steep, V-shaped gorges similar to those in mountains of the seasonal tropics. Petrovariance is sharply accentuated on the slopes, in contrast to the summit plain's complete indifference to lithology. This resembles the contrast between the smooth hammada surface above Tassili and the extreme dependence on lithology in the rugged canyons of the Tassili escarpment.

The larger valleys, particularly in their upper reaches, still have steep, lithologically controlled steps in their longitudinal profiles. The intervening stretches, however, have broad sandy floors, due to the dominance of sanding on the slopes. In these open areas Rognon (1967a) has found a system of terraces which he correlates with the Pleistocene pluvials. The older, coarser-grained terraces are less important than the familiar marl sand terraces, ascribed in the Hoggar to the Middle and Late Pleistocene. They evidently postdate the third (basaltic) eruption phase, for they nearly always overlie the valley lava streams. More importantly, their morphology differs markedly from that of ectropic river terraces, especially of ectropic mountains. Firstly, they are thin-bedded and fine-grained; secondly, they never seem to be erosional terraces; and thirdly, their association with the adjacent slopes is unusual (see Fig. 56), as can be seen particularly in wider spots in the Hoggar valleys. Rising gently from the terrace surface, the concave slopes gradually become steeper and assume an unusual parabolic profile as they climb toward the upper slope rim. We call these *parabolic slopes*. In the Hoggar, they nearly always connect at or just below the thick latosol soils under the basalt cap. According to Kubiena (1955), they sometimes connect with the brown loam remnants which he discovered on the Atakor. This association is also suggested by the kaolin content of the debris-mixed fine matter of the parabolic slopes and the marl sand terraces, while the greater carbonate content in the terraces suggests later reworking in an arid climate. Thus Mensching (Mensching *et al.*, 1970, p. 124), much like Kubiena and Büdel before him, concludes that here sediments "of a soil with some red loam and much kaolin deriving from beneath the oldest of the capping strata are included among the reworked material."

A fourth feature in the Hoggar, stressed by Büdel (1955b), has been given less weight by later observers. The valleys of the Hoggar, even when large, have long gentle stretches with broad sandy floors, separated by bedrock steps. These steps are very important for the Tuareg nomads, for here the groundwater, which otherwise runs under the sandy floor, is forced by the rock barrier to surface in the water holes.[47] The marl sand terraces march smoothly across these steps; some distance away, their normal height is 6-10 m. From here they rise gradually upstream, reaching a relative height of over 40 m at the step, and returning to their normal height of a few m at the next basin. This clearly shows that they have nothing to do with the cutting of these valleys through vertical erosion. Instead, not only did another kind of soil development take place during the pluvials, but also another kind of valley was developed, with thick transitional deposition far up into the valley. We have already described this marl sand accumulation as a pluvial depositional facies which contains material from pre-Pleistocene humid phases, but whose shape in the mountain valleys is to be correlated with more recent pluvials. This agrees with Grunert's observation (1972a and b, 1975) in the Tibesti, that the normal terrace height is 7 m, with a maximum height of 25 m. Here also the marl sand terraces are purely depositional, showing neither lateral nor vertical erosion.

Two Pleistocene phases of vertical erosion are clearly evident in the Hoggar: an older, energetic phase preceding the marl sand deposition, and a later phase which carved out the desert valleys and their tributary gorges to or nearly to their former depth. Grunert has found two similar phases of gorge development

---

[47] "Tilmas" dug in the sands upstream of the step produce clear, sand-filtered water, and are therefore preferred for human use. The camels are watered at the open water holes or "gueltas" located at the rock barrier itself. These holes contain a wealth of life, including water beetles, mosquito and dragonfly larvae, and all sorts of plankton. The porous sandstones of the Tassii contain larger, karst-like gueltas, such as the famous Guelta Tiguelguemine, 40 × 60 m in size. This very cold pool, over 10 m deep, is located in a steep-walled valley head 160 km SE of In Salah. An even larger guelta contains the famous pygmy crocodiles, diminutive pluvial relicts.

in the Tibesti Mountains as well, a more vigorous one before, and a less vigorous one after the deposition of the marl sand (there the Main Terrace or *Hauptterrasse*). Well-developed relief-covering soils from this pluvial depositional period have been found by Grunert (1972) and Sabelberg (1972), as well as by Kubiena and Büdel in the Hoggar,[48] and C^{14} dating of their organic contents yields an age of 7,000 and 14,000 B.P. This proves that a marked humid period with soil development, forest cover, and considerable valley deposition lasted from the Late Würmian into the Early Holocene, i.e., that the Saharan pluvials do not correspond with the ectropic cold stages (Jäkel and Schulz, 1972; Grunert, 1975).

Clearly the Saharan arid phases were times of valley-cutting, while during the pluvials the valleys were filled in with a fine-grained facies deriving from the concurrently forming soils. Older terraces were at the same time always reworked into the new terraces.

2.5.1.4 The Altitudinal Zones of the Highlands of Tibesti

Only twelve years ago the Tibesti Mountains qualified as the most remote and unexplored region of the Sahara. Thanks to the intiative of Hövermann (1963, 1972), who founded the Bardai research station there, and to his colleagues and students who worked there (particularly Hagedorn, 1968, 1971), the Tibesti is now one of the best studied regions of the Saharo-Arabic desert belt.

The forelands of these mountains have been discussed in works by Ergenzinger (1968, 1971-72), Kaiser (1972), B. Gabriel (1972a and b), and Busche (1972a and b, 1974). One of the many features studied by these and other authors is a broad fluvio-eolian sandplain resembling that of the Hoggar. This stretches S to broad plains, studied by Ergenzinger, which were inundated by the Later Pleistocene and Holocene transgressions of Lake Chad. As around the Hoggar, the sandplain, deriving from Quaternary phases of relative aridity or humidity, veils a Tertiary etchplain relief consisting of partially exposed basal knob relief, shield inselbergs, high moated inselbergs, inselberg ranges, intramontane plains, etchplain passes, etc., not to mention features now undergoing traditional development (disintegrating tors and wind-scoured rock surfaces). Remains of lateritic or latosol-like soils are common. Near the Enneri Dohou on the E edge of the Tarso Ouran,[49] Kaiser (1972, Fig. 35)

---

[48] In the discussion to Jäkel and Schulz (1972, p. 141) on pollen investigations of the marl sand terraces of the Tibesti, Sabelberg remarks: "the pollen analyses carried out by Schulz are characterized by a very well-developed vegetation, including even deciduous trees. It is clear, as Schulz has pointed out, that such a vegetation would require a climate with considerable precipitation. Yet one ought to find a well-developed soil in a position stratigraphically equivalent to the pollen-bearing sediments." Evidence of this has been collected by Grunert (1975).

[49] The local Toubou name for valleys, called wadis or oueds in Arabic, is *enneris. Tarso* is the Toubou word for a high plateau.

found fully kaolinized schists over 10 m thick preserved under basalt sheets. Vincent (1963) described several phases of intensive laterite weathering under a series of basalt sheets; these are frequently exposed in the mountains under conditions similar to those of the Hoggar. Hagedorn (1971, pp. 25 and 38-39) and Busche (1974, pp. 37-54) have made a number of similar observations, pointing out that these occurrences resemble older ones mentioned by Tricart and Cailleux (1961-64) and Michel (1967). Hagedorn (1971, p. 39) states that "the laterites are probably of Later Tertiary age," and even considers it possible that "during the Pleistocene, climatic conditions may also have permitted the development of laterites." He refers here to the brief periods of fine-material development which we have mentioned above. We must remember that hard laterite is far more durable than soft latosol, and that the Tibesti Mountains, being much nearer to the humid tropics, catch the inner tropical W winds far better on their westward upwind sides than does the Hoggar.

The inselberg-studded fluvio-eolian plain with its thinly veiled basal knob relief joins the Tibesti Mountains at a sharp piedmont angle, while growth apices of the plain reach into the mountain fringes as triangular reentrants and partly even as broad wadi beds. To what extent true pediments form the transition between the plain and the mountain edge is hard to determine in view of the great ambiguity of this term as used by some authors,[50] but the feet of the mountains and neighboring inselbergs often have deflated moats and notches, the latter frequently adorned with well-preserved rupestrian art. "It is therefore not inappropriate, on the basis of their wide distribution in the etchplains of the S and W mountain fringe, to associate the development of the notches and of the footslopes with the creation of the etchplain, the latter having demonstrably occurred under a different climate from that of today" (Hagedorn, 1971, p. 34). This highly characteristic etchplain cuts across a number of tectonic and structural boundaries and reveals a partially stripped basal weathering surface with basal knobs and inselbergs, etchsurface

---

[50] We define true pediments only as those flat forms radiating from a mountain edge, whose upper portion is carved slightly into the bedrock, and whose debris cover stretches with quickly diminishing grain-size as a glacis over the Recent basin sediments (see also Section 2.6 and footnote 44). Figure 53 shows this diagramatically, as defined by Mensching (1958a). Many authors have watered down the term to include all flat erosional features from concave footslopes, which can form in nearly any climate in a few thousand years, to continent-wide etchplains, which can only form through the unique seasonal tropical mechanism over tens of millions of years. A word which, disregarding formative process, is used for any flat erosional slope (which could have at least a dozen possible origins) obviously has no clear genetic definition. Even serious workers dealing with etchplain development have taken up this modish word. Busche aptly observes (1974, p. 55) that "frequently one need only change labels to make sense out of detailed and valuable observations." No science can afford to work without carefully defined concepts.

depressions, and etchsurface passes (the reflections in the bedrock of the wash depressions and wash divides of the former wash surface). The effects of today's climate continue traditionally to work this surface within the framework of the fluvio-eolian sandplain, disintegrating the basal knobs by sanding, washing, and deflation, and even partly by corrasion. On the whole, Hagedorn comes to the conclusion that "the volcanic deposits, laterite remnants, etc. show that the etchplain and its inselbergs are relict forms which attained their present shape during the Tertiary. . . . The . . . flatlands correspond in shape and character to recent forms from the Sudanese savanna; etchplain depressions and etchplain divides typify the Sudanese etchplains" (Hagedorn, 1971, p. 72). On the particularly prominent inselberg of Ehi Goudia (500 m relative height) it could be shown that the etchplain formerly ran right to the inselberg, forming a sharp piedmont angle. Most mountains in this altitudinal range (800-1000 m) have the same sharp feet. All observers agree that slope erosion has slowed considerably. At this height, slopes and scarps are dormant (*Ruheformen*) in the sense of Passarge (1912); this agrees well with observations by Mortensen (1956) and Bremer (1965a and b) in other arid regions. Even intramontane basins below 800 m, such as the basin of Zouar at 700 m, cut like etchplains "discordantly across the gently warped Cambrian sandstones and . . . pre-Cambrian schists" (Hagedorn, 1971, p. 67).

In the altitudinal zone between 800-1000 m and 2000 m, according to Hövermann (1966) and Hagedorn (1971), things are quite different. Here, in addition to large valleys and basins, we find a heavily dissected relief with "canyons, gorges, and steep-walled, V-shaped ravines. . . . The zone of maximum vertical erosion lies at present between around 1500 and 2000 m. Lower down vertical erosion decreases, and material is simply transported. Below around 1000 m, deposition sets in in the larger valleys" and intramontane basins (Hagedorn, 1971, pp. 121-122). Naturally these altitudinal boundaries fluctuate with the major climatic changes. A similar altitudinal structure was suggested previously by Hövermann (1963, 1967). Grunert (1972b, 1975) has distinguished a currently active "gorge region" (*Schluchtregion*), with particularly powerful active dissection between around 1500 and 2000 m, in the partially inactivated "gorge relief" (*Schluchtrelief*) at 800-2000 n. All this fits well with our picture in the Hoggar Mountains. The surface edges, particularly of the intramonatane basins, also change between 800 and 2000 m, and rocky piedmont surfaces or true pediments appear. Pediments, often tripartite, first begin to appear at 600 or 1000 m, and grow more marked with altitude. "The fairly thorough inactivation of these surfaces shows that they developed under a different morphodynamic system. . . . The pediments are relict and belong to various Quaternary stages, shown by their relationships to basalt flows" (Hagedorn, 1971, p. 71 and 110). It is important that true pediments first appear at an altitude where the slopes were heavily affected during the Pleistocene by regular periods of heavy frost. As already stressed, Höverman (1963, 1967) was the first to point out the dependence of true pediments on a winter-cold desert climate.

This is corroborated by Hagedorn's discovery (1971) of marked frost effects in the highest Tibesti above 2000 m. Though further S than the Hoggar, the greater height of the Tibesti explains the wider and more marked appearance of frost effects here than in the Hoggar. Relict solifluction sheets are also common here, and are still moving even today, though far less than during the Pleistocene cold stages.

2.5.2. The Central Australian Arid Region

The central Australian arid region is the earth's second largest area of tropical desert and desert steppe. As in the Sahara, the lowlands here are always warm and experience little or no frost. While the Sahara, and especially its central mountains, may seem even more forbidding than central Australia, this "land behind the back of beyond" forms a desolate region within the heart of an isolated continent.

Surrounded by a ring of low-lying dune-covered deserts, the interior of Australia consists largely of a low crystalline shield forming a rough square around 700 km to a side and climbing to a height of roughly 1000-1250 m. Broad basins running WNW to ESE segment this upland, which is further broken up by numerous intramontane plains and groups of inselberg ranges (see Fig. 47). Both the underlying rocks and the relief sphere itself are of enormous antiquity. The distribution of the parallel strips of highland and lowland was basically set in pre-Cambrian times, and has since been subject only to slight epeirogenic movement, e.g., in Variskian and Alpidic times. The lack of any subsequent tectonic, volcanic, or climatic breaks considerably complicates geomorphologic interpretation of this region, as does the fact that the successive relief generations developed within a far more limited range of altitudes than in the Sahara.

Geomorphologically speaking, it is Bremer (1965a, b, and c, and 1967a) who opened up this cheerless area. Apart from the works of Mabbutt (1963, 1965, and 1966), she had very little preceding work in Australia to go on, nor had the work in the Sahara been done at that time. We have already discussed Bremer's conclusions, and have relied on them in describing the Sahara, which, lying on the other side of the equator, forms a mirror image of Central Australia (Büdel, 1969c).

Lacking other evidence, Bremer had to concentrate on extant stages of soil development, crustal relics, slope weathering, valley networks, and above all, on the overall relief itself for clues to the relief development. This led to a new methodology for tropical geomorphology.

2.5.2.1 Etchplains and Soil Development

While latosol and laterite sheets in the Sahara are preserved only in protected areas, their relicts, according to Bremer (1971), are still widespread in arid Inner Australia. They occur, for instance, as remnant caps of massive laterite up to ten m thick. This laterite, consisting of fully indurated iron crusts, overlies a mottled zone, which in turn lies above a kaolinitic pallid zone. Above both Cretaceous and crystalline bedrock the basal soil surface is usually featureless and slightly uneven, clear characteristics of the former basal surface of weathering. The laterites are the remains of an Early Tertiary tropical planation surface. Yet they are not omnipresent; Bremer has clearly shown that the crusts originally accumulated only in the very gentle wash depressions, through convergent lateral seepage of iron-bearing solutions. Where laterite was lacking, the surface has since been lowered by 20-50 m. Today the laterite sheets cover sharp-rimmed mesas which rise by this amount over the new surface. Between the Early Tertiary and the Recent, then, planation has brought about an at least partial relief inversion. Siliceous crusts (silcretes) are often interbedded in the laterite crusts of the mesas, further hardening the mesa caps and sharpening their breakaways. Bremer rightly assumes that the development of laterite caps implies greater precipitation and probably also higher temperatures, while the interbedded silcretes may indicate intervening brief periods of slightly less humidity. The planation between the older and the younger surface nevertheless occurred over a long period of more or less constant climatic conditions and relative tectonic quiet, falling in the Early to Late Tertiary. This lowering would not have taken place over such a large area had not the Early Tertiary landscape already provided the necessary conditions for this. The later surface inherited and traditionally continued the development of the older, flat surface, at whose level many of the rivers continue to flow today.

The younger etchplain is covered by the second major soil here: a thick, plastic latosol, usually red in color (yellow-red, carmine, or brown-red). The fine sand fraction invariably present in such red loams ranges in grain-size almost into the clay fraction. According to Kubiena (1953, 1954), the red loam is frequently altered to a red earth, becoming looser, more porous, and highly mobile. The clays are mostly kaolinite and illite. The humid phases indicated by the younger latosol seem to have lasted into the Early Pleistocene (compare Rognon's data, 1967b, on concurrent humid conditions during the Villefranche in the Hoggar Mountains).

The younger latosols are already adapting to the present arid climate, though this is clearly wetter than that of the Central Sahara (the annual rainfall in an equivalent part of Australia is 180-200 mm, as opposed to 50 mm in the Sahara). The soils are accordingly not as altered as in the Saharan sandplains. The denser, clayey subsoil is often still extant, and sometimes shows gilgai patterning. The sandy topsoil from which the clay has been washed out is usually still red (unlike the white to pale yellow Saharan sandplains), even when it has developed a fluvio-eolian sandy upper layer 10-20 cm thick. Serir-like gravels of lateritic, siliceous, and calcareous concretions (pisolites, gibber, and calcretes) also occur. The polish on the gibber and pisolites shows no traces of transportation; the rounded forms are concretionary, and are easily distinguished from the angular fragments of the massive laterites and siliceous crusts found on the mesa slopes. Clearly the concretions developed early in the history of the second, younger latosols, and after washing and deflation in the top soil, now form a lag gravel. They are largely limited to wash divides and local watershed divides. Bremer rightly adduces them (along with the relief inversion) as evidence of long-term planation. The rivers have not actively induced lowering of the landsurface, but have been passively integrated in the overall erosion, as is invariably the case in active etchplains.

A few special or marginal areas of the younger surface are slightly dissected, partly through deepened wash depressions resembling wash valleys (see Sections 2.3.5 and 2.3.6 above), and partly through steep gullying. On the intervening wash divides and on all other interfluves, planation still proceeds undisturbed today. From this Bremer drew the important conclusion that etchplains are extremely durable as a relief assemblage even when climatic conditions controlling their relief-forming mechanisms have fully or partially changed. This was the first step toward formulating the law of "traditionally continued development" (see Section 2.3.9 above). In the tropics, with rarer and less thorough phases of valley-cutting, this law is far more important for the relief appearance today than in the ectropics.

"Much recent work on tropical geomorphology is wholly concerned with current processes and, projecting present processes (occurring passively and traditionally on the old etchplain) too far into the past, assumes that these were responsible for creating the etchplains from the start" (Büdel, 1969c, p. 225). In central Australia especially, where the younger plains still have much of their old latosol cover, current processes are still influenced by this latosol, and not just by the main inherited relief forms. The active processes today still resemble their predecessors which created the etchplains. Bremer states that "not only the previous [relief] forms, but also the [inherited] soils affect the current, younger morphological processes" (Bremer, 1967a, p. 128). This is far truer here than in the Sahara. Bremer believes that even the Australian soils have undergone "a long multi-phased development" (Bremer, 1967a, p. 128). According to her, only about one to three m of the latosol cover have been removed (largely in the latest Quaternary), producing the pisolite and gibber serir. What the water film washes away during the rainy season often becomes windborne during the dry season, and may even be blown back to its original location.

2.5.2.2 Inselbergs and Slope Weathering

The etchplain, for long ages lowered parallel to itself, is surmounted, much as in the Sahara, by inselbergs rising to relative heights of up to 350 m. They are concentrated in areas where the wash and etchsurface divides have remained fixed through both planation phases, which means that they existed in about the same size on the Early Tertiary surfaces as well. The consistent distribution of the inselbergs on the plain is an important observation of Bremer's, as is the fact that in their present sizes these sharp-footed mountains are by no means associated only with harder rock units. The famous Ayers Rock in the middle of Australia consists of the same steeply dipping arkose that has been beveled flat in the surrounding etchplain. The latosol cover on the etchplain continues in full thickness right up to the razor-sharp mountain foot (see Photo. 21), and old subcutaneously formed notches or concavities are now partially exposed at this piedmont angle. Above this the completely barren rock walls rise at over 60° to a flattened top.

The sharp piedmont angle shows that both the mountain's shape and the latosol-covered plain derive from humid periods probably reaching back into the Early Tertiary. The surface and the mountain flanks are extraordinarily stable in their present form. This is shown above all by well-preserved, inactive notches or concavities at the mountain foot. The stability of the overall mountain was proved on the basis of more recent, largely Pleistocene microforms on its walls. The last phase of Tertiary humid tropical activity is represented by rills, shallow in cross-section but steep in longitudinal profile, segmenting the rock walls. These are interrupted by large tafonis from a somewhat drier, probably Early Pleistocene stage. A later phase of highly characteristic flaking is assigned by Bremer to the arid period of the main dune activity. Today only extremely slow grus weathering (familiar from the Sahara) takes place. A very old, deep violet crust is preserved on the summit plain of the mountain. All this shows that the inselberg walls, having long ago achieved their barren steepness, were no longer weathered back. This already pointed toward Bremer's later formulation of the rule of divergent erosion (Bremer, 1971 and 1972; see also Section 2.3.3.2).

Ayers Rock is especially significant, as it lies in ancient sedimentary rock. Similar forms can be found 25 km to the E in the Olga mountains. Otherwise, the accepted tradition is that such classically dome-shaped inselbergs are found only in granites or other crystalline rocks.

2.5.2.3 Fluvial Work and Valley Development

The lag sands, grus, and pisolites on the basal latosol sheet indicate that it is gradually being destroyed. Yet surface wash is still considered to be the dominant wash process here. As long as the latosol sheet is widespread, surface wash will continue to behave as in a seasonally tropic savanna climate, without causing as much destruction and redeposition as in the much older fluvio-eolian sandplain of the much drier Sahara. The basal weathering surface, with its knobs and penitent rocks, is rarely exposed in Australia.

During the rare, one-to-three-day periods when streams traverse the plain, the low wash ramps (*Spülsockel*) surrounding some inselbergs (though not Ayers Rock) become covered by a dense network of small rills, 10-20 cm apart and 1-2 cm deep. Similar forms were described above as rainy season rivulets (Section 2.3.2.3[4]). Here again the rills branch and reconverge. "They are limited to the fine grus layer above the latosol" (Bremer, 1967a, p. 171). The rills flow off the ramp toward the surface, collecting into larger rivulets one m wide, which cut 1-2 cm into the dense latosol beneath. They tend to form sharp banks in latosol, and gentle banks in sandy material. *Erosion cuts no further into the surface*, and Bremer emphasizes that she usually saw such rills right after rains. Obviously they are quickly erased again by wind.

Of the further relief dominating these surfaces, Bremer names an example of "wash depressions up to three km wide, separated by divides five to eight km wide. The difference in height between the depression floor and the divide crest is only 10-15 m. The wash slopes have no rivulets, only broad wash surfaces" (Bremer, 1967a, p. 171). All drainage is areal in nature, and "leaves and twigs are sometimes washed into long bands." Clay and fine sand are not deposited in these bands, but carried on in suspension, as shown by the suspension-rich rivers. Even small rivers can flood several hundred m of their tributary wash depressions during rains. But the dissolved and suspended load remains small, and neither hinders erosion nor promotes it by providing mechanical weapons. Bremer's description makes it clear that erosion processes and their relief product on the surfaces of central Australia today resemble the seasonal tropics far more than do those of the Central Sahara.

But the analogies go still further. Gravels in Central Australia occur only in exceptional patches at most 100-200 m long in pure mountain valleys—never in water gaps. These gravel patches are stable, that is, they do not move downstream. A continuous gravel bed is never present, and was not even found by coring (Bremer, 1967a, p. 153). As in the seasonal tropics, gravel does not provide tools for vertical erosion. In this point also, then, despite the recent change to an arid climate, the surfaces of Central Australia still resemble the wash surfaces of the seasonal tropics.

Even rivers of the central mountains, with their rare, stable gravel patches, potholes, rock bars, and loads of sand and suspended material, are incapable of mechanical vertical erosion. Bremer has produced a series of new proofs for this, demonstrating that the rivers have cut down in rhythm with the plain.

Signs of lateral erosion are equally lacking. Tributaries bring at most a few pisolites or fragments dislodged from chemically disintegrating basal knobs or conglomerates. Undercut or slip-off slopes are absent here as in active etchplains, nor is there lateral undercutting even at narrows. The steep footslopes are often silicified or covered with violet crusts of iron or other material; at most they show the usual slow sanding, indicating an extremely long history of chemical development without fluvial mechanical attack.

Of very basic and far-reaching significance is Bremer's evidence that the Central Australian network of permanent and intermittent rivers shows no signs of mechanical headward erosion. Like Wadi Tamanrasset of the Hoggar Mountains, which in the pluvials flowed over 1000 km down to the Niger Bend, the Finke River, rising in the Central Australian dome, flows over 600 km from the heart of the MacDonnells to the Eyre Sea Depression (at minus 12 m altitude); at times it may even have run yet another 110 km S to Spencer Gulf. This drainage area, singularly enough, includes the entire central upland. The headwaters of the Finke River even cut through the northern chains and hogbacks of the MacDonnells across to the S edge of the Ngalia-Burt Plain (bounding this upland to the N, running NW to the undrained salt sea region of W Australia). The headwaters of the Finke River system thereby cut through the N part of the MacDonnells in numerous water gaps, slicing through a series of prominent hogbacks of nearly vertical quartzite. Crossing the central dome of Australia, the rivers then run S and SE. The only possible explanation for this is that the river network was *not* created by gradual headward erosion after the mountain uplift, but rather that it is older than the mountains. Yet this region has been a highland since the Late Paleozoic. The river network must therefore have already been established in the Late Paleozoic, and must be older than the entire central dome and older than the weathering-out of the hogbacks (some of which are several hundred m high). This weathering-out took place when the land still sloped E and SE. One may therefore call the many narrow water gaps through this hard region of the MacDonnells "epigenetic" only in the sense that at the time the river network was formed, the surface was not a depositional surface (for which there is no evidence), but an extremely ancient erosional surface (etchplain) with an overall corresponding slope.

Two extremely important dogmas of geomorphology, still lingering in the wake of Davis (1899), are thus heavily refined or limited: namely, the greatly overestimated role of headward erosion, and the concept of epigenetic water gaps.

2.5.2.4 Intramontane Basins

Finally, Bremer (1967a) succeeded in showing that intramontane basins and their uniting gaps were lowered into the central uplands at the same rate and by closely related mechanisms (see Section 2.3.3.3). Numerous surface apices extend as elongated triangular reentrants into the mountains, while the mountain area itself is so riddled with intramontane plains that it could almost be considered a collection of inselberg ranges. The intramontane plains are not necessarily found along main rivers (as required by all previous theories): independent of the main river network, they form broad, deep, and level ribbons cutting through the mountains, even across harder strata. The level gaps or gateways which unite basins or connect them with the circumjacent surfaces *often contain no river at all*. In such cases these openings (which in India I named etchplain passes or *Flächenpässe*) must have formed through planation itself.

A particularly striking example, found straddling a divide in the SW part of the MacDonnells, consists of Wild Eagle Plain (Bremer, 1967a, pp. 144 and 185), which was clearly autonomously lowered. It is shown in Fig. 47 and is the model for Fig. 48. This intramontane plain was formed at the headwaters of several rivers in the fluvial network. It is drained today by no fewer than four small rivers running in four different directions. Their headwaters are united within the plain by gentle etchplain passes, and leave the plain through separate gorges. Other basins have etchplain passes connecting them with the surface surrounding the mountains. Both the river gorges and the passes uniting the basin with the outside have the same very gentle gradients. One cannot say that lowering of either the surfaces or the rivers took place faster than general erosion. Rather, both occurred in full harmony with the general relief-developing mechanism of tropical deep weathering.

Both started off at the same level on a high ancient surface. What is notable is that during more humid times, the rivers with their linear chemical erosion cut down no faster than the intramontane plains with their chemical areal washing. Bremer justifiably cites Credner's words, written as early as 1931(!) that "the intramontane basins are lower remnants inherited from a former, more widespread flat relief" (cited in Bremer, 1967, p. 186). This was also aided by the fact that, once started, such an area would collect rain and seepage water from the surroundings (see Fig. 48 above).

2.5.2.5 Total Relief

The total relief of the Central Australian desert today is stamped far more clearly than that of the Sahara, for its climate has remained seasonally tropic for much longer. This climate has had but brief interruptions (involving both cool and arid fluctuations), after which the relief-developing mechanisms continued as before. No great tectonic-epeirogenic disturbances have occurred, and the relief-developing mechanisms which created the major features may go back to the Paleozoic. The major relief assemblage is therefore of extreme antiquity, far more so in fact than the unique biosphere or the mesolithic culture of the aborigines. The arid Quaternary has been brief by comparison, a fact connected with the very gradual expansion of the Antarctic continental ice (see Hoinkes, 1961; Büdel, 1963b; and Bremer, 1967a). In Australia, surrounded as it is by oceans, the trade-wind deserts have been desiccated far less than in the Sahara. Thus the humid tropical relief inherited in Australia has been far less altered.

2.6   The Winter-Cold Arid Zone of Surface Overprinting (Transformation), Largely through Pediments and Glacis

Traveling in the previous sections from the humid tropics toward the trade-wind deserts, we have distinguished several stages of traditional development and/or transformation of the unique etchplain-and-inselberg relief created so long ago.

The least change is found in Central Australia, which has only become moderately arid, and that in fairly recent times. This region is a classic example of traditional development. Not only are nearly all of the major relief forms preserved here, but even widespread laterite and latosol sheets in the subsoil as well.

The second stage is shown in the Central Saharan lowlands. Here the bedrock frame of the former relief is still extensively preserved, with its etchplains, etchplain escarpments, abruptly protruding inselbergs, partially stripped basal knobs, etchplain passes, and intramontane plains. Yet the level surfaces are covered by a fluvio-eolian sandplain, and the clayey red to yellow-red sheets of subsoil can only be described as sand-mixed latosol derivatives. Unaltered remains of latosol sheets are rare, and are preserved only under old basaltic flows. Where the sandplain cuts into the mountains as broad wadi floors, terraces show that the arid erosional processes have involved a range of altitudes. The fact that the mountain terrace system commonly contains a fine-grained marl sand terrace (the only terrace of the fluvio-eolian sandplains in the forelands) shows that arid periods having little soil cover and charac-terized by valley-cutting in the mountains alternated with humid periods of stable soil sheets and thick plant cover. In the humid phases valley-cutting was replaced far into the mountain valleys by aggradation, though the upper slopes contributed no coarse debris to this. Outside the Saharan mountains, the landscape of basal knobs is gradually being stripped, and its cover replaced by a fluvio-eolian sandplain which is only limitedly capable of traditionally lowering the old etchplain.

Traditional development grades even further into actual transformation or overprinting of relict surfaces in the small areas of the frost-prone altitudes, particularly between 800-1000 m and 2000 m in the Tibesti Highlands. Here the etchplain escarpments and the walls of the intramontane plains are fringed by true pediments, for much periglacial solifluction rubble was produced above 2000 m, especially during the Pleistocene cold stages. This agrees with the observation (first mentioned by Hövermann, 1966, 1967) that true pediments occur largely along mountains reaching into the periglacial altitudinal zone.

This conclusion is fully supported when, leaving the tropical lowland deserts, we cross into higher latitudes to the winter-cold deserts. This is evident even on the N fringes of the Sahara, e.g., in Tunisia (Mensching, 1964b), the NE end of the High Atlas, or the E edge of the Middle Atlas toward the Moulouya Basin, where Mensching (1953, 1958b, 1968) and Mensching and Raynal (1954) investigated very extensive pediment systems. Many pediments have been found along the mountain fringes in the Iberian Peninsula and other Mediterranean countries, by, among others, Mensching (1964a), Seuffert (1970), Wiegand (1970, 1972), Stäblein (1973), K. Fischer (1974), and Wenzens (1976). We will discuss some of their work later, but it should be stressed here that the Mediterranean and the rare Central European pediments are inactive, relict forms. The highest glacis or pediment can usually, where datable, be assigned to the Earliest Pleistocene (equivalent to the Villafranche) (Seuffert, 1970). K. Fischer (1974), has emphasized that the earliest pediments around the Montes de Toledo in Central Spain are even older, and must date back to the Pliocene. Their basal relief surface, therefore, was not formed during the cool arid climate generally assumed for the ectropic Villafranche, but rather under a warm humid climate with a kaolinitic-illitic weathering zone over 15 m deep. The pediment surface, then, developed on the stage set for it by an etchplain, and like the sandplain, is a type of traditional development of the old etchplain. The later glacis surfaces cutting into these pediments are of Pleistocene age, though K. Fischer doubts that they can be attributed to specific cold or warm stages. Wenzens (1976) has corroborated this picture, and is elaborating on it with investigations (still in progress) in many other parts of the Iberian Peninsula. Even the most recent of these Mediterranean forms are inactivated, and are covered by Holocene soils and vegetation.

Pediments are much better developed in the winter-cold deserts of NW Argentina, the SW Rocky Mountain basins of the United States (Colorado, Utah, Nevada, and the Colorado Plateau in Arizona), and most especially in Central Iran and Tibet. The flat forms radiating from the mountains and filling the intervening broad basins are covered by fresh arid debris rather than by weathered soils. Considerable debris movement occurs on these piedmont slopes, shown by the facts that (1) debris is still being produced by the barren front slopes of the mountain block, (2) it is deposited across the pediment by gravel-rich rivers, and (3) it is (in some cases) then reworked by rill erosion (Weise, 1970, 1974). Movement is especially rapid when the piedmont surfaces are still undissected. The debris on the upper part of the piedmont surface is a coarse fanglomerate, and usually forms a thin veneer over the bedrock (see Figs. 53b and d). Only at this upper margin of the surfaces does the term "pediment" actually apply. Further downslope, the grain size becomes finer (Weise, 1974), and overlies recent basin fill. The accepted term for the piedmont surface's gently sloping periphery is "glacis." These grade directly into alluvial fans. The entire system usually feeds a playa (also called a kavir, sebkha, or salar), or it may empty into the bed of an intermittent stream.

This produces the impression of "mountains drowning in rubble," as described in Iran and Tibet by many of the earlier investigators, from Sven Hedin (1910, 1916-22), von Niedermayer (1920, 1940), Trinkler (1932), Hörner (1933), Machatschek (1938-40), and A. Gabriel (1934, 1942, 1964), to Bobeck (1934, 1961, 1969). In Iran many of these mountain ranges towering above the desert are crowned by green arable heights. These have served since the time of the Arian migration from W Eurasia into India (around 1500 B.C.) as a passage for countless migrations and invasions, the most famous being that of Alexander the Great (334-325 B.C.), and the most recent that of O. von Niedermayer (1915-1916) in the First World War.

In the arid SW of the United States, numerous American investigators, following McGee (1897), Waibel (1928), and Bryan (1935), have investigated pediment development. These have lately been joined by many non-Americans, including above all Birot and Dresch (1966), Tricart, Raynal, and Besançon (1972), Oberlander (1974), Werner (1972), Wenzens (1972), and Mensching (1973). This New World region has thus become the classic area in the literature for this question. The interior of Iran, however, has even more outstanding examples of this phenomenon, and has recently been studied by Weise (1974), who relies largely on the theories developed in the United States.

The basin-and-range structure produced in both regions by recent tectonic movements makes it all the easier to find analogies. Wide basins filled with fairly recent sediments are fringed by recently uplifted and freshly dissected mountain chains of various sizes. The ranges often have fairly straight piedmont angles, from which the gently inclined surrounding piedmont surfaces radiate in all directions, as shown schematically in Figs. 53 and 57. In larger mountains whose piedmont surfaces already bear larger rivers, broad surface embayments (triangular reentrants) have begun to eat in the manner of triangular reentrants into the interior of the mountain (see Photo. 49 and Fig. 53c). These highlands are often pitted with intramontane plains whose floors are also covered with arid debris (compare Fig. 48, IV). Larger inselbergs often protrude through the arid debris of the "pediments" around such mountains (Photo. 50), and smaller sharp-footed inselbergs (Fig. 57) are often scattered along the rises of the gently undulating flat landscape. Other rises lack remnant inselbergs.

The constant movement of debris across these surfaces, particularly on their steep upper portion at the mountain foot (which may reach 10° in Iran), has naturally led to the assumption that not only the debris sheet, but also the

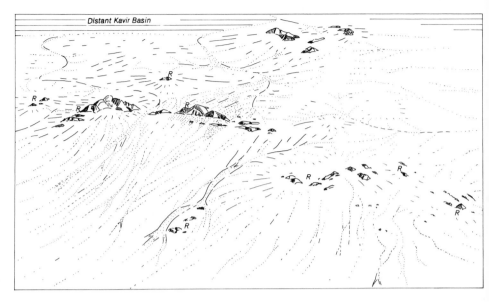

Fig. 57. Inselberg ranges worn down to remnant mountains (R) in the S Lut (Central Iran, N of Bam). Drawing by Büdel from an aerial photograph of February 22, 1964. Glacis fans cover large areas of the winter-cold inland deserts having basin and range structure. In the interior of the basins they cut across downwarped, unconsolidated, Recent sediments, and their peripheral pediments cut across older rock along mountain ranges (now often broken up into inselbergs and inselberg mountains). These are partly traditional pediments (Fig. 53b) and partly basal pediments (Fig. 53d).

rocky substrate and the mountain setting as a whole are all active features produced by current processes. Some researchers take this assumption as an unquestionable axiom. The concluding, key statement in one very exhaustive work on the subject states that seasonally tropical parent forms are not present in Iran (Weise, 1974).

This hypothesis is based on two assumptions: firstly, that backwearing of the mountain front slopes (or the walls of intramontane basins) has continued at constant slope angle over long distances, in some cases, over tens of km; secondly, that the plains left by this backwearing process are automatically and autonomously lowered.

Objections can be raised to both assumptions. The most important of these, again, involves the total relief picture carved into the bedrock. Like the framework of the Saharan and Australian arid regions, this relief, with its plains, inselbergs, etchplain passes, triangular, reentrants, intramontane plains, and stepped mountain V-shaped valleys corresponds so closely to the relief assemblage of the seasonal tropics that it seems unlikely that it could have been produced by extremely different relief-forming mechanisms under

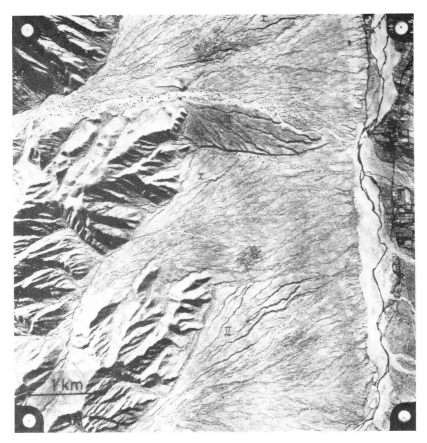

49. Mountain fringe deeply scored by fluvial embayments filled with broad alluvial fans (I). Above this, two generations (II and III) of older lateral pediments. 13 km S of Santa Maria, 80 km W of Tucumán, Pampine Sierra in Argentina. (Photo: Werner, 1972, Plate I. Aerial photo number 160/ 1.5.)

a totally different climate. Today's arid climate in Iran has probably only existed there since the turn of the Plio-Pleistocene (with slight fluctuations of humidity), and reaches back at most only slightly further into the Pliocene than that of the Sahara. Arid conditions can have prevailed in Iran for at most only a few million years. Yet we know that an etchplain-inselberg-relief in the humid tropical climate requires several tens of millions of years to mature.

The "pedimentologists" rationalize by declaring all these features to be typical of a highly arid climate. They take it for granted that bedrock under an arid debris cover can be planed off under fully arid conditions exactly as (and even faster than) under tropically humid conditions. The nomenclature painstakingly established for the *chemically* produced features of the season-

50. Inselbergs on an inactivated etchplain on the S edge of the Shir-Kuh mountains, 60 km SW of Yazd, Central Iran. White area to the left marks where old kaolinitic latosol sheets above the basal surface are being quarried. (Photo: Busche, October 1973, Nr. IR 1296.)

ally humid tropics is then sweepingly adopted for the features produced by *mechanical* weathering, by the simple expedient of prefacing each word with the term "pediment," e.g., pediment inselberg, pediment pass, intramontane pediment basin, etc.

This does not impress me as acceptable procedure. When a unique relief assemblage such as the etchplain relief has been established as the product of a very carefully investigated dynamic relief-forming mechanism, one cannot simply reconstrue all the major features of this assemblage as products of a completely different active dynamic system when they happen to appear in a currently arid climate. Despite different endogenic bases (old basement in S India, recent basin-and-range structure in Inner Iran), the similarities in the relief are really major. No relief-developing mechanism other than that of the seasonal tropics can accomplish the planation of huge erosional surfaces, and at the same time weather back and sharpen the slopes of small inselberg ranges and large mountain blocks, as shown even in Fig. 57.

Slight backwearing of slopes is possible in winter-cold deserts. Many large rivers emerging from the mountains into a sediment-filled basin form conical valley mouths filled with alluvial fans. These valley mouths may often be eroded laterally in the sense of von Wissmann (1951) and Johnson (1932). Yet Photo. 49 (from Werner, 1972) shows that this process quickly ceases when the anastomosing river on the alluvial fan, eroding gradually downwards, builds terraces (I, II, and III in Photo. 49) separating it from the mountain edge. This creates *lateral pediments* (Fig. 53c).

Further evidence of this is provided by Weise (1974). He describes several intramontane plains of partly tectonic origin (e.g., that of Sheytur) which contain alluvial fans of varying heights and ages. As in Werner's example above, Weise found three (sometimes even four) depositional surfaces in vertical succession. These covered a height of at most 50 m and lay on bedrock which had a shallow microrelief. Despite their obviously different ages, the highest and oldest join the wall in the same way as the younger: the wall has not weathered back since the deposition of the oldest fan. During its entire slow infilling the basin has stayed the same size.

Many things suggest that even elsewhere the upper edges of a broad glacis have only a very narrow pediment carved into the bedrock, i.e., that even in winter-cold arid climates, only a very modest backwearing of the backslope has taken place. Figure 53 shows one such possibility. Where pediments have developed on the stationary edge of a downwarped etchplain, we speak of a *traditional pediment* (Fig. 53b). This is not uncommon, for the downwarped surface provides an erosional base for the mountain block behind. But even when there is no relict surface at the foot of the mountain, even when no lateral erosion has taken place, and even at a distance from large rivers entering the foreland, a narrow *basal pediment* may develop in the bedrock at the upper edge of the glacis fans simply through pronounced backwearing of the mountains's front slopes (see Fig. 53d). Such cases often resemble the sharp piedmont angle of frost mid-slopes (see Fig. 31, right). Photograph 54 shows horizontally straight-running piedmont angles which may be produced in this way. Certainly backwearing may sometimes take place faster. But no precise investigations have been made on where and through what mechanisms this does or does not occur.

Given the generally very slight backwearing of the front slopes of desert mountain ranges (or the walls of intramontane plains), we must greatly restrict the old assumption that broad pediments might have been formed by slope backwearing. Where an erosional bedrock surface underlies a fanglomerate sheet or glacis fan (shown in exposures or by protruding inselbergs) we are dealing simply with an incipient traditional pediment (see Fig. 53a). Naturally, coarse debris from the hinterland will attack the weathering blanket on the basal surface, will transport it, and even attack and remove the bedrock itself far more vigorously than a Saharan fluvio-eolian sandplain. Yet chemically derived paleosol relicts are occasionally preserved beneath desert debris sheets (see examples below). These indicate that current transformation processes are merely performing on the stage set by inherited etchplain environments. The older, sometimes very broad surfaces carved into the bedrock were not created by today's processes. Current processes have only destroyed much of the old soil sheet, replacing it with the present sheet of arid debris, and have attacked the bedrock in many places, altering it as a whole. This can be even more marked than in Photos. 44 and 45 from the Sahara. On the other hand, ancient kaolin-rich soil sheets have sometimes remained preserved beneath the fanglomerate, as in Photo. 50, and on the traditional pediments, as in Fig. 53b. The various degrees of transformation provide a wide field for future research.

It is above all the traditional pediments along the upper edges of these piedmont surfaces, which, even when the kaolin-rich soil sheet itself is largely gone, most frequently have roots of old disintegration layers preserved in a few deeply weathered pockets in the bedrock. I observed such a case on a field trip in New Mexico, at the E foot of the San Andreas Mountains toward the Tularosa Basin. Here, on a pediment sloping at 6°, beneath about 50 cm of fanglomerate, a loamy red grus zone 2-3 m, in places even 6 m thick was present in the basal granite. This cannot have developed in the arid climate in which this pediment was formed, but must derive from an older warm humid climate. The upper edges of many pediments only reach into the bedrock to the extent prepared for them by a pedosphere or decomposition sphere created under humid warm conditions (see Fig. 53b).

Guided by Székely (1969, 1970), I saw a particularly striking example of this on the upper edge of a very broad piedmont surface on the S side of the Matra Mountains in Hungary. Most of this footramp, which reaches over six km into the foreland, consists of a glacis cutting across soft Neogene sediments: only the upper 150-200 m of the surface cuts as a pediment into the andesite of the Matra Mountains. This occurs only where the rock has already been chemically decomposed. We have already mentioned the corresponding conclusions of K. Fischer in the Montes de Toledo in Central Spain. Even Mensching (1973, pp. 133-134, 147-148, and 152) admits to previous Tertiary deep weathering in pediment development, and says (p. 148), "one must agree that a morphogenesis of very effective Tertiary deep weathering may have formed a relief generation preceding today's pediments, as Büdel has often stressed."

The sub-Recent to current transformation of the older form is aided by the substantial supply of debris from the mountain hinterland in the winter-cold desert climate. Rubble is supplied in three ways. Firstly, debris falls from the front slopes directly onto the pediment. Secondly, debris falls from the many interior slopes into the valleys and intramontane basins and is carried along the valleys out onto the pediments and glacis. Since the valleys only occasionally experience powerful flushing, the debris is moved intermittently and remains poorly rounded.

The debris on the piedmont surface itself is transported even more slowly by anastomosing channels. Where the mountain foot does not contact a major river (as in Photo. 49), but falls directly to the pediment (as in Fig. 53a, b, and d), further transport of the fanglomerate usually takes place as described by K. Fischer (1974, p. 10). The debris veneer (called "raña" there) is "poorly sorted and shows little stratification. Only on the outer edge of the pediment and on the adjoining glacis does stratification begin; the clasts are rounder, and sorting into lenses becomes clearer. The orientation of the components shows the increasing importance of fluvial transport with greater distance from the mountain. The thin raña on the middle or upper part of the pediment, on the other hand, shows orientation diagrams which closely resemble those of mudflows in the mid-latitude highlands." Mudflow deposits in W Argentina show nearly exactly the same orientation, as described by Werner (1972), who connects them with the development of the piedmont surface.

These comparative observations suggest that pediment growth, especially laterally, is closely connected with mudflow-like movement of loose material. Water here is far less a transporting medium than a triggering and lubricating agent. Particle roundness, stratification, and above all, fluid behavior of the mass increase downslope, indicating that water aids increasingly in its transport. Blackwelder (1928) and Beaty (1963) among others, have described greater fluidity at the noses of mudflows.

This provides a better explanation of a few important features of glacis development: namely, the fact that surfaces and their debris cover are sometimes enormous; the thinness of the cover on their upper portions; and the unconformity with which they overlie their substrate. The large size of these surfaces in intramontane plains or in basins between more or less isolated mountain ranges implies two things: firstly, slow, non-fluvial debris movement *which does not lead to valley-cutting*, and secondly, the presence of an older flat surface, whether this be several km of horizontally stratified Neogene basin fill, or a low-lying etchplain.

The intermittently moving mudflows of winter-cold deserts contain more material of a coarser size than do the Saharan sandplains, which are fed by sanding of inselbergs and inselberg ranges. The sand can be washed immediately even into the flattest of forelands by the sudden rains, but has little ability to attack the bedrock, being at most capable of reworking and stripping the soil sheet on the lower relict etchplains. The coarse rubble emptied as glacis fans from the mountain slopes, e.g., of the dasht plains of Iran, can attack the substrate more vigorously. In Iran, the debris at the mountain foot collects up to the gradient (usually 6-8° or slightly more) at which the mudflows can move the debris downslope. Lower down on the piedmont surface, as Weise (1974) has stressed, the grain size in the fanglomerate or raña rubble sheet is quickly reduced. The gradient becomes gentler, for, as K. Fischer's observations (1974) have shown, the material can be washed fluvially even on nearly horizontal terrain. Transitional deposition begins, leading eventually to semipermanent deposition. The strata of fine material in the basin center become thicker, and their anastomosing channels resemble those of a very flat alluvial fan.

This asymptotic profile is typical of all glacis, and accords with their manner of growth, spreading out from a mountain foot toward a lower foreland basin. The Neogene basin fill usually dips slightly from the mountain edge toward the basin interior (Fig. 53b, and d), and is cut by the glacis. This means that the glacis must perform some superficial erosion. To what extent this occurred through previous areal erosion by double planation, as K. Fischer's observations suggest, and to what extent the inner edge of the glacis sheet areally attacks the loose basal strata, as Weise (1974) assumes, or to what extent it may be due to varying combinations of the two, is hard to determine. In any case, in its inner portion, the lower surface of the fanglomerate or raña is erosionally slightly unconformable with the more steeply dipping strata of the basin fill beneath. Downslope of this follows a transitional area, where transportation and temporary deposition balance each other out, until eventually the latter gains the upper hand. The increasingly fine-grained debris sheet becomes thicker, and the unconformity at its base becomes a mere gap in stratification. Finally the glacis runs out into a salt clay flat (kavir, sebkha, or salar). *The entire glacis consists of a very broad, gently concave fan of transitional deposits.*

When such a glacis fan is bordered by an intermittent stream running along the lower edge of the whole glacis or dasht surface, and when epeirogenic movements or changes in climate cause this stream to cut downwards, younger, narrower glacis fans will begin to dissect the older glacis, working inwards from the stream. The older fan becomes a high glacis (*Dachglacis*, Seuffert, 1968, 1970), or initial surface (*Initialfläche*, Mensching, 1955, 1958a), below which the younger glacis stages then develop.

Such glacis fans form a very specialized case of areal transitional deposition tied to specific epeirogenic conditions. In active form they are typical of winter-cold deserts, and in inactive form they have a wide distribution in the etesian subtropics, and a restricted distribution in the drier mid-latitudes.

Let us return to the development of a basal pediment (Fig. 53d). The processes unfolding here are best understood when compared with the growth of a haldenhang at the base of a talus slope below an Alpine rock cliff (see Fig. 36), though the gradients are gentler on the arid front slope and its adjacent piedmont surface. In the unvegetated desert, the upper slope (or arid mountain front, *Trocken-Fronthang*) remains barren even on less stable or gentler slopes, and is fully exposed to exogenous attack. This attack consists of insolation, salt wedging, and slight chemical attack, but above all of frost-shattering. Because of the latter, the disintegrational products produced there are coarser than in the frost-free deserts, where sanding prevails. We are dealing here, so to speak, with a frost-intensified sanding and weathering. The material accumulates at the slope foot to an angle of 6-8° (at most 10°), permitting further transportation in mudflows. At the same time, the front slope above can weather back slightly where not protected by the rubble heap at its foot. Here the debris accumulation and its bedrock base gradually grow back and upwards at the expense of the slope above, just as the scree and haldenhang accumulate at the cost of the rock cliff. In active glacis development, a true pediment may gradually form in this fashion at the base of a glacis which is growing into the mountain. This we may call a basal pediment. But this occurs only where the glacis is actively developing, where its upper edge is *not* attacking the substrate, and where the accumulation of rubble from above outweighs any attack below. A basal pediment can never become very large. On inactivated pediments, and particularly on inactivated traditional pediments, as K. Fischer (1974) has rightly pointed out, no further pediment growth occurs at all. The only processes occurring on the surface here are simply those of traditional development.

Even where glacis surfaces are actively forming, basal pediments, when present, grow only very slowly. Naturally, they lead to the lowering of the arid front slope, attacking small remnant mountains on all sides and ultimately reducing and destroying them.

The longitudinal courses shown in Fig. 53 a-d of glacis and pediments vary greatly. Where the mountain edges are broken up by old triangular reentrants or by larger river valleys, lateral pediments predominate. Where the mountains are straight-edged, due to fresh epeirogenic uplift, lateral pediments occur at the valley mouths, separated by traditional and basal pediments. The valley mouths here form the local base level of erosion. Along fairly straight mountain fringes, the incipient glacis edges and pediments are also straight and form one level (see Photo. 54). This is because the adjoining glacis fans all aggrade toward the interior of the large bolson, toward an erosional base that is either nearly flat (as a salt-clay flat or kavir), or slopes only very gently (where drained by an intermittent river running through the basin).

The glacis fans are dissected in varying ways, working back from the base level of erosion. Young, narrow glacis fans may cut triangularly into older ones, the latter becoming the initial surface (*Initialfläche*, Mensching, 1958a, or upper glacis, *Dachglacis*, Seuffert, 1968, 1970) for the lower one. The upper glacis ceases to be active and undergoes at most traditional development.

Investigations by Hagedorn and his co-workers (1975) in the higher mountains of Central Iran (the Shir-Kuh massif, around 4000 m high), indicate that the Pleistocene snow line was probably lower than hitherto assumed, reaching well below 2500 m. The Ice Age moraines came down to about 1300 m. Terraces (one broad upper one and two narrower lower ones) from higher and younger moraines at around 2500 m on the S run down into the narrow mountain valleys and enter the foreland well below the level of the bordering glacis surfaces (Busche, personal communication). This agrees with observations by K. Fischer (1974) in the Montes de Toledo in Spain. He found no evidence that the individual glacis levels could be correlated with the Ice Ages, as often assumed elsewhere. Rather he found that the Pleistocene river terraces invariably lie below the levels of the glacis (see Wenzens, 1976).

3

Climato-Genetic
Geomorphology

3.1 Investigating the Relief Generations:
The Core Problem of Geomorphology

All continents have several relief generations, due to changing climatic conditions in the course of the earth's history. Particularly marked climatic changes have occurred in the mid-latitudes and in the etesian subtropics since the end of the Tertiary, and here we therefore find the richest variety of relief generations. At the same time, the relatively weak relief-forming mechanisms of the brief Holocene period have had little effect here. Thus, 95% or more of the total relief in these latitudes may consist of forms inherited from earlier relief generations.

But the relief-forming mechanisms which created these assemblages are now long gone, and cannot be observed or measured today. We can therefore only deduce their existence from the traces which they left behind. The most important evidence consists of the actual assemblage of forms, for these forms, stamped firmly into the bedrock, prove the existence of extremely long-lasting stable climatic conditions. The second most important clue consists of paleosols, including inconspicuous remnant gravels, pisolites, concretions, iron rinds, ventifacts, permafrost traces, etc. Finally, a third clue, overrated in the earlier literature, consists of correlated sediments. Fossils in such sediments may be useful for climatic studies and for dating. But their record, particularly near places of former erosion, is likely to be incomplete.

Once the entire inventory of clues (particularly the relict landforms of a specific relief generation) has been identified, one can proceed to the main task: comparing the basic traits through space and time (chrono-geographic comparison). This is done by searching for a morphoclimatic zone today in which the actively developing assemblage of forms is the same or similar to those landforms painstakingly identified as belonging to a single relief generation. Here their developmental process can be studied in full.

In this manner we have found six morphoclimatic zones today which serve as patterns for the relict assemblages of the mid-latitudes and etesian subtropics. One of these modern zones, the perhumid inner tropics, is as yet poorly studied, but relict examples of it have not been found at other latitudes. Another, the glacial zone, with its subglacial and circumglacial assemblages, is exceptionally well defined, and has been so extensively studied since the time of A. Penck and E. Brückner (1909) that its major features are familiar to most readers. So far our main emphasis has been on the remaining four subaerial zones, whose relict analogs dominate the relief of the mid-latitudes and etesian subtropics.

The first of these is the polar zone of extreme valley-cutting, or the active periglacial area. Here permafrost and scant vegetation have permitted such powerful Recent mechanisms to unfold that the previous relief generations which undoubtedly existed here have been decisively overprinted during the brief Holocene epoch.

The second is the peritropical seasonally wet zone of extreme planation, which along with the similar monsoonal subtropics covers about 80% of the humid tropics. Here the relief-developing mechanisms, which under stable conditions create inselberg-studded etchplains, have been almost uninterruptedly at work for tens of millions of years, continuing through the Ice Age hiatus up to the present. The earth's crust here has accordingly been stamped with a characteristic, mature relief assemblage. Any extant older relief generations owe their existence more to epeirogenic causes than to climatic change.

In the Early Tertiary this relief type extended even into the polar areas, and its traces can still be found in all climatic zones. For due to their flatness, relict planation surfaces at all altitudes are the most durable of all terrestrial relief forms. Regardless of the climatic regimes, their flatness continues to force traditional planation until their final transformation and destruction. Flat surfaces are particularly durable on karst.

Etchplains and etchplain stairways are still visibly the basis for the relief assemblage of the frost-free arid regions, where traditional planation is carried on by the fluvio-eolian sandplains. We find them in more veiled form in flatlands of the winter-cold arid zones, where surface overprinting has been brought about by glacis surfaces and pediments.

The first prerequisite to climato-genetic geomorphology is therefore to examine and differentiate today's morphoclimatic zones.

Figure 13 (see endpaper) shows the earth's present morphoclimatic zones (excluding the high mountains), where climatically controlled and relatively uniform relief-forming mechanisms are at work. The processes involved are analyzed by "dynamic geomorphology" (*dynamische Geomorphologie*), while their effect on the relief is studied by "active geomorphology" (*aktive Geomorphologie*), and the manner in which current processes transform relict features is the concern of "synactive geomorphology" (*synaktive Geomorphologie*). We have already covered much of this in the preceding sections, and a more thorough discussion of the three methodological branches will be presented in the third (German) edition of this book, where we will also examine the sub- and circumglacial relief, and sub- and perimarine relief. At present we limit ourselves to discussing the development of the dominant continental subaerial relief assemblages. We turn next to a brief consideration of the conditions necessary for the development of relief generations.

To transform a piece of crust into the stabilized product of a particular morphodynamic system, extremely long periods of geologic time are required, even in the zone of extreme valley-cutting. On the order of 10^4 to 10^5 years are necessary for a relief to become even roughly adapted to a new climatically controlled mechanism, and for the minutiae of the relief to become adjusted to the overall relief style, an average of 10^6 to 10^7 years may be necessary. This naturally varies according to the vigor of the exogenous processes, petrovariance, and epeirovariance.

But during such long periods of time, the climate usually changes, and in the ectropics this has in fact happened several times. This means that the processes of the current generation are suspended; a new mechanism becomes active, and begins to work toward a new relief type. But the inherited features remain, stamped into the bedrock, and often it takes much time for them to be destroyed. Flat features often undergo prolonged traditional development, and are only gradually transformed before being replaced finally by the new relief style.

3.2. The Role Played by
Older Relief Generations Within
a Given Morphoclimatic Zone

The zones discussed in Sections 2.2 through 2.6 are dominated by the features actively being created there today, yet traces of older relief generations are still present even there and require investigation.

In the polar zone of extreme valley-cutting, features inherited from earlier generations still build the framework of the present relief, a framework of Tertiary etchplains and Pleistocene glacial valleys. Yet Holocene processes working on the slopes and valleys have clearly overprinted 15-20% of this relief within the short post-glacial period. This clearly demonstrates that this zone is the most effective of all the subaerial valley-cutting zones.

In the perhumid, and especially in the seasonal tropics, excessive planation has dominated the lowlands throughout the entire geomorphic era without any Pleistocene hiatus. Hygric fluctuations in climate have of course occurred, at times intensifying the relief-forming mechanisms, at other times weakening them to the point where only traditional planation could continue. But as long as the relief-forming soil guaranteed the survival of the relief-forming mechanisms, planation could proceed again after each hygric fluctuation. In the mountains, the development of joint-controlled, steeply-stepped V-shaped valleys has also continued throughout the geomorphic era. Relief generations preserved here in the form of uplifted and partially or fully inactivated etchplains are due more to endogenous than exogenous processes.

In the tropical trade-wind deserts, the same system of ancient etchplains and etchplain stairways has been inactivated to a greater degree, and is now being eaten away by dissection. In mountains below around 2000 m, this dissection proceeds much as it did in the Tertiary humid periods (no Pleistocene hiatus here), forming steep, markedly stepped V-shaped valleys. In the lowlands, the inherited etchplain-and-inselberg relief is undergoing traditional planation by the fluvio-eolian sandplains or by gravel-covered deserts, without their significantly attacking the bedrock framework.

In the winter-cold inland deserts of the subtropics and the mid-latitudes, the inherited etchplain-and-inselberg relief of the lowlands is being transformed more rapidly by glacis surfaces. Where affected by recent tectonic movements, i.e., in areas of basin and range structure, the basins are frequently downfaulted and filled with Neogene sediments, often overlain by wide glacis fans. Pediments along the upper glacis edges cut into the bedrock. Often these are traditional pediments, developing in the weathered zone along the edge of the former etchplain. Where the former weathering sheets have been stripped away, the pediment then resembles a basal pediment (see Fig. 53d and Section 2.6), which, of course, is very narrow. Lateral pediments forming where rivers emerge from the mountains are broader.

The broad glacis fans dominating the basins of all winter-cold inland deserts were largely formed during the Pleistocene, which was more marked here than in the lower latitudes. The high basins of Iran experience heavy winter frosts even today (minima in Teheran are $-16.4°C$, in Isfahan $-19.4°C$). The highest mountain ridges were glacially altered during the Pleistocene, in places quite markedly. Periglacial frost-shattering was significant even at medium altitudes, and still continues weakly today. Solifluction sheets, solifluction-smoothed slopes (Spreitzer, 1960; Klaer, 1962), and intensified valley-cutting with terraces are widespread in the mountain valleys. But in the lowlands the Pleistocene hiatus produced only a broad glacis covering either Neogene sediments or preserved etchplain remnants. Since permafrost was absent, it cut no valleys. The relief, therefore, though heavily veiled, still resembles the Tertiary relief generation.

This is not at all the case in the fully to semi-humid subtropics and ectropics, regions which today comprise 45% of the continents in both hemispheres (the tropics occupy 32%, the ectropic inland deserts around 7%, and the polar regions slightly over 16% of the continental masses). In this region of the non-tropic relief-covering soils (*Ortsböden*), the relief consists of several, usually vertically distributed, relict generations. But the book of the past is not tidily opened for us page by page. Remnant features and traces of a relief generation may be widely separated or preserved only in hidden places. Remains of different generations often overlap in a confused mosaic. The difficulties of correlating features which historically and developmentally belong together may be imagined if one considers the many relief generations intertwined in one lithologically varied and glacially overprinted Alpine valley, where the weak relief-forming mechanisms of the Holocene have contributed but little to a complex of ancient features. *In the great ectropic regions of the continents, wherein lie the major centers of world civilization, the climato-genetic viewpoint is more important than in any other continental zone.*

There are distinct differences between the ectropic zone of retarded valley-cutting and the subtropic zone of mixed relief development (the etesian region, see Fig. 13 on the endpaper). In the ectropic zone of retarded valley-cutting, the Pleistocene Ice Age was very marked both in the small glacial region and in the large periglacial region. In this zone we may distinguish four major relief generations, of which the three oldest can be divided into several subgenerations (see Figs. 61 and 62).

The first major generation consists of Tertiary etchplain systems, the oldest of which reach back into the Late Cretaceous, as we will show using an example from the Franconian Alb. To what extent the great faunal break at the Cretaceous-Tertiary boundary (see Fig. 1) was accompanied by changes in climatic features (e.g., radiation), soil development, and morphodynamics remains for future climato-genetic investigation. In any case, several sets of Tertiary etchplains, dating from the Eocene into the mid-Pliocene, crown the heights of everything from shield areas and Mesozoic tablelands to hill country in Early Tertiary bedrock. Fragments of this etchplain system still exist in the central Alps, particularly on the broad relict surfaces of the Rax landscape on the karstified limestone massifs of the NE and SE Alps. Here as well as on the Swabian and Franconian Kuppenalb, they are broken by strips of an extraordinary hummocky relief, tropical analogs of which may be found in parts of the Highlands of Semién (see Fig. 10 and Photo. 29).

Many of the other relict surfaces can be classified as etchplain stairways. Where they underwent rapid Tertiary uplift, valleys of the tropical mountain type could for a short while peripherally dissect the flat-topped blocks. Sometimes these valleys were then filled in again by transgressions. Renewed planation often set in across both the filling and the surrounding bedrock surface. A typical example would be the lignite-bearing Tortonian and Sarmatian strata, which fill the pre-Tortonian valley system of E Bavaria (in the S Oberpfalz or Upper Palatinate between Regensburg and Schwandorf-Naaburg; see Fig. 68). This valley system was part of the original Naab system.

The highest regions of the Alps were similarly dissected during the pre-Pleistocene uplift, but no subsequent refilling occurred (see Section 3.3.2.3, where this is described for the Hohe Tauern). At all altitudes in Central Europe *the surfaces from this oldest generation bear no relation to the present valley network, and cut across the present divides.*

The second relief generation is from the Latest Pliocene and Earliest Pleistocene. Peripheral dissection of the higher relict surfaces intensified in the highlands. The planation which had cut across today's divides ceased in the lower areas, giving way to weak traditional planation and valley-cutting. This produced the shallow "Broad Terraces" or *Breitterrassen* (see Figs. 14 and 62). These broad valley floors laid out the lines followed by the valley systems today, even in the high mountains. In the central Alps (the Glockner group), which had been undergoing greater uplift since Tortonian times, Späth (1969) has found a subgeneration (the lowest of four) consisting of a "High Valley System," or *Hochtalsystem*. "In contrast to the older relief generations, the High Valley System is closely linked to the Tauern valleys. Nowhere does it cut across divides" as do the older plains (Späth, 1969, p. 132; see also our Fig. 80). Späth and others assign the High Valley System to the Pliocene, without further specification. Possibly it belongs to the Late Pliocene. But it is also quite possible that traditional planation of the older surface was narrowed down to the broad ribbons of later valley-cutting earlier in the Alps than elsewhere.

In the Rhine Plateau (*Rheinisches Schiefergebirge*), two sets of planation features, the "Trough Surface" or *Trogfläche*, and the "Principal Terrace" or *Hauptterrasse*, have been assigned to the Late Pliocene (see Fig. 61 and Brunnacker, 1975). In the Franconian Gäuland along the Middle Main Valley, Körber (1962) and Büdel (1957d) have dated the oldest (bipartite) Broad Terrace which follows the network of the present valleys to the Earliest Pleistocence (see Figs. 14 and 62).

The edge of the Upper Rhine Valley, the E edge of the Alps, and the inner edge of the Carpathian Arc all have terraces of this age in the form of piedmont slopes running from narrow pediments into broad glacis surfaces. Fink (1973) dates this "pediment period" to the Late Pliocene, but stresses that the first forerunner of the Danube (which aggraded heavily without eroding) must be set at the Plio-Pleistocene transition. Only subsequently did the lower Ice Age Danube terraces in the Krems-Vienna-Pressburg region form. Pécsi expressed himself similarly (1959) when he dated the first predecessors of the Danube to the period after the Levantin Transgression in the "Late Pliocene" (or better, the Latest Pliocene). At the same time he admitted that "those terraces composed of coarse gravels are Pleistocene in age" (p. 309). Thus the transitional period from the Late Pliocene to the beginning of the actual Pleistocene Ice Age is well defined in the E Austro-Hungarian region, while represented on the Main River by the two closely set Broad Terraces.

This transitional period is the turning point between unlimited traditional planation (which at medium altitudes in Central Europe lasted into the Middle or even Late Pliocene) and excessive valley-cutting. The boundary can be set fairly accurately, as it coincides with the start of the Günz Ice Age (corresponding to the Nebraskan Ice Age in N America) at around 800,000 B.P. The size of the Alpine ice cap and of the N American and Scandinavian continental ice sheets is uncertain, but they were probably smaller than in the Würm. On the other hand, traces of permafrost are found in the periglacial region of Central Europe, such as the earliest ice wedges in the Middle Rhine area (Brunnacker, 1975). The end of the Günz Ice Age coincides roughly with the Matuyama-Brunhes magnetic reversal, which has been dated to 700,000 B.P. in the Neuwieder Basin (see Fig. 61).

While the end of the second relief generation may be clearly datable, its beginning is not. The Late Pliocene changes which introduced this new relief generation set in at different times in the various companion spheres (see Fig. 61). The mammalian changes which constitute the majority of the "minor faunal break" (Schindewolf, 1954) occurred, according to Brunnacker, between 2.8 and 2.5 MY B.P. in the Middle Rhine area. This scant half a million years was the transition between the Late Pliocene (the Astian and the Levantine, perhaps also Reuver A), and the Uppermost Pliocene (Reuver B and D and the Lower Villefranche). The boundary in the Middle Rhine sediments between predominantly "Tertiary" sediments (independent of the Rhine) and dominantly "Rhenish" heavy minerals (deriving from the Rhine drainage area) lies at 2.4 MY B.P. I take this to indicate not only that a change in weathering type took place, but also that the Rhine began to develop its own character. The era of trough development may have begun at this time. This corresponds with striking accuracy to the geomagnetic reversal from the Gaus (positive) to the Matuyama (negative), which occurred at 2.45 MY B.P. The decisive floral boundary, however, the best evidence of further cooling, lies at 2 MY B.P. It is here, therefore, that we set the beginning of the Earliest Pleistocene, which opened with Schaefer's (1951) Biber cold stage (equivalent to the pre-Tiglian and the mid-Villefranche). The Earliest Pleistocene then continued as shown in Fig. 61 up to the beginning of the Günz Ice Age.

The transition to the second major relief generation therefore started with a Latest Pliocene preparational period lasting from around 2.8 to 2 MY B.P. By this time the earliest hominids had already developed, as indicated by the australopithecine remains from the Tiglian warm phase of the Late Villefranche.

The main period of the second major relief generation, consisting of the Earliest Pleistocene, extends from 2.0-0.8 MY B.P. The second major relief generation as a whole lasted a full two million years. It was in this period that the revolutions shown in Columns 4 and 5 of Fig. 61 took place. During this time the ancient etchplains were first dissected and the main areas of today's rivers became fixed at medium elevations in Central Europe[51] and in the corresponding latitudinal zones in the rest of the world. This occurred in the form of the so-called Broad Valleys or *Breittäler*, which, though shallow, were many km wide. Broad glacis fans along the edge of the Pfälzer Wald and the Vosges ran toward the Upper Rhine Valley, and similar glacis fans developed along the mountains encircling the Danube lowlands.

What is important is that in areas of uplift traditional planation became discontinuous, while Ice Age valley-cutting, with permafrost and ice rind effect, was not yet possible.

It is hard to tell what climatically governed relief-forming mechanisms operated during this transitional period. Certainly they must have been unique in character. The broad glacis surfaces at least indicate a brief period of drier, cooler climate. This would fit well with the Biber-Donau-Eburon cold stages (or Schaefer's Biber-Donau cold stages, 1950 and 1951). The Trough Surfaces and Broad Terraces are more plausibly assigned to the Latest Pliocene and beginning Earliest Pleistocene. Here widespread planation still continued, although now fixed to the ribbons of the later valley network. Planation was certainly traditional, that is, it would not have existed were it not for the previous Tertiary relict surfaces, which the traditional planation processes lowered only slightly. To what extent this was produced by continued decomposition at depth during the humid-warm interphases, along with river braiding (due to excess bedload with reduced water supply) and lateral erosion during the cold and arid phases, and to what extent other factors played a part is difficult to determine, for these Trough Surfaces and Broad Terraces bear little in the way of sediments today. On a Broad Terrace on the Main River below Miltenberg (in N Bavaria), Körber (1962) found a large polished ventifact.

Both the correlated lowland sediments and the faunal and floral history (shown in Fig. 61) provide general data on the climatic changes of these two million years. But they provide little evidence regarding the concurrent pedological and morphodynamic processes. Information on the latter can only be derived by studying the relict forms themselves.

This could be done without too much difficulty for the first and third relief generations (those of the tropicoid paleo-earth and of the glacial and periglacial cold stages), for in both cases we can find similar active regions for detailed study, namely, the seasonal and the subpolar regions. But it is not so easy to find an equivalent active morphoclimatic zone for the relief forms of the

---

[51] In the high mountains, the older etchplain had already started being cut up by younger ones (through tropical mountain type dissection), and broad dissection like that of the second major generation had already established the drainage network of today.

second major mid-latitude relief generation. One of the most important problems facing climato-genetic geomorphology concerns the relief-forming mechanisms of this generation.

The third major relief generation, the Pleistocene Ice Age, completely suspended the previous relief-forming mechanisms in two major ways. In the northern parts of N America, in the Scandinavian mountains, the Baltic countries, the Alps, the High Tatra, and the Caucasus, continental ice sheets, or, in the more southerly mountains, ice fields (all, strikingly enough, equivalent in size) developed. Through erosion and aggradation, these created a diverse but well-studied landscape in which the previous relief was variously accentuated, transformed, and covered with a host of minor new forms. This process began with the Günz, was repeated in the Mindel and two or three Riss stages, and concluded in the Würm.

The periglacial region away from the glaciers was affected quite differently by the Ice Age hiatus. Here thick layers of permafrost developed at depth. The forests disappeared from nearly all of W, Central, and E Europe, so that the ice rind effect produced a zone of excessive valley-cutting. The valley courses which had been laid out but not incised during the Earliest Pleistocene were powerfully and rapidly cut down in the approximately 350,000 years of all the cold stages added together. This created the characteristic box-shaped valleys through which wander the rivers of today. The individual cold stages are marked by pronounced gravel or erosional terraces on the slip-off slopes of these valleys (see Fig. 14). The influence of endogenous movements on valley-cutting will be discussed in Section 3.3.1.2. Valley-cutting also dissected the glacis surfaces along the mountain fringes of Austria, Slovakia, and Hungary. The pleasant, meadow-covered valley floors which lend the German Mittelgebirge and hill country such a familiar and friendly appearance, are actually nothing more than Würmian braided riverbeds (see Fig. 14 and Photos. 10 and 52). These riverbeds, which cut evenly and at constant width deep into the core of the mountains, have contributed greatly toward the widescale cultural development of the European midlands, and have helped them remain at the forefront of history from the time of Charlemagne to the Second World War. Europe may also thank the periglacial Ice Age hiatus for having loosened the soil through alternating winter frosts and summer thaw above the permafrost table. Fragments deriving from the ice rind were mixed by cryoturbation and incorporated into the Holocene relief-covering soils. Lastly, Europe has the Ice Age hiatus to thank for its blankets of loess.

The fourth major relief generation is that of the Holocene. The warm postglacial climate entered the central belt of Europe, covering the former glacial floors, cryoturbated frost-debris, and loess blankets of the periglacial regions with thick non-tropical relief-covering soils (Fig. 6 shows the gray-brown podzolic soil as an example) and with dense forest and wooded steppe. Both the soils and the vegetation retarded slope erosion, reducing fluvial loads, while the melting of the ice rind and permafrost further reduced the river's capacity for lateral and vertical erosion. Only above the forest line and the alpine turf zone are periglacial conditions still present. In the Mittelgebirge and in the lowlands, fluvial activity is limited to today's floodplains, narrow strips running tamely through the broad gravel beds of their braided predecessors (see Photo. 10). These broad Würmian gravel beds are indeed sometimes filled with exceptionally high floods, but this has only come about since the time of the Neolithic deforestation. The gravel of the valley floors is rarely picked up and moved along now. Even today many valley floors show the gently arched shape produced by the series of elongated, coalescing gravel fans which characterized the former Ice Age valley floors (see Fig. 14 and Büdel, 1944). This gravel, often quite coarse, is covered today by Holocene alluvial loams and sands, the majority of which have only been deposited there by increased flooding since the Neolithic deforestation (Mensching, 1950)

3.3. The Ectropic Zone of Retarded Valley-Cutting

3.3.1 THE RELIEF GENERATIONS IN NON-ALPINE EUROPE

The glacial relief of N Europe and the Alps and the much larger periglacial relief of the mid-latitudes have been covered during the scant 10,000 years of the Holocene only by a stable blanket of relief-covering soil. The Holocene period has not only done remarkably little to change the inherited Würmian features, but has even failed to disturb the mechanically formed weathering sheets on slopes of less than around 27-33°. Rather the new relief-covering soils have incorporated and stabilized the old weathering sheets. The development of a new soil sheet was aided by the forest, wooded steppe, and steppe vegetation which grew with and on it. If one imagines today's landscape minus its present soil and vegetation, 95% of the resulting relief would be a relict Ice Age relief, and would appear as though the Ice Age climate had only just taken a turn for the better.

Two exceptions to this are well known, one being the few mountain peaks which reach above the alpine tundra zone. Here we find active glaciers, rock cliffs, scree slopes (see Fig. 36), mudflows, and localized solifluction. The other exception is in the present coastal regions, where dune islands, tidal flats, marshes, and open and filled-in lakes are forming (the latter particularly in the areas of former glaciation). Much of what we regard as filled-in Alpine trough valleys, scree slopes, and alluvial fans were formed in the first few hundred years after glacial retreat. A second period of locally increased soil mobility was brought about by anthropogenic deforestation, which has been taking place since the Neolithic (in Central Europe, since 3500 B.C.) We have already mentioned the increase in flooding since that time. This has been accompanied by an increase in local surface wash (e.g., in steep vineyards), and by occasional soil slips and slides, especially in clayey pasturage. Brunnacker (1958) has shown that long-cultivated farmland on certain gentle slopes in Lower Franconia has lost nearly a third of its soil profile: this profile is still fully preserved under the adjacent forest, with the result that a bank about half a meter high has developed at the forest edge.

All these events together have not stripped much of the relief-covering soil sheet, let alone the bedrock. The wall of the Roman *limes*, for example, which since its erection 1800 years ago has lacked the protection of primary forest, runs straight and practically undisturbed, even where it crosses particularly unstable concretionary Keuper marls (*Knollenmergel*) on slopes of 12-13° in the Welzheimer Wald (Wagner, 1960, Plate 65). In both the humid and the semihumid ectropics the only visible traces of general relief activity consist of slight linear erosion along today's river courses. These, where not correctively interfered with by man, gradually rework their Würmian valley floors through slow displacement of their free meanders. Large rivers do this more vigorously than smaller ones. In this manner the rivers, though far less effective than their Pleistocene predecessors, still form conveyor belts for an extremely reduced erosion. We therefore call this zone the "ectropic zone of retarded valley-cutting." Here the total relief is almost completely dominated by remains of older relief generations. While the effects of the first (Tertiary), second (Earliest Pleistocene), and very weak fourth (Holocene) relief generations were fairly uniform throughout this region, the third generation shows marked differences between the glacial areas of circumglacial and subglacial effects, and the periglacial regions, where the relief was attacked subaerially. The latter region includes nearly all of W Central Europe, which we will discuss first in various subsections of 3.3.1. In Section 3.3.2 we will then examine the relief generations of the Alps, as well as a few examples from the Danube Basin adjoining it to the E.

3.3.1.1 The Erzgebirge, the Leipzig Embayment, and the Harz Mountains

The first work on etchplain stairways in Central Europe was done by W. Penck (1924), who investigated these in the Fichtelgebirge under the name of "piedmont stairways" or *Piedmonttreppen*. At this time the mechanisms of etchplain development and the concept of the four major relief generations were not yet known. Yet we owe much to Penck. He was the first to see the etchplains of the Mittelgebirge in a new light. Disagreeing with the cycle and peneplain theory of Davis (1899, p. 112), he developed the idea of a *primary planation surface* (*Primärrumpf*), a surface which does not necessarily imply the wearing down of a previously uplifted mountain belt, but rather can develop at low altitudes, as long as erosion keeps pace with simultaneous uplift. The mechanism responsible for such erosion was left unexplained. Penck's view that stepped surfaces could be produced even when uplift proceeded at a constant rate was based on an assumption which, though theoretically possible, was very unlikely, and it seems probable that he himself would have modified his views on this had he lived longer. Criticism of this point has prevented some successors of Penck's from accepting the advanced concept of a primary planation surface, although such a concept was already in the air, as expressed by Sölch, Scheu, O. Lehmann, and others.

To continue work on Penck's ideas, I studied the W Erzgebirge at the beginning of the 1930's. Here the preservation of Tertiary remains under basalts and the clear correlation of the relict surface with the Tertiary sediments of the Leipzig Embayment provided a convenient natural test site for dating the surface.

The result of this study was a fourfold system of etchplain steps, as shown in Fig. 58. This system developed on the gentle NW slope of the tilted Erzgebirge block, and is rudimentary on the steep SE drop toward the Eger Graben. Later workers (H. Richter, 1963; Gellert, 1958) have generally substantiated this view. A central upland (*zentrales Bergland*) is found around the Keilberg (1244 m) and the Fichtelberg (1214 m), and drops with a marked etchplain scarp to the "Gottesgab Surface." This is preserved without breaks to a maximum width of 14 km, and slopes evenly from 1050 to 901 m. Tertiary remains beneath its basalt sheets date it as pre-Late Oligocene. The next lowest "Schöneck Surface" could be dated in the same fashion to the Late Oligocene to Mid-Miocene. The fragments of this surface are up to 7 km wide and slope from 850 to 750 m. This is adjoined by the lowest and broadest surface, the "Vogtland-NW Saxonian etchplain," which slopes from 600 m at Schwarzenberg (S of Zwickau) toward the area N of Zwickau (in some places with slight steps) to the edge of the Leipzig Embayment near Grimma, where it is slightly under 200 m high (see Figs. 58-60). According to one description, "the correspondence of the relief assemblages strongly supports the idea that the ancient Central European etchplains and etchplain stairways were formed by the same climatically controlled erosional processes active in the etchplain stairways of the seasonal tropics today. In order to interpret the European etchplain stairways as climatically formed relict landscapes, the first step was to prove, firstly, that they are inactive relict forms, and secondly, that they do indeed derive from Tertiary times. For the Tertiary climate at these latitudes corresponded largely to that of the regions in which etchplain development is occurring today" (Büdel, 1938, p. 231).

W. Penck (1924) considered that dissection of each surface proceeded up from the next lower surface, which formed a local base level of erosion. This is the case in active etchplains of the seasonal tropics today, but it is not true of inactivated ones in the ectropics. I was able to show even then that planation and valley-cutting here belonged to two climatically quite different and chron-

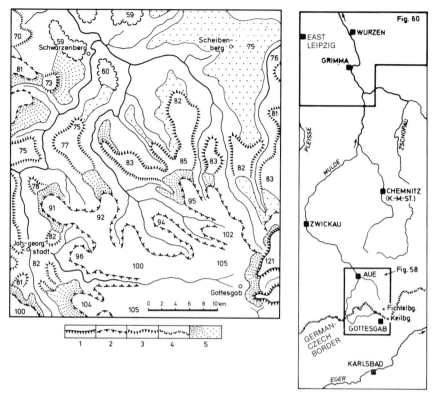

Fig. 58. (at left) The stepped etchplain of the W Erzgebirge (Büdel, 1935). For position of map see Fig. 59; height shown in tens of m. (1) Central Upland of the Fichtel Mountain Group. (2) Pre-Late Oligocene Gottesgab Surface. (3) Schöneck Surface, Late Oligocene to Mid-Miocene. (4) Vogtland-NW Saxonian etchplain, Late Miocene to Late Pliocene. (5) Etchplain escarpment and/or heavily eroded and remodeled surface fragments. The steep Pleistocene dissection along the rivers has been left blank.

Fig. 59. (at right) Position of maps shown in Figs. 58 and 60 in the Erzgebirge.

ologically widely separated relief generations. "The V-shaped valleys cutting deep into the etchplain stairways are nearly all completely adjusted *to a base level of erosion at the mountain foot.* Their longitudinal profiles do indeed show nickpoints, but these are regularly associated with resistant rock formations, especially contact zones of Variskian granite intrusions. Nowhere does a stream, falling from a higher surface, flow along at the level of the next surface before cutting into it, as would be required by W. Penck's theory. . . . Even the highest surfaces are directly attacked by the form assemblage of the steep erosion valleys, which attack both the highest and the lowest surfaces. . . . Since this valley-cutting began, *no* further planation has taken place, only progressive destruction of the surface through linear dissection reaching upwards from below and adjusted only to the general base level of erosion at the edges of the entire domed area" (Büdel, 1938, p. 230).

Mentally reconstructing the inactivated surface steps above the present valleys, one is struck by their resemblance to active etchplain stairways, having triangular reentrants, and even intramontane plains. Inselbergs, on the other hand, except where capped by basalts, are low and rare. Whether the shallow swales or dells[52] crossing the relict surface today are remains of ancient wash depressions, or are Pleistocene features, remains questionable. Naturally the younger dells often follow older wash depressions. But it has been shown that in their present shape they were formed at the latest during the Würm, and have not been further developed in the Holocene, for in the Erzgebirge they are filled with Holocene raised bogs. In the Riesengebirge on Zobten Mountain in Silesia, similar methods have proved marked Würmian movements in the solifluction sheets, which in many places form blockmeere. In the Erzgebirge, blocks 0.5 m in diameter (their upper surfaces often covered with a pale white, clayey weathering rind) were transported to distances of two km, sometimes across slopes of only 2°! The blockmeere are concordantly overlain by extensive raised bogs, without any traces of disturbance or jamming. This shows that the boulder sheet and the overlying bog have not moved since the peat began forming. Accurate palynological investigation of the raised bogs has shown that they encompass the entire post-glacial stage, including the later parts of the Late Glacial. This proves that movement in the boulder mass must have ended at the latest with the disappearance of the Ice Age tundra climate, and that no further movement or development occurred in the post-glacial period. This, then, was a very convenient "natural test site" for determining the age of the boulder movements. The undisturbed sedimentation of even the oldest, slow-growing *Phragmites* peat shows that the movement of the blocks had already ceased by the time they became covered with a very thin humus layer, no thicker than the common humus forest soils today (Büdel, 1936). As we know today, this means that solifluidal movement of the boulders was brought to a halt not by the regrowth of forest, but by melting of the permafrost and ice rind substrate.

This also throws new light on the wearing down of the inselbergs. The heavy periglacial erosion produced by combined solifluction and surface wash as observed in modern Spitzbergen (see Sections 2.2.9 through 2.2.14 above) were at work during the Central European Ice Ages, and while preserving the overall character of the relict surfaces, traditionally continued their development. This even included wearing down inselbergs, making the surfaces

[52] This concept was developed by Schmitthenner (1926) with regard to the structural scarp landscape (*Schichtstufenlandschaft*). There "swales" are defined as the shallow, uppermost portions of Recent valley systems, reaching back into the adjacent summit plains. They also occur on relict surfaces of shield areas or plateaus, in Tertiary hill country, and (as "dells") on all pre-Würmian Pleistocene deposits (for more details, see Section 3.3.1.7).

even flatter. On the other hand, as Hövermann (1953a) in particular has emphasized, the deeper Tertiary grus zones or decomposition spheres were in places exposed by periglacial erosion, providing core stones for the block-meere, having been moved by solifluction during the cold stages. Most of the castellated tors (*Felsburgen*) in the Mittelgebirge are simply resistant relics of the old decomposition spheres, which have become prominent now that their surroundings have been removed. Observations leading to these con-clusions were made by Mortensen as early as 1932.[53] At the same time, erosion in the old wash depressions caused greater sculpturing of the surface itself. The flattening of the inselbergs can be regarded as a restricted form of "cryoplanation" (see Section 3.2.9 above), but the erosion of the wash depressions cannot be so regarded, for it sows the seeds for surface dissection. Both processes can be included under the term *traditionally continued de-velopment of the relict surface* (with incipient dissection and transformation).

The lowest surfaces of the Erzgebirge dip gently northwards under the Tertiary and Quaternary fill (largely Oligocene in age) of the Leipzig Em-bayment (the *Leipziger Bucht*). Lignites near Geiseltal bei Halle, recently dated as far back as the Eocene (Pietzsch, 1962), are replete with a "Tertiary faunal paradise" of fully tropical species, wonderfully preserved even to the soft parts of the cell structures. The whole of these sediments covers an old etchplain-inselberg landscape. Below the Oligocene strata, a deep 15-20 m thick kaolin sheet may in some places be found (Gellert, 1967). On the whole, of course, the etchplains preserved here are older than those of the neighboring Erzgebirge, which continued to be lowered even during the Late Tertiary. The etchplain buried in the Leipzig Embayment has far steeper and more numerous inselbergs (some with slopes of over 25°), for they were protected from Ice Age influences. Many of these inselbergs protrude through the sedimentary sheets in the foreland zones of the Erzgebirge, studied in detail by H. Richter (1963) and Gellert (1967) E of Leipzig. Richter's map is reproduced here as Fig. 60. Gellert stresses that such inselbergs also occur in the E in the Silesian Sudetic Forelands; they are also widespread between Leipzig and the Harz mountains (H. Richter, 1963).

The counterpart to the Erzgebirge W of the Leipzig Embayment consists of the Harz, whose W portion, set off by fracture lines, rises like a ship's prow over the scarplands (*Schichtkammlandschaften*) of the Lower Saxonian Hills (*Niedersächsisches Bergland*), patterned by the weak phase of "Sax-onian" folding. Here Hövermann (1949, 1950) has described an etchplain stairway corresponding in many features to that of the Erzgebirge. Previously it had been dogmatically maintained that etchplains and etchplain stairways

[53] A very elegant proof of this has been provided by Louis (1934). The many castellated tors of the British Isles are found only in the unglaciated S, and are completely lacking further N in Ireland, England, and Scotland. Even where the rock type was particularly prone to the devel-opment of such features, they were totally destroyed by ice scour.

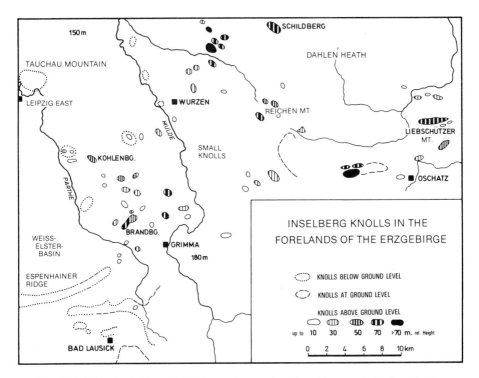

Fig. 60. Etchplains buried beneath Tertiary sediments, through which protrude inselberg knolls. Forelands of the Erzgebirge, S edge of the Leipzig Embayment. (After H. Richter, 1963, and Gellert, 1967. Location of map shown in Fig. 59.)

could only exist on crystalline rocks, while only bedding-plain surfaces, structural scarp reliefs, and scarplands (*Schichtflächen, Schichtstufen-landschaften*, and *Schichtkammlandschaften*) could develop on these subhor-izontal Mesozoic strata of the geologic superstructure. Following suggestions by Mortensen, Hövermann was the first to follow the peripheral planation surfaces of the Harz over onto the plateau of the N Eichsfeld, a step which pointed the way for further research.

3.3.1.2 The Rhine Plateau, the Wetterau, the Vogelsberg, and the Rhön

Tertiary relict surfaces are splendidly developed on the *Rhine Plateau (Rheinische Schiefergebirge)*. This plateau is dissected mainly by the Rhine, Moselle, and Lahn valleys, all three of which take root in basins along the mountain edge and join in the middle of the plateau at Koblenz. Their courses must therefore predate the mountain uplift. During active Tertiary planation this block must have sloped gently toward today's Middle Rhine; this zone was then vigorously uplifted, so that the contrast between the gently stepped Tertiary relict surfaces and the sharply cut Pleistocene valley is particularly striking (see Photo. 51).[54]

This simple picture has in some cases been refined, in others confused by detailed research. The problems arise from two sources. For one thing, uplift did not take place uniformly in time or space. Many pre-Pleistocene valley courses and old tectonic depressions have been found, filled with Tertiary sediments such as the Vallendar and Hofheim gravels. The explanation for this is probably similar to that described above (Section 3.2, p. 256) for the pre-Tortonian valley N of Regensburg.

A second problem consists of the fact that we have here a particularly well-preserved system of flat relief features, clearly associated with the climatic

51. Middle Rhine valley near St. Goarshausen, with three relief generations. High Tertiary relict surface; Latest Pliocene to Earliest Pleistocene Broad Terraces (*Breitterrassen*, the Trough Surface or *Trogfläche*, and the Principal Terrace or *Hauptterrasse*); Late Pleistocene valley incision, with marked terraces in wider areas such as the Neuwied Basin. Compare Figs. 61 and 62. (Cekade Aerial Photo. Crämers Kunstanstalt KG Dortmund, released under No. NRW Cr. 5616.)

[54] Unlike smaller rivers such as the Tauber (Photo. 10) and the Nagold (Photo. 52), the gravel-rich Rhine has in the Holocene reworked its Würmian gravel bed thoroughly enough that in the Middle Rhine gorge (see Photo. 51), where the river fills nearly the entire valley, settlements and communication routes are limited to narrow banks and partly truncated alluvial fans of smaller tributaries.

interphase between the Early Tertiary period of etchplain development (which continued at least traditionally into the Mid-Pliocene) and the later abrupt down-cutting of the Ice Age Pleistocene. This intermediate system consists of the "Broad Valleys" or *Breittäler* (also called the "Broad Terraces" or *Breitterrassen*). These are terraces, often many km wide, perched along the recent valleys 15-30 m (sometimes even less) below the crowning etchplain surface. Like the etchplains, the Broad Valleys are erosional surfaces, but are associated with the valley network of today. The rare gravels found on these surfaces are totally different from those of the later terraces. We will return again to this special relief generation belonging to the Latest Pliocene and Earliest Pleistocene. This system is shown as a single Broad Terrace in Figs. 14 and 62.

At least two very broad such systems are cut into the older etchplains of the Middle Rhine Plateau. At first glance these may be confused with the relict surface (see Photo. 51, with sharp rims above the Pleistocene gorge), yet they form the shallow predecessors of today's valleys. The higher, very broad surface, called the "Trough Surface" (*Trogfläche*) has been dated to the Latest Pliocene, while the lower "Principal Terrace" (*Hauptterrasse*) has been dated to the Earliest Pleistocene (see Fig. 61, after Brunnacker, 1975). Photograph 51 clearly shows the gentle transition from the older etchplain to the Trough Surface and the Principal Terrace. Neither of these would be so wide, had not etchplain development preceded them. In the sharp, vineyard-covered drop down into the Pleistocene Rhine Valley, the four later "Middle Terraces" (*Mittelterrassen*) corresponding to the Günz, Mindel, and Riss Ice Ages, are barely visible. These Middle Terraces are better developed at wider areas in the valley, such as in the Neuwied Basin (the *Neuwieder Becken*).

Quitzow (1969) has very impressively described the Tertiary etchplain landscape in the area of the Middle Moselle, with the Trough Surface/Broad Valley and the sharply contrasting Pleistocene valley system. Here the highest relict surfaces carry kaolin-rich weathering horizons and basin clays, which supply several ceramic industries. The raw material for the Upper Franconian and Bohemian porcelain industries is also taken from the thick kaolinitic weathering rinds on the relict surfaces of the Frankenwald and the Fichtelgebirge. The actual Trough Surface of the Middle Moselle area carries only a cover of alluvial loams (*Decklehme*) with little or no kaolin, a sure sign that other mechanisms were already at work here. "One cannot conclude from this that these are Recent or Pleistocene weathering features, for this interpretation is refuted by their position on the Trough Surfaces, outside the area of Recent dissection" (Quitzow, 1969, p. 44). Below this Trough Surface, Quitzow's map shows a second broad (slightly tripartite) system of terraces, Late Pliocene to Earliest Pleistocene in age, perched above the steep cliffs of Pleistocene dissection. Due to the greater distance from the main axis of

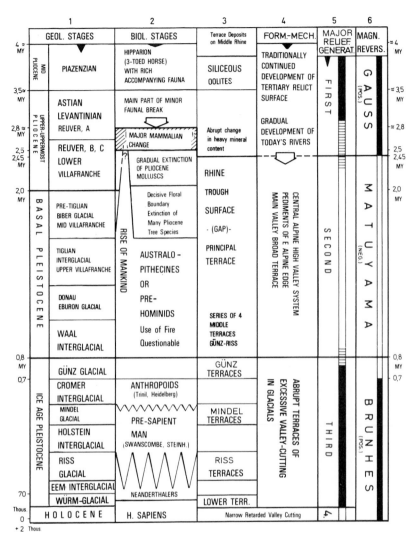

Fig. 61. Sequence of relief generations in Central Europe (Columns 3, 4, and 5) along with the corresponding changes in the companion spheres (Columns 1, 2, and 6) during the last four MY.

upheaval on the Rhine block, and due to the base level of erosion of the Middle Rhine, Heine (1970a) found only one undivided Late Pliocene Broad Terrace on the Upper Lahn, while Körber (1962) and Büdel (1957d) found only one (slightly doubled) Broad Terrace on the Main River below Würzburg. This is shown as the Latest Pliocene-Earliest Pleistocene terrace in the schematic profiles of Figs. 14 and 62.

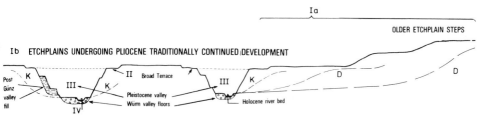

Fig. 62. The four main relief generations in the periglacial region of Central Europe. (R) ravines, (S) swales. (Ia) Pre-Pliocene etchplain and etchplain stairway. (Ib) Etchplain undergoing traditionally continued development during the Pliocene. (II) Broad Terrace (*Breitterrasse*), Latest Pliocene to Earliest Pleistocene. (III) Valleys and Würm valley floors incised by excessive valley-cutting during the Ice Age Pleistocene. (IV) Holocene riverbed with floodplain.

Despite some disagreements on detail,[55] we may take it as assured today that the high surfaces of the Middle Rhine Plateau, like those of the Erzgebirge, derive from an etchplain stairway whose period of development lasted (with some interruptions) from the Oligocene into the Mid-Pliocene. It is important to realize that the first laying-out of the very gently incised Trough Surfaces and Broad Terraces in the Latest Pliocene and Earliest Pleistocene would not have been possible without the preceding Early Tertiary etchplain. This is true regardless of what mechanisms may actually have created these Trough Surfaces and Broad Terraces. They represent a type of traditional development of the older surface along narrow ribbons of terrain, which at the same time were the first forerunners of today's valley network.

The discoveries made by Bibus (1971, 1973, 1975) on a geomorphologic cross-section from the Taunus via the *Wetterau* to the W edge of the Vogelsberg (see Fig. 63) throw much light on the development of the later etchplains in this area. Bibus was able to show that the same surface cuts both the old crystalline Taunus and the recent volcanic mountains. The profile crosses the Wetterau Depression at its highest point near the Main-Lahn divide. This area was so little disturbed by Pleistocene erosion that the surface cutting across it could be dated with exceptional accuracy by its thick Pliocene latosol sheets, which contain bauxite concretions in commercially minable quantities. This surface is cut by recently down-faulted grabens, such as the Nauheim Graben with its lignite-bearing Late Pliocene sediments.

At the edge of the Taunus (W in Fig. 63) Bibus identified the Usingen Basin as a very well-preserved intramontane basin, sunk 400-420 m into the

---

[55] From the plethora of literature on this subject, some of the important works, besides that of Quitzow (1969), are: Stickel (1927), Mordiziol (1927), Gurlitt (1949), Louis (1953), Macar (1954), Kutscher (1954), Kremer (1954), Körber (1956), Panzer (1965, 1967), Hüser (1972), and Birkenhauer (1972). The last of these, despite all tropical experience, attributes the development of etchplains in large part to lateral erosion.

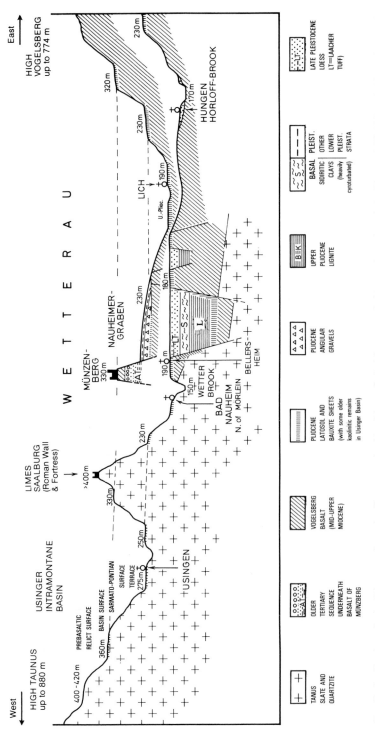

Fig. 63. Profile through the Wetterau from the Taunus to the Vogelsberg. Early Miocene etchplain on the edge of the Taunus (400-420 m high), with several intermediate surfaces, largely Pliocene, in the Usingen Basin (an inactivated intramontane basin) and on the W edge of the Vogelsberg. The youngest of these surfaces (180-190 m high) carries the thickest latosol sheets of this age in Central Europe. These are now being mined for aluminum.

pre-basaltic (i.e., pre-Middle Miocene) relict surface. Bibus calls this relict surface a Trough Surface, due to its shape, but it is of course earlier than that on the Middle Rhine. This relict surface was last active in the Aquitanian (Early Miocene).

As in some active intramontane basins, remains of older surfaces are preserved on the slopes of the Usingen Basin. One of these is the "Basin Surface" (*Beckenfläche*) of Sarmato-Pontian age, at a relative height of 360 m, and the other is the "Surface Terrace" (*Flächenterrasse*) at a relative height of 250-275 m. Numerous cores and exposures provided Bibus with evidence that these levels still carry deep Tertiary weathering sheets, consisting of "pale gray to whitish decomposed rock, in places with reddish or even yellowish tinges. . . . All zones of decomposition are clearly dominated by kaolinite" (Bibus, 1973, pp. 12-13). Thus kaolinite development, as Semmel has stressed (1972, p. 18) was still possible in the Central European Early Pliocene.

The thickest weathering sheets, however, are found lower down at 190 m on the basalts of the *Vogelsberg*. Here the volcanism belongs largely to the Tortonian (Burdigal to Late Miocene). The clayey weathering profiles above this are 35 m thick near Nauheim, 33 m thick near Gambach, and at places even 50 m thick. The Early Pliocene etchplains covering tens of square km on the basalts along the headwaters of the Wetter Brook near Lich are particularly impressive. A bauxite pit located 1.5 km ENE of Lich shows the following profile: above lies a latosol sheet, several m thick, containing bauxite concretions. Below this is a decomposed mass, totaling eight m in thickness, in which the former basalt structures are still recognizable. Unweathered basalt occurs 20-30 m below the surface.

This proves the presence of a seasonally wet, fully tropical climate in the Early to Mid-Pliocene, which is corroborated by fossils of the same age (Bibus, 1973). Of course it is true that the basalt substrate was particularly suited for latosol development even under the already cooling climate of the Central European Early to Mid-Pliocene. This is shown by comparison with soil blankets of Mid-Pliocene age on the basalt sheets of E Styria (Winkler-Hermaden, 1951, 1957). Nevertheless, it is generally true that a temperature drop in the Pliocene, shown by the reduced number of tropical plant species and possibly caused by colder winters, would have had little effect on soil development and morphodynamics, as long as a monsoonal climate with warm rainy summers provided for a seasonally continued development or at least preservation of such latosol sheets and thereby of the underlying relict surface.

According to investigations by Mensching (1957a, 1960a) and Rutte (1957), the tuff layers and basalt sheets of the high *Rhön* are of the same age or perhaps younger than those of the Vogelsberg, and are probably Tortonian-Sarmatian to Early Pliocene (see Fig. 64). The Rhön had not yet been elevated at that time, and the volcanics spread out over a series of lignite-bearing, fine-grained sediments of Helvetian to Lower Pliocene age. These overlie Muschelkalk

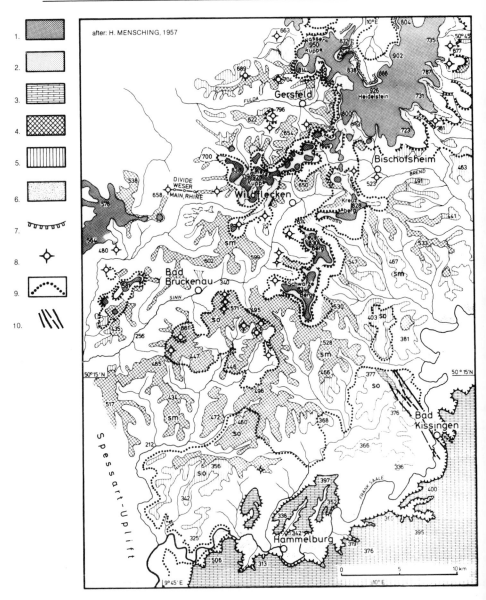

Fig. 64. Geomorphological map of the Rhön Mountains and their southern foreland. (1) Early Pliocene basal (2) Remnants of a single, prebasaltic, Miocene surface. (3) Muschelkalk (Triassic limestone). (4) Early to M Pliocene post-basaltic surface. (5) Late Pliocene level of the Pliocene erosional surface. (6) Fluvial terraces the Franconian Saale River, undifferentiated. (7) Scarp bordering the Muschelkalk base of the High Rhön. Cone-shaped mountains, many formed around resistant volcanic pipes. (9) Boundary between the Middle a Late Triassic Sandstones. (10) Morphologically significant faults.

(shelly Triassic limestone) which is preserved under the basalt flows and which in turn rests on a broad base of Bunter Sandstone. Shortly after the eruptions, the entire series was uplifted and broken.

The geomorphic development here has been unraveled by Mensching. Besides the basalt-covered surfaces, some exposed remnants of the pre-basaltic relict surface have been preserved at heights of 650-800 m. These were the parent surface for the entire subsequent development. Below the escarpment of the Muschelkalk (which is usually very marked) spreads a very wide Mio-Pliocene foreland surface (Mensching, 1957a, p. 61), which slopes from around 550 m below the Muschelkalk scarp of the Rhön, SE to around 350 m at the marginal depression of the Streu River and Franconian Saale River (between Mellrichstadt and Kissingen, see Fig. 64) to the SE. Traces of a valley floor found on this N-S marginal depression have been dated by a series of yellow to bluish-white sandy clays and clays containing mammalian remains from the Mio-Pliocene. This valley floor had a gentler gradient than do today's valleys. Its sediments, despite proximity to the Rhön, contain no basalt cobbles. Clearly, under the climatic conditions of the time, the High Rhön produced no coarse weathering products. The surface of these Mid-Pliocene fine-grained deposits is covered with coarser gravels, containing frequent clasts of basalt. The clay-rich Mid-Miocene fine sediments probably belong to Mensching's Mid-Pliocene foreland surface, while the overlying gravels belong to Pleistocene river terraces.

But the foreland surface and the Pleistocene terraces are separated by yet another interesting surface, the *Ausraumfläche*, lying about 70 m below the foreland surface between the Rhön and the Streu-Saale Depression. Although Mensching notes its extensions onto the gravel-covered terraces of the Streu and the Saale rivers, he nevertheless attributes it to the Latest Pliocene. This intermediate surface displays several features which deviate from those of the Pleistocene terraces. "The entire drainage system of the Rhön was surely laid out as early as the Late Pliocene. But this certainly does not mean that the valleys existed at that time as deeply eroded lines even approximating those of today" (Mensching, 1957a, p. 68). Gravels are lacking on the Ausraumfläche, which Mensching regards a Quaternary glacis. We believe that it may be equivalent to the Late Pliocene-Earliest Pleistocene Broad Terrace on the Main River (Figs. 14 and 62).

3.3.1.3 The Upper Rhine Valley, the Black Forest, the Swabian Neckar, and the Swabian Alb

We will discuss the *Upper Rhine Valley* only in terms of the edges of the Pfälzer Wald and the Vosges. Along the W edge of the valley the relief generations (where not too disturbed by step-faulting, as around Molsheim) are preserved in their entirety from the summit of the adjacent mountains down to the broad Rhine floodplain. This is not the case on the E edge of the valley, where Late Pleistocene terraces of the Rhine and its tributaries cling to the mountain slopes, the Early and Earliest Pleistocene features having been swept away by lateral erosion or buried by subsidence.

The best sequence of relief generations has been investigated by Stäblein (1968) along the edge of the Pfälzer Wald from Durkheim in the N to Weissenburg in the Alsace to the S (see Fig. 65). The highest etchplain (Oligo-Miocene) of the Pfälzer Wald is pocked with intramontane plains or "Dahner Surfaces" (*Dahnerflächen*), which according to Ahnert (1955) and Liedtke (1967a) are of Pliocene age. Along the fault scarp facing the Rhine Plain, the "Haard Piedmont Plain" (*Haardrand-Fussfläche*) forms a typical fanglomerate-covered glacis surface with a narrow lateral pediment along its upper edge and a lower Principal Terrace at its foot. The glacis surface has been assigned by Stäblein to the Late (or Latest) Pliocene, and the lower Principal Terrace to the Earliest Pleistocene. Stäblein was able to trace it as a continuous ramp for 50 km from N of Neustadt to S of Weissenburg. It reaches a maximum width of 15 km just S of Landau. This is the largest such feature in W Central Europe, and resembles one I found in 1974 on the edge of the Vosges near Reichsfeld-Ittersweiler, 35 km SW of Strasbourg, which deserves closer investigation. Further N in Stäblein's working area the ramp is cut by Pleistocene rivers (the Lauter, the Klingbach, the Quiech, and the Speierbach) debouching from the mountains. The rivers then quickly spread out into wide alluvial fans running into the Rhine. The difference between the erosional glacis fans and the predominantly depositional alluvial fans is very clear here. Counting the Lower Terraces along the Rhine, then, all four relief generations are displayed here in exemplary and vertically ordered fashion.

The *Black Forest*, much like the Erzgebirge, the Harz, and the Rhine Plateau, is crowned by a system of etchplains of Early Tertiary age, whose lower steps, however, may still have been active in the Early Pliocene. The old problem of whether any of these surfaces cut smoothly from the crystalline Black Forest in the W to the Bunter Sandstone Black Forest in the E (a brief discussion may be found in Semmel, 1972), was a question which only made sense as long as one considered the crystalline relict surfaces to be exogenically formed etchplains and the Mesozoic surfaces to be structurally controlled, and until Bremer's discovery (1971) that etchplain scarps may grade longitudinally into structural scarps in the seasonally wet tropics of today. In many places a quite uniform surface cuts both the crystalline and the Bunter Sandstone, while in other places a scarp has formed at the Bunter Sandstone. But this does not contradict the basic fact that the remarkably even relict surface on the Bunter Sandstone represents an etchplain which in many places is only very loosely adapted to bedding-plane surfaces (see Photo. 52). Weise (1967) has proved in great detail that, looking across the later valley incisions, one of the lowest of the Early Pliocene surfaces runs from the Bunter Sandstone with a scarcely visible rise of 2% onto the Muschelkalk.

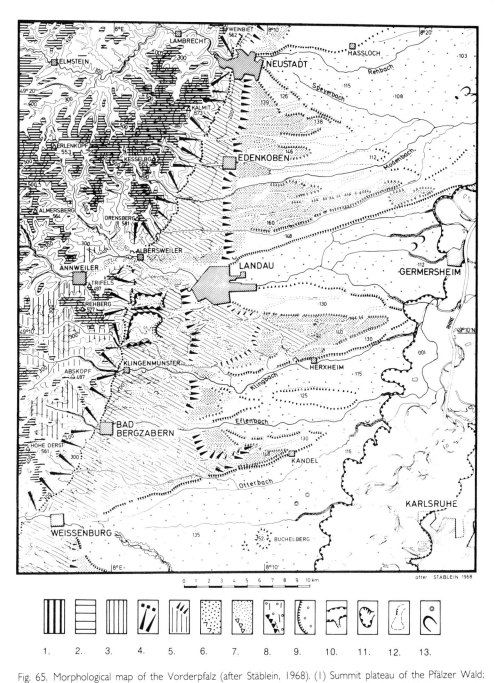

Fig. 65. Morphological map of the Vorderpfalz (after Stäblein, 1968). (1) Summit plateau of the Pfälzer Wald; Oligocene/Miocene etchplain remnants. (2) Dahner Surfaces. Pliocene intramontane basins. (3) Kraichgau margin. (4) Graben fracture rim: Oligocene, with Pleistocene slope debris sheets. (5) Haardt margin piedmont surface, including Late Pliocene galcis remnants, the Principal Terrace (Earliest Pleistocene) and fault scarps (pre-Mindel) along the central valley lowland. (6) High Terrace (Mindel). (7) Talweg Terrace (Riss). (8) Lower Terrace (Würm). (9) Frankentahl Terrace (lower Lower Terrace). (10) Holocene floodplains, floodplain scarps, and small V-shaped valleys. (11) Areas of Pleistocene erosion. (12) Dells and gullies. (13) Dune fields and pingo scars.

52. Nagold Valley near Hirsau, N Black Forest. Etchplain adapted to stratum of Bunter Sandstone, cut by a Pleistocene valley. Valley floor formed by Lower Terrace. (Commercial photograph.)

Between the gentle eastern slope of the Black Forest on the W and the steep scarp of the Swabian Alb to the SE lies the low northward-facing triangle of the *Swabian Neckar* (*Neckarschwaben*) or the ''Württemberg Lowland'' (*Württemberger Unterland*), as it is called locally (see Fig. 68). This region enjoys a very favorable soil and climate, and is characterized by three geographic units: the earliest settled, loess-covered Gäu surfaces on the Muschelkalk and the Letten Coals (*Lettenkohle*); the heavily forested, recently settled Keuper mountain country at medium altitude, and finally the Neckar and its tributaries, which cut through the other two landscapes. Only in the Pliocene did this river system begin draining to the Rhine instead of the Danube. The valleys today form a nexus of highly important industrial corridors.

Rarely has an area been geologically so well investigated as this one. This may be due to fossiliferous facies whose horizontal uniformity and rapid vertical change make them easy to recognize across fault breaks. The beds of the Mesozoic substructure are nearly horizontal, dipping very slightly toward the S. Geology was taught here in many of the lower schools, so that a knowledge of the subject became exceptionally widespread in this region.

The stratigraphic sequence is characterized by astoundingly rapid changes in morphologic hardness, changes which were emphasized by the Ice Age Pleistocene of the third relief generation. On steep slopes each hard bed crops out as an erosional or structural bench (*Schichtstufe*), and the resulting agreement between structure and morphology is often extremely detailed. Since there are many such slopes, particularly on the Upper Neckar upstream of Stuttgart (some are valley slopes, some are "free" slopes below ledges of Keuper, Schilf, and Stuben Sandstones, Lias-alpha, or Malm Limestones-beta to gamma of the Alb), the geologist, who after all looks for stratigraphic structure, will see the correspondence between structure and shape so often that he will readily formulate a general theory of a structural scarp relief (*Schichtstufenlandschaft*).

If one climbs from the NW up the three major steps described above, one will cross over the last sharp escarpment onto a broad plain which is structurally adapted (i.e., follows the SE dip of the beds) for at best a few hundred m length and a few km width. Overall, however, it rises gently toward the SE. This means that it cuts across increasingly younger lithic units, cutting smoothly from Muschelkalk to Letten Coal, from Schilf Sandstone to the variegated marls, from Stuben Sandstone to the concretionary marls, from Lias-alpha to Lias-beta through Lias-gamma, and finally onto the Opalinus Clay (Dogger-alpha). Faults are frequently crossed with little or no evidence of their vertical displacement. The plateaus of the various Gäus and the many relict surfaces of the Keuper upland and the Alb are true etchplains both in shape and genesis. This will be shown in detail below, using Franconian examples.

As Wagner indicated (1950, pp. 574-575), the Upper Miocene basalt vent at Scharenhausen near Stuttgart contains Malm blocks 23 km away from the Alb edge. A volcanic pipe of similar age on the Katzenbuckel in the Odenwald contains remains of Dogger whose nearest outcrops are now 90 km further S in the Hohenstaufen, the nearest part of the Alb. This does indeed prove that the lithic boundaries have receded this far during the last 12-14 million years, but it does not prove that this occurred in the form of scarp retreat. According to Wagner, about 700-800 m of rock were removed in the region of the Katzenbuckel. The broad surfaces (the Gäu surface, the plateaus of the Keuper mountains, the Lias surfaces, and the Alb surfaces themselves) which cut across the faulted strata make it highly probable that this enormous amount of erosion took place areally during the first relief generation, under conditions completely different from those of today. Certainly there is not a shred of paleoclimatic or other evidence to the contrary. Erosion kept pace with concurrent phases of uplift in this portion of the crust. Alternating epeirogenic uplift and planation gradually weathered out the large scarps whose beautifully etched out treads and risers reveal strata which are little if at all evident on the surface itself. We know now from studies in the seasonally wet tropics that etchplains may cut perfectly smoothly across different rock units, and we also know how erosion there will sharpen the piedmont angle at the foot of an inselberg or etchplain escarpment so that the slope above becomes steepened. On such precipitous barren slopes, petrovariance will be minutely revealed, even while it is completely suppressed on the latosol-covered surface (see Sections 3.3.1.4 and 3.3.1.5 for a discussion of the paleosols). This corresponds to Bremer's rule (1971, 1973b) of divergent erosion in the savanna countries of today. In Central Europe, weathering out of the harder strata was accentuated by the abrupt Pleistocene valley-cutting. This happened to an even greater extent in the Swabian Neckar than in Franconia, for the former was uplifted and dissected more than the latter.

Using examples from the Swabian Neckar, Dongus (1970) draws very basic distinctions between the relict surface in the crystalline rocks, which he accepts as etchplains, and the surfaces on the horizontal superstructure, which he regards as structurally controlled. Endogenous structure, however, is of no importance to the active exogenic processes involved in the relief development. Naturally the relief created will here and there adapt to differences in hardness of the underlying rock. But although the rock units (e.g., horizontally bedded structures) may remain the same, the exogenic processes will work in entirely different ways, varying geographically according to morphoclimatic zone, and chronologically according to relief generation. In the tens of millions of years of humid tropical conditions present in Europe during the first major relief generation, the very slight uplift with dominant planation under thick blankets of red loam hardly provided the conditions for such a structural adaptation,[56] except on steep slopes. During the following brief Earliest Pleistocene relief generation, conditions were not very different. In the third relief generation, on the other hand, morphologic structural adaptation became increasingly clear. Thus the problem of the "structural escarpment landscape" (*Schichtstufenlandschaft*) could not have been solved without the help of climatic and climato-genetic geomorphology.

The high plateau of the *Swabian Alb*, which crowns the sequence of surfaces, is, as I have stressed before (Büdel, 1951), an etchplain surface. It is well known that this surface is longitudinally bipartite. The *Kuppenalb* (hummocky alb surface) to the NW is surely far older than the Villefranche remains in the Bärenhöhle (a cave near Erpfingen on the alb surface) have led some authors to assume. Its relief still shows features of an Early Miocene tropical karst adapted to a knobby surface of limestone reefs and later cryogenically

---

[56] When, during the first major relief generation, an etchplain is lowered parallel to itself, approaching the surface of a hard horizontal stratum, it may come to coincide partially (but never wholly) with this layer. This is still a case of etchplain genesis: one cannot call this a bedding-plane surface in the sense that the hard stratum created the surface, for it was created by climatically controlled, exogenic processes. When such a surface becomes dissected after a climatic change, the hard stratum at the upper edge of the resulting valley will interrupt erosion (Späth, 1973) and help to preserve the surface. It does not create it, as Dongus contends. Examples are common in limestone. Broad, well-described relict surfaces have been found on karstified limestones in the NE and Dinaric Alps, cutting smoothly across contorted folds and nappe structures. It is clear that the preservation is due to the permeability of the stone and lack of surface dissection, not to stratification.

flattened. The Kuppenalb is separated from the Plateau Alb or *Flächenalb* to the SE by the well-known straight marine "cliffs," investigated so thoroughly by U. Glaser (1964). Although the fossils in the shore deposits were previously ascribed to the Burdigal Sea, recent reinvestigation of the cliffs has interpreted them as remains from the Helvetian Sea of the alpine forelands, which at that time were much wider. The limestone reef knobs are absent in the Plateau Alb to the SE, but this does not mean that the Helvetian Sea did not transgress over the older etchplain surface, traditionally developing it and accentuating the old etchplain scarp (now the marine cliff) with Helvetian shore deposits. Glaser's conclusion that the Plateau Alb continued to be flooded (along with the cliff and parts of the Kuppenalb) during the Tortonian and Sarmatian is particularly important. By means of well-substantiated Recent tectonic movements, he was able to show why the knolls of the Kuppenalb are so much better preserved in the W than in the E. His conclusion on the Pliocene paleo-Danube is also important: "its gravels lie distributed today over an area up to ten km wide on the remnant surfaces, not in the valleys. That it [the paleo-Danube] must have had a true etchplain as its parent relief is shown by the fact that the gravels are found at the same level on Malm-epsilon, Malm-theta, Oligocene, Early Miocene, and Late Miocene sediments. Usually they cut smoothly across rock boundaries.

"The tendency toward plantation must have dominated the Plateau Alb throughout the entire Pliocene, for even the northernmost gravels lie completely at the level of the surface. Down-cutting must have begun in the Villefranche, for the gravels appear a few tens of m downslope toward the Danube" (Glaser, 1964, p. 80). Thus the sequence of relief generations in this area corresponds to those in the rest of Central Europe.

3.3.1.4 A W-E Cross-Section through Franconia from the Spessart to the Frankenwald

All the features described in the scarp landscape (*Schichttafelland*) of Swabia reappear even more clearly in Franconia. Here there was less uplift and dissection, and the strata lie even more horizontally, though faulted heavily near the Rhön, the Thüringer Wald, and the Frankenwald (Rutte, 1957). These faults had no influence on planation, as is shown by the cross-section made by Cramer in 1937 (at a time when he can hardly have been influenced by theories of etchplain development) through the Heustreu disturbance zone (*Heustreuer-Störungszone*) shown here as Fig. 66. The cross-section clearly shows how the Gäu surfaces of the faulted zone cut smoothly across all strata from the middle Bunter Sandstone through to the Letten Coals (including the slip faults in these) without regard to their petrovariance.

The old cross-section shown here in Fig. 67 (taken from Büdel, 1957d) has not only remained valid, but has even been further clarified, e.g., with respect to the Muschelkalk escarpment in the W, on which the supporters of

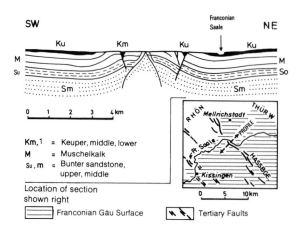

Fig. 66. Section through the "Heustreu Fault Zone" in the NE of the Pliocene Gäu surface of Lower Franconia (after Cramer, reproduced in Rutte, 1957, Fig. 48). The etchplain, carved largely into the soft lower Keuper, cuts smoothly across folded and upended Mesozoic strata. In no way can this be called a "bedding plane surface."

the structural scarp landscape theory placed so much weight. According to Fugel (1975), this escarpment is developed in the entire region between the Neckar near Mosbach and the Franconian Saale near Bad Kissingen (see the map in Fig. 68) only where a Pleistocene valley has cut down into the Röt, creating a steep Muschelkalk slope on one side and a gentle slope on the Röt on the other. *Except along such valleys, the Muschelkalk escarpment does not exist at all.* Rather the smooth Gäu surface cuts evenly without scarps from the hard lower Muschelkalk across the soft Röt to the main Bunter Sandstone. This observation of Fugel's, like that of Weise (see Section 3.3.1.3), clearly supports the theory (see footnote 56) that where valleys and dells of the third relief generation in the mid-latitudes cut into softer rock on the inherited relict surface, they formed structural escarpments along the valleys. This has been observed in all phases of development in the Swabian Neckar and in Franconia. Where these valleys did not reach far into the softer units, *the smooth surface from the first major relief generation, which cut evenly across all rock units, faults, and later-developed watershed divides, is fully preserved.*

This is obvious even when comparing the two W-E cross-sections through Franconia (Fig. 67). The upper cross-section represents the older view, still maintained in some quarters (see footnote 56), according to which all harder strata are weathered out as scarps both on the surface and in the slopes. The surfaces, according to this view, are bedding-plane surfaces. The lower cross-section depicts the true situation. Here petrovariance is revealed only along the edges of some of the larger escarpments, as the W edge of the Steigerwald

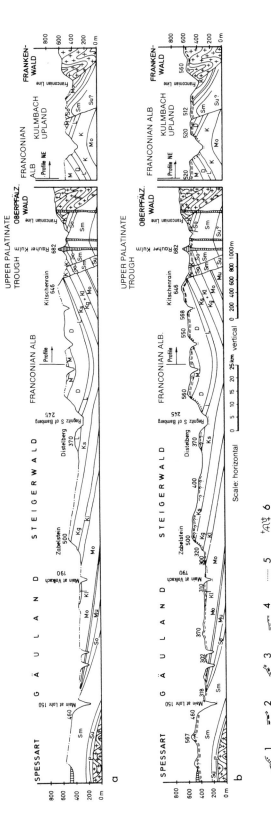

Fig. 67. E-W Profile through Franconia from the Spessart to the Frankenwald. (a) Traditional, and (b) revised interpretation (unaltered, from Büdel, 1957d). (1) Cuestas. (2) Miocene or pre-Miocene etchplains and remnant mountains. (3) Sarmatio-Pontian etchplain (with remnant mountains). (4) Late and Mid-Pliocene etchplains (3 and 4 are both inherited from older surfaces). (5) Pleistocene terraces and valley floors. (6) Crystalline units of the geologic infrastructure.

and the Franconian Alb (the *Frankenalb*). Petrovariance is not at all visible
on the undissected surface, and the Muschelkalk escarpment disappears completely between Lohr and Karlstadt (N of Würzburg). The scale in the figure
does not permit the depiction of the benches present on the slopes of the
Pleistocene valleys.

The appearance of the old etchplain is best shown by the slightly karstified
surface of the Gäuland. As one looks out across the valleys and dells which
cut and dent the surface, the impression is truly one of a flat plain. This is
especially true of the Ochsenfurter Gäu, covering around 80 km² between the
Main River near Ochsenfurt and the Tauber River near Mergentheim (see
Photo. 10). The airfield of Giebelstadt is situated here, and nearby was fought

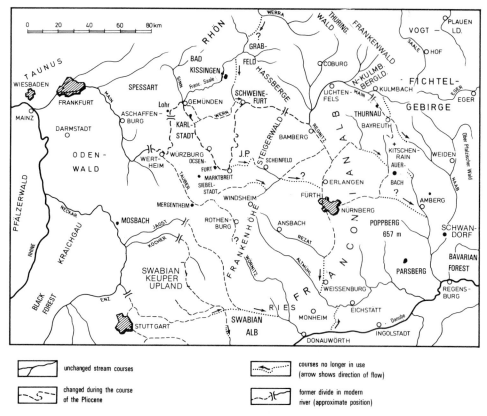

Fig. 68. Stream diversion in Franconia during the Pliocene: encroachment of the Rhine drainage
onto the Danubian drainage (Büdel, 1957d). Such diversion occurred easily on the Tertiary etchplain until the Middle or Late Pliocene. By the Earliest Pleistocene and certainly in the Ice Age
Pleistocene the rivers were locked into narrow valleys, after which changes in course by spilling or
piracy were rare exceptions. I.G.: Ifphofen Gap (*Iphofer Pforte*).

the tragic Battle of Sulzdorf (1525), in which the army of the Franconian peasants, led by Florian Geyer, was cut to pieces by the aristocratic cavalry, who could be deployed freely on the open plain. Even this, the most even plain in all of Franconia, is not a bedding-plane surface. It bevels the Upper Muschelkalk (which here contains the particularly hard Quader Limestone) just as evenly as it bevels the Letten Coal of the Keuper. Above all, it cuts right across faults which show displacements of up to 30 m (these faults were not yet known in 1957). Finally, the upper edge of the surface at the Main Valley near Ochsenfurt is at the same altitude (around 320 m) as its upper edge at the Tauber Valley near Mergentheim, 22 km away, the plain between being interrupted only by a single low shield inselberg.

The Gäuland surface (*Gäuflache*), last active in Late Pliocene times, is covered by a thick sheet to relict frost-patterned ground. A noticeable non-conformity is often visible (see Photo. 5) where the ice rind melted away from between the cryoturbated frost debris zone and the undisturbed substrate beneath. Above the frost debris zone lie 0.3 to over 2 m of loess, on which the wealthy farmers grow wheat and sugar beets.[57]

The Gäu surface remained largely undissected, even though, as Rutte (1957) and Körber (1962) have stressed, both the Middle Main and the Lower Neckar cut down twice during the Pleistocene, probably due to tectonic movements in the Upper Rhine graben. The first (Günz) phase cut nearly to the present valley floor. The valley was then refilled with sediments containing, among other things, the famous mandible of Heidelberg Man in an old Neckar loop near Mauer. Between the Mindel and the Würm, the Main and the Neckar rivers cut down again, reexcavating their former valleys, though cutting off a few of the old meander loops. All this took place below the Gäu surface, and influenced it only slightly through tributary valleys.

East of this region the fairly continuous Keuper scarp of the *Steigerwald* rises about 150-200 m over the Gäu surface. This is joined S of the Iphofen Gap (the *Ifphofer Pforte*, through which the Main River formerly drained to the Danube; see Fig. 68) by the similar forms of the Frankenhöhe, and N of the Main Gap by the *Hassberge* between Bamberg and Schweinfurt. The Hassberge, tectonically complexly faulted and shot through with narrow basalt vents, have been investigated in detail by Späth (1973). His most astonishing conclusion was that seen from the faulted zone of the Gäu surface, the Hassberge form a uniform scarp with a sharp upper rim. This, like any other "structural escarpment," is lithologically controlled. In its northern section, this escarpment stays at around 500 m altitude for over 15 km. But the rim is not formed everywhere by the same hard stratum of sandstone. Though the Keuper strata on the scarp show displacement along faults, these faults are

---

[57] Fairly early redistribution and consolidation of landholdings gave the farmers a considerable advantage over their neighbors, who had to struggle much longer with the former system of subdivided inheritance.

not morphologically evident on the rim. *The continuous straight rim is formed in some places of one stratum, in other places of another.* Späth (1973, p. 36) has described this as follows: the scarp of the northern Hassberge "towers like a wall over the faulted zone, as can best be seen from the W. Nowhere does the upper course of a ravine or dell reach as far up as the upper rim of the scarp. Thus the silhouette of the scarp rim runs perfectly straight for kilometers. . . . Yet, *like the Hassberge plain, it is composed of very different lithic units.* These range from the Heldburg stage (Middle Keuper) to the Angulaten Sandstone of the Lias-alpha-two; yet neither the scarp nor the summit plain changes in altitude." Of the Hassberge surface adjoining this to the E at a height of around 500 m, Späth says (1973, p. 30): "besides other features of size and form, it cuts flush across geologic strata of differing hardnesses (from the Burg Sandstone to the Lias, inclusive, a distance corresponding to over 100 m in the stratigraphic sequence), cutting without a trace across faults with nearly 200 m displacement." Thus the Keuperland is another case presenting telling evidence against the classical structural escarpment theory. When segments of the etchplain surface are isolated by progressive Pleistocene dissection, and are attacked and lowered from all sides, erosion will, understandably enough, be retarded by the first harder stratum. This produces surfaces which are secondarily adapted to harder strata (this is usually peripheral, but may occur cryogenically from above). The resulting surface, which lies slightly below the former etchplain surface, may be called a "bedding-plane surface" (*Schichtfläche*), though even here the inheritance from the relict surface is involved (see footnote 56).

The Hassberge surface (around 500 m high) has been dated by Späth to the end of the Early Pliocene (see Fig. 71). The relict surfaces of the Spessart (460-480 m) and of the Franconian Alb (520-550 m) were also traditionally active until about this time. Even the 300 m high Gäu surface which was active into the Late Pliocene was inherited from this Franconian parent surface. East of the "Franconian Line" (the *Fränkische Linie*, an ancient boundary separating the crystalline and the metamorphic rocks to the E from the younger, subhorizontal sediments to the W), renewed uplift seems to have raised a surface along the W edge of the Frankenwald (560-600 m). This surface cuts smoothly across the crystalline infrastructure (metamorphosed and heavily folded in Variskian times) and may have been inactivated somewhat earlier than the Hassberge and the Franconian Alb. Thauer (1954) has reinvestigated the etchplain stairway (studied by W. Penck, 1924, as a piedmont staircase) in the Frankenwald and in the Fichtelgebirge (see Section 3.2.1.1 above).

The importance of Ice Age periglacial slope erosion in Franconia, as in all the mid-latitudes, is well known. We will mention here only a few aspects of this which have come to light since the work in Spitzbergen. At the INQUA Congress in Vienna in 1936, among all the papers on glacial phenomena, only one presentation concerned itself with periglacial research (Büdel, 1936).

Since then knowledge in this area has broadened considerably, a few of the main workers including Washburn (1973) in America, Journaux (1972, 1974) and Tricart (1963) in France, Pissart and Macar (1968) in Belgium, Fink (1973) in Austria, Furrer (1954) and Bachmann-Vögelin (1966) in Switzerland, Sekyra (1960), and Demek and Kukla (1969) in Czechoslovakia, the Biuletyn Peryglacialny in Poland, and Gerasimov and Velitschko (1969) in Russia. The most recent discoveries on the type and manner of these processes in Franconia, particularly of those in the clayey Keuper, are shown in Figs. 69 and 70. Figure 70 is a profile drawn from an actual exposure, while Fig. 69 is an idealized profile with the unconformity accentuated by a heavy line. Equivalent layers in both profiles bear the same numbers.

Zones 1 and 2 consists of Holocene soil, and along with zone 3, represent the upper part of the solifluction sheet, which formerly moved turbulently downslope. These zones contain large clasts, usually pointing slightly upward

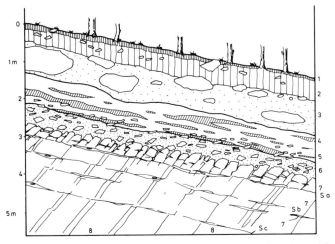

Fig. 69. Idealized section of a solifluction sheet on a gentle slope (12-15°) in the periglacial region of Central Europe. (1) Holocene soil, horizon A. (2) Holocene soil, horizon B. (3) Turbulently moved upper level of solifluction sheet, not affected by Holocene soil development. Fine material matrix with subangular fragments, usually allochthonous, oriented with the slope. Enrichment of coarser matter below. (4) Shear plane movement along the bottom of the solifluction sheet. Mostly fine material of various colors and grain sizes, thinly stratified parallel to slope. Formerly referred to as "pseudostratification." (5) At base of 4, zone of coarser, angular to subangular fragments deriving almost exclusively from the bedrock. During the Glacial Maximum this was moved along with zone 4. Upper zone of "barbs" or "downslope tipping" (Hakenschlagen) according to earlier authors. (6) Angular fragments of bedrock (no allochthonous matter) moved a few dm to at most one m, largely during the Late Glacial upon melting of the permafrost. This is the compressed remains of the former ice rind. Lower zone of "barbs" of earlier authors. (7) Bedrock with scattered shear plane movement (Sa, Sb, Sc) where episodic solifluction during the Late Glacial melting of the permafrost caused movements of a few cm. Deepest traces of such movement at four to five m depth. Zone of ice wedges, particularly in unconsolidated substrates. (8) Unaltered bedrock.

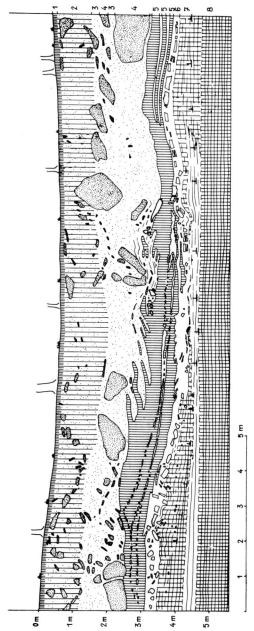

Fig. 70. Profile of a solifluction sheet on a 20° slope on marly Keuper. Lehrberg stratum on the W slope of the Schwanberg in the Steigerwald. View looking upslope, 435 m altitude. (Drawing by Büdel, August 1962.) (1) Holocene soil, A horizon. (2) Holocene soil, B horizon. (3) Turbated upper section of the solifluction sheet with many allochthonous fragments from upslope. (4) Fine-grained matrix of (3), containing floating boulders of Blasen Sandstone, transported for distances of 50-150 m. (5) Shear planes of stratified movement in lower part of solifluction sheet, consisting of alternating layers of (3) and (6) ("pseudostratification"). (6) Clayey basal layer, containing many medium-coarse fragments (usually transported only a few dm), oriented approximately parallel to the slope. This layer is the compressed and slightly displaced remains of the former ice rind (barbs and barely indicated in this profile). (7) Zone of episodic solifluction. Scattered shear surfaces in the barely disturbed strata of the marl bedrock. These formed with the Late Glacial disappearance of the ice rind and permafrost. Upside down T's show the lowest level at which slight movement along the shear surface is evident. The movement involved only a few cm.

and oriented with the slope. Much of the material here comes from higher upslope, the larger blocks having sunk with gravity toward the bottom of zone 3. Any slope wash was solifluidally mixed in. Zones 4 and 5 have usually been called the zones of pseudostratification: they are not sheets of slope wash, but traces of movement along shear planes (compare Fig. 32 from Spitzbergen). The material in this zone is almost exclusively derived from the bedrock and is often layered in sheets of overthrust material. Below this stratum is a marked unconformity, beneath which lies the barbed zone of downward tipping (*Hakenschlagen*), zone 6, containing what remained of fine and coarse material when the ice rind melted away. The barbs occasionally run up into the zone of shear movement (4 and 5) above it. Below is the bedrock, whose upper portion (zone 7) is slightly dislocated along the shear planes Sa, Sb, and Sc. The cracks here formed during the late glacial when the permafrost melted (Büdel, 1959), releasing the tensions which had slowly collected over the tens of thousands of years of permafrost locking. The lowest shear plane showing traces of such tearing and dislocation is frequently found at depths of four to five m in Franconia (compare Figs. 28, 29, and 32, as well as Photo. 5 taken from the Ochsenfurter Gäu).

3.3.1.5 The Relief Generations of the Franconian Alb

The Franconian Alb is as complex as it is informative, and deserves special consideration. Investigations on it are still in progress.

The Franconian Alb (the *Frankenalb*) is lower than the Swabian Alb (the *Schwabenalb*), but it is crowned by about a dozen inselbergs reaching heights of 600-657 m, the highest points on the Franconian plateau. The overall tectonic structure is the reverse of the topographic structure. The Mesozoic beds rise gently in the low "Franconian Arch" (*Fränkischer Sattel*), whose crest runs along an axis from the Spessart through Würzburg to Weissenburg (see Figs. 68 and 71), then plunges to the SE. The tectonic Franconian trough runs along a NE-SW line from the Grabfeld depression to Regensburg, its greatest area of downwarping lying directly under the N Franconian Alb. It is here, therefore, that the youngest members in the stratigraphic sequence are preserved, including the Malm and the lower section of the Upper Cretaceous (Cenomanian, Turonian, Coniac, and Santonian). The two wings of the Franconian Alb (the N wing running from the Upper Main to the Altmühl Valley and the S or Danubian wing running along the Danube from the Altmühl Valley to Ries) form an arc open toward the NW and embracing the Franconian Arch. The Cretaceous is found only in the tectonically lowest portion of the N wing, which, however, is topographically the highest. Further E in the region of the Upper Franconian-Upper Palatinate trough (*Oberfränkische-Oberpfälzische Senke*), older Mesozoic sediments (Bunter Sandstone, Muschelkalk, and Keuper) are heavily faulted and bent sharply upward along the Franconian Line at the edge of the basement. A small arch, formed of

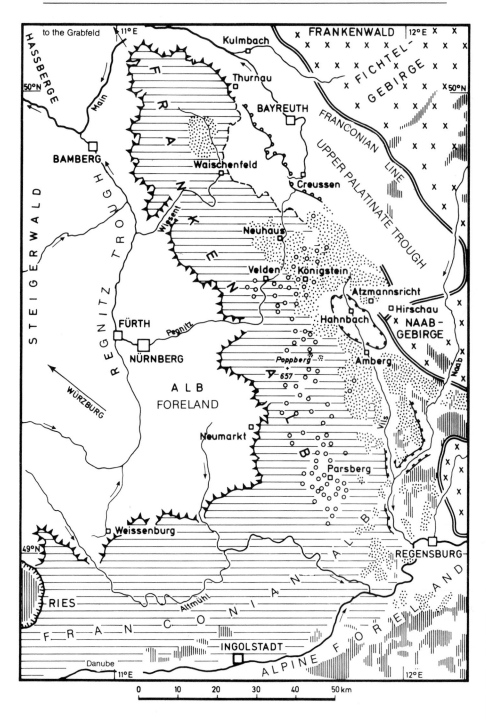

Fig. 71. The margins of the N Franconian Alb. The uplifted Malm plateau of the Franconian Alb is bounded by a steep escarpment (1) of resistant limestone overlying soft Dogger beds. It has generally been argued that the resistant limestone must have *generated* this escarpment. On the plateau's NE flank S of Thurnau, however, the escarpment disappears completely, without significant lithologic change, the plateau surface cutting almost imperceptibly (3) across the Malm/Dogger boundary. A gently rising escarpment (4) is found in the Dogger. This proves that both in the W and NE, etchplain escarpments were created under an ancient tropicoid climate. These have been accentuated in some places where brief retreat toward the uplifted plateau has sharpened a hard outcropping stratum. Such conditions are often met with in the tropics today. During the Pleistocene, S German escarpments like those of the Franconian Alb experienced only minor displacement and accentuation.

the crystalline Naabgebirge and the Keuper Hahnbach Arch, runs at right angles into the E flank of the N Franconian Alb.

This covers the older tectonic structures up to the Cretaceous-Tertiary boundary. The Danubian wing of the Alb sinks southwards, plunging below the Tertiary sediments of the Alpine forelands. Oligocene to Miocene marine and freshwater sediments overlap the S edge of the Alb, and run up into former valleys such as that N of Regensburg (see Section 3.3.1 above).

Looking from the Regnitz trough in the W, the N wing of the Franconian Alb looks just as structurally controlled as does the slightly higher Swabian Alb when viewed from the N. There are, however, four major differences. The first we have already mentioned, namely, that etchplains of almost the same height as that of the Alb surface occur to the W in the Keuperland (the Steigerwald and the Hassberge) and at a slightly higher altitude to the E in the Frankenwald.

The second difference is perhaps more striking. Along the W side of the N wing, the Dogger crops out (as along the edge of the Swabian Alb) as a soft bench-forming layer far below the cliff-forming Malm Limestone. On the E side, however, the surface of the Alb passes in many places without a break from the Malm to the Dogger. Along the W and N sides of the Alb edge, the Malm scarp maintains a height of ± 550 m, while the bench of the Dogger below is usually at a height of around 350 m. Local watershed divides always run along the Malm scarp. The morphologic behavior of these two stratigraphic units changes radically slightly S of Thurnau (see Fig. 71), though no geological change whatever occurs. This extraordinary behavior, deviating so greatly from the classical structural escarpment theory, struck Brunnacker (oral communication) even while mapping the geology on the topographic sheet of Mistelgau, 1:25,000. Just S of Thurnau the Dogger scarp rises to the same height as the Malm scarp. At the Vogelherd (11 km W of Bayreuth) the Malm scarp has disappeared. A plateau here measuring 300 by 800 m contains a barely visible depression 11 m wide separating the Malm on the W of the plateau from the Dogger on the E. Both units are at exactly 574 m relative altitude. Here at least they occur side by side. South of the Vogelherd, the Malm scarp moves four km westwards, while the Dogger scarp jumps further E. Eleven km S of Bayreuth, the two scarps are 15 km apart. The Malm scarp, now barely noticeable, rarely exceeds 500 m here, while the Dogger, which has now become the actual E edge of the Alb, is 600 m high (Büdel, 1957d, pp. 21-22).

The third major difference is that the Alb summit plain crosses at a uniform height in the center of the N wing from the Malm onto the Upper Cretaceous, proving that it is a gently rolling etchplain or system of overlapping etchplains.

The fourth major difference from the Swabian Alb deserves closer consideration. As in the Swabian Alb, one may distinguish between the smooth summit plain of the plateau alb, and the locally very energetic relief of the Kuppenalb. The latter has been interpreted by Höhl (1963) as a relict karst landscape from warm humid times. This is the same conclusion to which I came in the Swabian Alb. But the Franconian distribution of the Plateau Alb and the Kuppenalb is different from the Swabian. The plateau relief in Franconia occurs mainly in the Danubian wing, being largely replaced in the N wing by a more energetic Kuppenalb. The most striking example of a Kuppenalb runs S from the "Hersbrucker Schweiz" on the Upper Pegnitz River in the center of the N Alb through the center of the N Alb W of Amberg via the Poppberg (whose peak, at 657 m, forms the highest point in the Franconian Alb) S to the equally vigorous relief around Parsberg. The astounding thing, however, is that the Late Cretaceous sea flooded the entire N wing of the Franconian Alb from the NE, its shallow sea deposits twining intricately through the small basins between the knolls. The relief bears erosional traces both above and below this Cretaceous fill (see Fig. 72 below). It is therefore older than the Cretaceous transgression, and is *the oldest relief generation in Central Europe whose overall form is still present on the land surface.*

These conclusions are the more notable, as they help explain the manner in which this relict surface was formed. The closely set rounded knolls, some of which are quite steep (steeper than those of the Swabian Alb) have a diameter of 300-600 m at the base, and rise up to 200 m above the surrounding depressions of Malm Limestone (see Fig. 72). As in Swabia, many of these knolls are associated with sponge reefs. But again, just as with the inselbergs of the tropical etchplains, sponge reefs are also found here which do not crop out as knolls. As in Swabia, we find that on the Plateau Alb they tend to be dolomitized and well weathered out on the valley slopes, while morphologically scarcely visible on the crowning relict surface. Just as mogotes in tropical karst topography are by no means always tied to particular variations

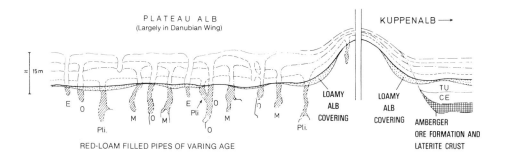

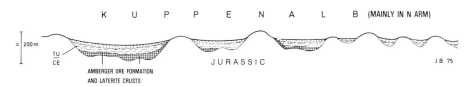

Fig. 72. Sketch of the relief generations in the Franconian Alb. Upper Profile: The *Plateau Alb*, covering the Danubian wing of the Franconian Alb and neighboring portions of the Swabian Alb, was investigated by Dehm (1960, 1961b), who found nearly 4000 red-loam filled karst pipes having a very uniform depth of 7-15 m. Two hundred of these contain rich Tertiary faunal remains (largely mammals), the fossils in each pipe being highly period-specific. The pipe contents can thus be dated to relatively brief stages of the Eocene and Early Oligocene (E), three further stages of the Oligocene (O), and to the Miocene (M), and Pliocene (Pli). A total of fifteen stages of Tertiary pipe development were identified. All are similar in depth and contents, and indicate similar climatic conditions. Other than a few concavities which were later filled in and reexposed, all start at about the same height. This proves that the surface has been lowered parallel to itself by only a few m (at most 10-15 m) under uniform climatic conditions (with varying humidity) since Eocene times. Lower Profile: The *Kuppenalb*, largely distributed in the N wing of the Franconian Alb, was completely covered by the Late Cretaceous Sea, whose sediments still cover the narrow depressions between the knolls. Below these lie the Amberg Ore Formation of the Lower Cenomanian (CE), a thick, humid tropical soil containing rich lateritic and iron crusts. This is overlain by younger Late Cretaceous sediments, largely Turonian (TU). This shows that the Kuppenalb formed subaerially in partial adaptation to sponge reefs under humid tropical conditions before the Late Cretaceous transgression. The reexposed upper portions of the knolls still dominate the relief today. *The Kuppenalb is the oldest relief generation of Central Europe which still dominates the landscape today.*

of limestone, and may be surrounded by depressions filled with thick kaolinitic latosol soils, so in Franconia, the depressions on the Jurassic limestone beneath the Upper Cretaceous are filled with thick kaolinitic weathering sheets. In places these become thick enough to form the Amberg Ore Formation, which contains enough high-grade iron to be mined commercially (see Fig. 72). Particularly rich deposits of this are found N and S of Auerbach, as well as in the elongated paleo-basin of Königstein. This basin, six km long and one km wide, could be regarded either as a pre-Late Cretaceous polje or as an intramontane basin. Reasons for considering the two concavities to be closely related will be discussed in Section 3.4.3.

The laterite sheet of the Amberg Ore Formation fills only the bottoms of the basins and the highly intertwining system of depressions running between the knolls. Associated weathering sheets run beneath the Cretaceous cover up the sides of the hills as far as their protecting cover. The knolls are usually steeper below, where covered, than above ground (see Fig. 72), due to Pleistocene cryogenic processes. The Pleistocene "loamy alb covering" (*lehmige Albüberdeckung*), which covers the Cretaceous fill of the depressions in thicknesses of a few dm to over 10 m, contains many interbedded paleosols, which, like the paleosols in loess, can be dated to several glacials.

Where this loamy alb covering is absent, extensions of basal latosol weathering can be seen running right up the knolls. Often this sheet consists of bright yellow ochre, which until recently was dug up in small scattered pits for use as pigment in wall paint. Along the northern edge of the Naabgebirge N of Hirschau, the coarse-grained Cretaceous sandstone overlies the fine-grained Dogger Sandstone in a continuous sheet. In large exposures N of Atzmannsricht, along the boundary of the two, are stratified sheets of hard laterite up to 40 cm thick. Journaux and Pécsi (oral communication during a field trip in the Franconian Alb on July 10, 1974) interpret this as a "*cuirasse ferrugineux*," created by lateral supply of dissolved iron in a swale of the tropical paleo-relief. A short way S of these sheets, near Hirschau, the Bunter Sandstone cropping out in front of the Naabgebirge is kaolinized from the level of the same relict surface downwards to great depth. The kaolin content rises eastward toward the higher crystalline mountains of the Oberpfälzer Wald, where it is industrially extracted from the sand in great quantities. The purified quartz sands are piled up on the so-called "Monte Kaolino," which is now used as a summer ski resort, for the sand is as loose as powdered snow. It would be a very rewarding task in the future to examine systematically all the variants of this thick, widely spread, pre-Cenomanian tropical pedosphere and decomposition sphere. Such a study might provide insights on whether the tropical soil had a different composition before the angiosperm conquest of the land. At present such a climato-genetic approach to soil science is in its infancy, although some work points in this direction (see Brunnacker, 1970; Heine, 1972; Pfeffer, 1968).

The Plateau Alb on the Danubian wing is not protected by any sediment sheet, and the pre-Cenomanian relief has been planed off, if indeed it ever was hilly. By good fortune it is possible to follow its subsequent erosional history from the Eocene to the Early Pliocene step by step, and to show that this proceeded in planar fashion. The key is provided by paleontological remains found by Dehm (1961a and b). In the Danubian wing of the Franconian Alb and in neighboring parts of the Swabian Alb he has found nearly 200 karst pipes filled with tropical red loams and very precisely datable remains of Tertiary mammals. Most of these are small (shrews and small rodents), but some remains of larger mammals have been found (deer, hippopotamuses, and carnivores). Twelve pipes are of Eocene and Early Oligocene age, the majority derive from various Oligocene and Miocene stages, while the latest ones are from the Early Pliocene. The older pipes are often shallower than later ones (see the upper profile in Fig. 72). What is particularly striking is that no pipe contains a mixed fauna, except where an Oligocene pipe has accidentally been tapped laterally by a Pliocene one, as shown in Fig. 72 (eighth pipe from the left). Most of the pipes run only 7-15 m deep and end abruptly. There are also many thousands of narrower pipes filled with red loam but lacking fossil contents.

The explanation for this phenomenon is shown in the upper part of Fig. 72. Since Eocene times the relict surface has been worn down parallel to itself through countless stages, each forming a thick red loam blanket which reached down into the Malm Limestone in karst pipes. These pipes grew downwards only for the given, strikingly uniform distance, before they were washed full of impermeable red loam, and could no longer serve as karst pathways for the seeping water. Soon the water found new cracks nearby and formed itself new pipes, so that from the Eocene on through the Tertiary, the entire process repeated itself over and over again. This is why each fossil-bearing pipe contains faunal remains of only one geologically brief period. Dehm was able to distinguish around fifteen stages for the given period of time. We have here, then, a specialized form of karstification which remained extremely stable during this period of the tropicoid paleo-earth. The deep weathering sheets must have filled these pipes quite quickly, sealing them off. The sinkholes which presumably formed just over the pipes then became waterholes sought out by animals (mainly small mammals, but also some larger species), until such time as the seeping water found a new path downwards, usually somewhere nearby. With time the new path was enlarged, washed full of loam, and sealed off, whereupon it became a water hole for the animals of the next period. The fact that the pipes always reached a uniform depth before being sealed up is probably due to the repeated sequence of events in a uniform climate, rather than to a stable water table, although the two explanations are not mutually exclusive. Dehm (oral communication) has rightly pointed out that the Tertiary alb karstification, lacking any larger caves, cave systems, or dripstones, differs markedly from Quaternary alb karstification, which has very large branching cavern systems with dripstones. This again points to the great change in soil and relief-forming conditions which accompanied the transition from the relief generation of the Latest Pliocene to that of the Earliest Pleistocene (see Fig. 61).

Since the latest Pliocene pipes are never deeper than 15 m, and since the tips of one Eocene and several Early Oligocene pipes have been found, the Alb etchplains can have been lowered no more than 15 m between the Eocene and the Pliocene. The gentle Alb relief (of which considerable portions are preserved on the Malm Limestone interfluves between the Pleistocene valleys and dell systems) was therefore *lowered a maximum of 10-15 m*, and in some places can have been lowered *only a few meters*.

In at least two parts of the Danubian Alb (in the upper Regnitz and N of Regensburg), local upheaval, partly connected with marine regressions in the Alpine forelands, led to Mid-Miocene dissection. The resulting valleys were refilled in the Helvetian, Tortonian, and Sarmatian. Lignite mining in the neighborhood of Schwandorf (see Fig. 68) has laid the former valleys bare nearly in their entirety, for the last stages of flooding had made the upper valley reaches particularly rich in coal and peat. It was revealed that these steep-headed valleys were completely different in character from today's periglacially formed Alb valleys, but correspond fully to the steeply stepped tropical mountain valleys. This and the fossil contents of the lignites are yet more evidence of a warm and humid climate. The conditions portrayed in the upper profile in Fig. 72 furthermore show that once these valleys were filled up, the slow planation and/or traditional planation of the S Alb surface set in once more, and continued undisturbed into the Early Pliocene. As long as the valleys remained open, of course, pipes such as those described above could form in the valley itself. These, however, have not been depicted in the figure.

3.3.1.6 The E Sudetes and the Mühlviertel

Leaving the triangle of the SW German tableland, we return to the Bohemian basement, an area we have already mentioned in discussing the Erzgebirge and the Frankenwald. The Riesengebirge, with its high relict surfaces, has been studied by (among others) Büdel (1937, pp. 22-45). Based on the monograph by Hassinger (1914), Czudek (1971, 1973b) has recently published a thorough analysis of the Nizky Vesenik (the *Gesenke*). Here a broad ancient etchplain cuts across the Kulm strata (folded in Variskian times); though deeply dissected now, the old, slightly delled surface is well preserved. Between 500 and 560 m, spots of kaolinitic weathering 1-2 m thick may be found beneath the Ice Age solifluction sheets. These old weathering zones derive from the surface's last active phase, which, though dated by the marginal sediments to the Tortonian and Early Pliocene, must surely have started earlier. In one place, the Tortonian marine Tegel Clay, cored by Czudek to a depth of over 20 m, reaches up to a narrow 100 m broad valley to the level of the Gesenke, a sign that here, as N of Regensburg, the plateau was deeply dissected in tropical fashion in pre-Tortonian times. Subsequent flooding of the S portion of the plateau by the Tortonian Sea must have been very sudden, for the Tegel Clay lies directly on the Kulm strata of the valley floors, without any intervening shore facies. The remainder of the plateau then underwent at least traditional planation, governed by a new base level of erosion formed by the Tortonian Sea. This continued until Pleistocene dissection set in.

The Later Tertiary epeirogenic movements in these mountains cannot have exceeded a few hundred m, and were not nearly as marked as the concurrent subsidence in the surrounding basins (including the Olmütz Basin, a series of Moravian basins extending via Brunn into the Lower Austrian Weinviertel, and the Vienna Basin). An interior graben of the Vienna Basin has subsided by nearly 5000 m since the Helvetian-Tortonian transition, an interesting geophysical fact stressed as early as 1914 by Hassinger. We can elaborate on this today, now that Pécsi (1964) has found downwarping of up to 1000 m during the geologically brief Pleistocene in the Alföld and Wallachia. If this downwarping has occurred at a uniform rate, that would mean over one m per thousand years. One may ask whether subsidence in such basins should not be geophysically reinvestigated in the light of taphrogenesis of the large graben systems.

Another area of this region (which forms the largest block of the Central European basement), the *Upper Austrian Mühlviertel*, has been described by H. Fischer (1963-64). This plateau of nearly pure granite is broken up into three tilted blocks, whose steep sides face the N and whose gentle slopes fall to the S. Tectonically this forms a stairway with steep steps and gently rising treads. Fischer's work area covered the S part of this, running from the Danube and its swampy floodplain between Mauthausen and Grein (altitude 230-250 m) up to the watershed area, 50 km away in an isolated central upland just over 1000 m high. The Late Oligocene to Early Miocene Schlier Clays of the Alpine forelands in places lap over the lower block, running in the depressions between the two lower blocks. The tectonic structure is therefore pre-Oligocene.

On the S slope Fischer counted seven Danubian terraces reaching 400 m in height. These correspond in number and sequence to Fink's terraces (1973) in the Vienna Basin. The highest terrace, between 350 and 400 m, is dated by Fischer to the Earliest Pleistocene or Latest Pliocene. Above this the granite block is topped by an etchplain stairway, whose S edge is covered by both the Pleistocene terraces and the Schlier Clays. "Beneath the sheet of sediment, the basement is covered by weathering crusts which in places are over 20 m thick. These have not been preserved everywhere, for the pounding Early Tertiary Sea destroyed them in exposed places. In basins and hollows they were protected from the surf, and even preserved by sediments deposited over them. In such cases the weathering crusts have a kaolin horizon sometimes over 20 m thick. This gradually turns to kaolinized grus, which grades via loose and more solid grus into the granite itself. Since kaolinitic weathering takes place only in a warm humid tropical climate, we may conclude that this climate was present here during Miocene times'' (H. Fischer, 1963-64, p. 65). Thus all three major pre-Holocene mid-latitude relief generations are found here in their characteristic form.

3.3.1.7 The Shape and Age of the Dells, as Illustrated in the Neukirchener Feld

It has already been stated (Section 3.3.1.1) that the highest branches of the Pleistocene valleys run up onto the old etchplain in the form of gentle, usually dry, spoon-shaped valleys, often several km long. We have suggested that these swales or dells (*Dellen*, named by Schmitthenner, 1926) may join old wash depressions on the relict surface. This is particularly likely where, after running a long way at a very gentle gradient, they suddenly plunge into deep Pleistocene valleys. On the other hand, we were able to show in the Erzgebirge that the last phase in the development of the dells took place during the Würm Glacial. We now wish to investigate how and to what extent the Würm Glacial formed such dells independent of any spoon-shaped predecessor.

A suitable "natural test suite" for solving this problem consists of the dell systems on the Riss terraces, which provide a gently sloping setting of uniform, generally slightly indurated, material. Dell systems on these terraces must of course postdate the gravel platform. As an area of investigation, I chose the Riss platform of the Neukirchener Feld in the Alpine forelands S of Altötting and Mühldorf am Inn (Büdel, 1944). This area had the advantage that its dell system ran N to the Principal Lower Terrace, or PLT (the *Hauptniederterrasse*, or HNT) of the Inn, for the Lower Terraces along the Inn piedmont glacier have been extensively studied by Troll (1924).

The Neukirchener Feld is the largest Riss gravel platform in the Alpine foreland. From Kraiburg am Inn in the W to S of Altötting in the E, it measures 25 km, while from N to S it reaches a maximum breadth of 12 km. From the Riss moraines at the SW, the Feld slopes in a gentle curve NE and then E from an altitude of 492 m to 437 m, or at a gradient of 2.75‰. With its thick loess blanket over permeable Riss gravels, the Neukirchener Feld 40 years ago was almost solidly covered with wheat fields in a countryside otherwise devoted to pasturage, and its many breweries still make beer from wheat instead of barley. Rich single farmsteads are scattered around the villages of Ober-Neukirchen and Unter-Neukirchen. Today, however, the land is being increasingly used for pasturage.

An exceptionally striking system of parallel dells runs at a slight angle to the general slope directly north to the Inn PLT (Fig. 73). The dells, of which there are a good dozen, are mostly 6-12 km long, and cut through the entire platform. The two longest have a total length of 25 km and cut back through the older moraines to the Würm moraines of the Inn-Chiemsee glacier. They can have received but little water and gravel from the outwash plains of the latter, however, for the outwash plains have only been active a few thousand years, in the last phases of dell growth. The long dells furthermore have exactly the same shape as the shorter ones. They are all completely periglacial or autochthonous in origin.

In their lower courses the dells have symmetrical, broad-floored, box-shaped valleys running evenly out onto the PLT of the Inn River. In their upper courses, however, they show a very consistent asymmetry. The explanation which I found for this thirty years ago has been supported by Helbig (1965) as well as by experience in Spitzbergen and by renewed investigation of the Neukirchener Feld and of the Lower Bavarian Tertiary hill country in 1973-74.

Basically the W-facing valley slopes are steep, while the E-facing slopes are gentle (see Photo. 53). The steep slopes hardly ever face WSW, but commonly WNW, NW, and in a few cases even NNW. This excludes any significant involvement of sunlight as proposed by some authors. The main cause must have been wind-driven snow, the wind direction, as shown by the loess and dune distribution in Europe (Louis, 1928), and particularly in the W-E trough of the Alpine forelands, being clearly from the W and NW. The snow drifts collected in eddies to the lee of obstructions, or on the E-facing W slopes of our dells. Here it lay longest, causing the greatest saturation, solifluction, and slope erosion in summer. Most importantly, greater solifluction and slope wash on these slopes forced the streams in the valley floor over to the E. The W-facing E slope was thus steepened, remained dry, and resisted solifluction. The birth of the dells was also aided by a third, hitherto ignored factor. The dells run N to NNE toward the Inn River (cut down during the Würm) at an angle to the original ENE slope of the Neukirchener Feld. This means that they cut across the flow of meltwater (as it flowed on the undissected Riss gravel fields) at an angle of 40-60°. The water supply of the dells (at a time when the substrate was locked in permafrost) therefore came largely from the W, so that the streams must have been forced eastwards from the start. We will see an even more marked effect of this kind when we consider the E Stryrian Grabenland.

Recent investigations from 1973-74 (in which I was vigorously aided, particularly in the drilling, by my colleagues Späth, U. Glaser, and Busche) elaborated on this. The most important new insight was that the dell system was not entirely a product of the Würm Maximum, but developed in three phases (see Fig. 74). Of the two Riss stages shown by other remains in the N Alpine foreland (Graul, 1952; Weidenbach *et al.*, 1950), we found that the gravel platform of the Neukirchener Feld dates exclusively from the first

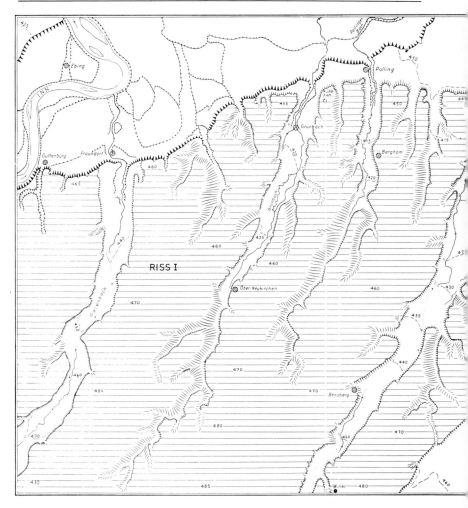

Fig. 73. Autochthonous valleys on the Riss-I gravel platform of the Neukirchener Feld, formed during the Riss-II and the Early Würm. Northern Alpine foreland, E of Munich. Arrows indicate sites of the cross-sections shown in Figs. 71-74.

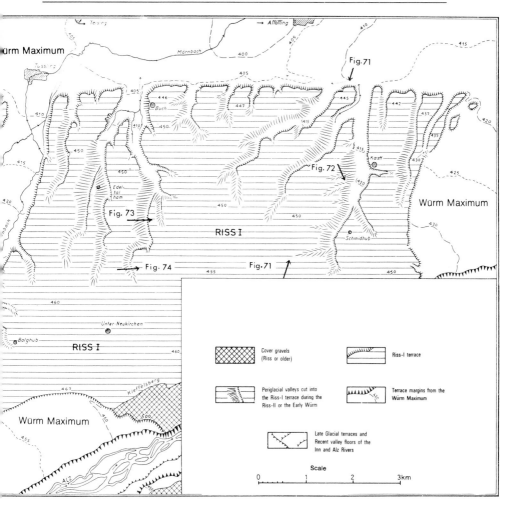

Cover gravels
(Riss or older)

Riss-I terrace

Periglacial valleys cut into
the Riss-I terrace during the
Riss-II or the Early Würm

Terrace margins from the
Würm Maximum

Late Glacial terraces and
Recent valley floors of the
Inn and Alz Rivers

Scale

0   1   2   3km

53. An asymmetrical dry dell in the Riss-I gravel platform of the Neukirchener Feld, formed in three phases (Riss-II, Early Würm, and Würm Maximum); see Figs. 70-74. (Photo: Büdel, July 1962.)

(roughly equivalent to the N German Saale). The first shallow dells developed in Riss-II times (roughly equivalent to the N German Warthe) and were then covered with an Eemian interglacial soil, no further incision taking place during the warm period. The Eemian paleosol, a tough, clayey red layer called *"Pechlehm"* or *"Brenner"* by the farmers, frequently crops out near the upper edges of today's dells (see Figs. 75, 76, and 77, drawn on the basis of numerous corings). The presumed cross-section of the Riss-II dells is shown in all three figures by line 7. A very important point is that these incipient dells already extended across the entire width of the gravel platform, shown by the fact that the Pechlehm or Brenner is found even in the uppermost branches of the dells (see Fig. 77). Even this first system of dells was not formed by headward erosion from the N (and in any case the Riss-II Inn was not nearly so deeply incised as today's Inn, see Fig. 74), but was created uniformly across the entire surface by vertical downworking at a time when the entire substrate was locked in permafrost.

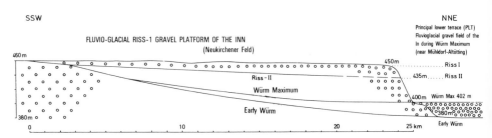

Fig. 74. Dissection of the Riss-I gravel platform of the Inn in three generations of cold stage periglacial valley-cutting. (N Alpine foreland E of Munich.)

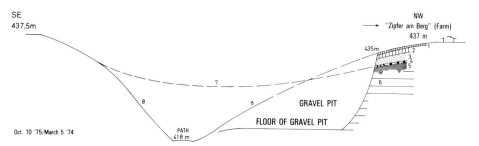

Fig. 75. Cross-section of an asymmetrical dell on the Neukirchener Feld near the "Zipfer am Berg" farm (3.5 km S of Altötting). (1) Ap, plow layer. (2) Brown forest soil. (3) Reworked loess. (4) Solifluidal transition from 3 to 5. (5) "Pechlehm" or Eemian soil, with gravel, reaching lower in pockets. (6) Riss-I gravels. (7) Hypothetical cross-section of dell during Riss-II. (8) Asymmetrical dell profile from the Würm Maximum, nearly unchanged today. (Drawn by Büdel, Späth, and U. Glaser.)

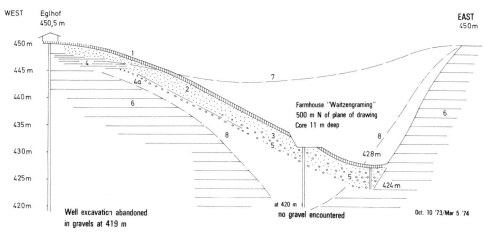

Fig. 76. Profile of the middle stretch of the Kiefering dry valley near Eglhof (a farm 5.5 km SW of Altötting), based on two deep cores and eight 4-m cores. View downvalley, toward the N. The last phase of development of today's asymmetrical valley occurred in the Würm Maximum. (1) Holocene soil. (2) Pure loess. (3) Solifluidally transported loess with gravel. (4) "Pechlehm" or Eemian soil, *in situ*. (4a) Solifluidally transported pechlehm. (5) Solifluction mass of gravels with loess, the latter partly calcified, partly sandy. (6) Riss-I gravel. (7) Hypothetical dell profile in Riss-II times. (8) Approximate dell profile during the Early Würm. (Model by Büdel, Späth, U. Glaser, and Busche.)

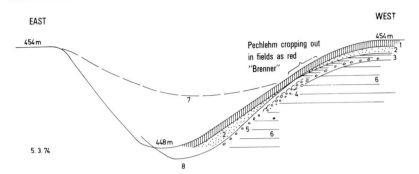

Fig. 77. Gentle slope of an asymmetrical valley in the upper Kiefering dry valley, 6.5 km SW of Altötting. View looking S. Last phase of development occurred in the Würm Maximum. (I) Ap, base of plow layer. (2) Loess and reworked loess, partly flecked with red. (3) "Pechlehm" (Eemian soil). (4) Solifluidally transported pechlehm with gravel. (5) Solifluction sheet of reworked loess with gravel from 6. (6) Riss-I gravel. (7) Hypothetical dell profile during Riss-II times. (8) Approximate dell profile in the Early Würm. (Drawn by Büdel, U. Glaser, and Busche.)

The major entrenchment of the dell system occurred in the cold and wet Early Würm Glacial. With the help of the ice rind effect, the Inn (which at that time was largely periglacial, for the Alpine glaciers did not pour onto the forelands until late in the Würm Maximum, as we will show in Section 3.3.2) cut down 20 m below today's level in the neighborhood of Altötting (down to 380 m, while the PLT is at 400-402 m; see Fig. 74). The dells simultaneously cut down to the same level, a process which, again, occurred "from above downwards" even in the highest branches, some of which cut through the entire Riss moraine system. At this time the Würm foreland glaciers had not yet pushed forward into the upper courses of these dells. The asymmetry of the dells developed due to heavy Early Würm solifluction and slope wash processes, but vertical down-cutting still predominated.

This was followed by the third phase during the Würm Maximum. The glaciers advancing from the S reached the upper courses of the longest dells and formed small outwash plains there. The Inn Valley to the N was filled by its heavily aggrading fluvio-glacial river, deposits S of Altötting reaching a thickness of at least 20 m (380-402 m; see Fig. 74). These deposits, which now form the PLT of the Inn, reach into the lower stretches of the dells, giving them their symmetrical box-shaped profile. The dell floors run evenly out onto the PLT of the Inn, but if one were to walk out toward the middle of the PLT, one would notice a rise of about two m. This illustrates how the elongated fluvio-glacial gravel fans of the Inn PLT near the moraines choked the lower courses of the dells. At the same time it shows another fact, which we recognized as early as 1940 in the Adelegg[58] (unpublished), namely, that

[58] The Adelegg (whose Schwarzer Grat is 1126 m high) is a small molasse upland in the Allgäu (SW Alpine forelands) which was nearly surrounded during the Würm by the Rhine glacier on

the periglacial valleys began cutting in the Early Würm and carved below the level of the later fluvio-glacial outwash plains and gravel fields of the Würm Maximum. Deposition and raising of valley floors during the Würm Maximum reached from the Inn River several km upstream along the dells of the Neukirchener Feld, favoring lateral erosion. In the middle stretches of the dells, greater water supply from the W caused undercutting of the W-facing slopes. This was what perfected the asymmetrical picture, best shown by the corings sketched in Figs. 76 and 77.

Further proof that the dell system had been completely laid out prior to the Early Würm is found in the lower right corner of Fig. 76. Here the fluvioglacial PLT of the Alz River has eaten a large loop into the Neukirchener gravels, and cuts across a dell forming the upper branch of the periglacial Mörnbachtal. This convenient "natural test site" shows that the PLT of the Alz found the entire dell system in completely developed form.

The extreme lopsidedness of these valleys was conserved by the loess of the Glacial Maximum, which covered the E-facing gentle slopes while leaving the W-facing slopes free. The latter therefore remained steep, while the re-worked loess on the E-facing slopes shows that cryogenic erosion was enhanced there.

On the whole, the analysis of the Neukirchener dells shows the general vigor of periglacial down-cutting during the cold stages, and particularly during the wet and cold Early Glacials. Dell and valley development ceased entirely during the interglacials (the Eem and the Holocene), when permafrost and ice rind effect, especially on this platform of permeable limestone gravels, disappeared.[59]

3.3.1.8 Dells and Ravines:
Local vs. Generalized (Climatically Controlled) Headward Erosion

The pattern of dell development on the Neukirchener Feld clearly shows that the wet, cold climate of the Early Würm Glacial greatly favored this autochthonous form of valley-cutting. This period was longer than has hitherto been assumed (as we will show in Section 3.3.2.2), lasting around 70,000 to 25,000

the W and the Iller glacier on the E. Running out from this semi-nunatak are deep periglacial valleys, whose lower courses were filled in during the Glacial Maximum by glacial outwash coming from both sides.

[59] In the Riss and Würm moraines of the Inn piedmont glacier, fragments from the Limestone Alps form 75% of the gravels, while the green Julier Granite so characteristic of the Upper Inn often forms less than 1‰ of these deposits. The fraction of Julier granite may rise to 5‰ in the Loisach forelands of the former Ammersee glacier (Büdel, 1957e, 1962), showing that much of the Engadine Inn ice came over the Fernpass and the Seefelder Sattel/Porta Claudia into the Alpine forelands during the Würm maximum, while a new Inn ice stream formed below Innsbruck in the Stubaier, Tuxer, and Zillertal Alps. It would be worthwhile to expand the investigation of the gravels in the areas of Alpine glaciation, a task which has long been completed in N Germany.

B.P., while the Glacial Maximum, characterized in the Alps by the last great glacial advances through the longitudinal valleys and out into the forelands, can have lasted only around 7,000 years (from around 25,000 to around 18,000 B.P.). It can be shown along the edge of the Scandinavian ice shield that in crossing what is now the German Baltic Coast, the glaciers of the Würm Maximum found a permafrost substrate already 200-300 m thick (chalk slabs of this thickness on the island of Rügen can only have been taken up and incorporated into the ground moraine in a solidly frozen condition). The permafrost must have developed gradually during the course of the Early Glacial, and so we can assume that in the Neukirchener Feld it only developed fully during the second half of the Early Glacial, and only then exerted its full influence on valley-cutting by the ice rind effect, and by blocking karst seepage in the limestone gravels (see footnote 59), increasing surface runoff, particularly from the snow banks. This in turn led to increased solifluction and slope wash on the E-facing slopes.

We further showed that these autochthonous valleys were already completely laid out in Riss-II times, probably even in an early stage of this. Their shape was fully preserved under the Eemian Pechlehm soil blanket. In the climate of the Early Würm Glacial, these valleys and their branches were scooped out back into the old moraines impinging to the S. This was permitted by concurrent down-cutting of the Inn River (which at this point was still largely periglacial), which lowered the local base of erosion. But the *active valley-cutting agent* consisted of the Early Glacial climate, as the shape of the valleys shows. We therefore cannot speak here of headward erosion undercutting and gradually wearing back a steep headwall; rather we must view this as a largely climatically controlled autonomous down-cutting, which occurred along the entire length of the valley, working more or less simultaneously from above downwards.

At any rate these valleys succeeded in cutting up the Neukirchener Feld in the Riss-II, and then in the Early Würm. During the Eemian and the Würm Maximum the valleys were stabilized, and their shape marked by the Pechlehm of the Eem (which developed under a forest), and by the loess of the Würm Maximum. From this we conclude that the down-cutting could not have taken place through mere lowering of the local base of erosion, had not a climatic change occurred at the same time.

This is proved along the undercut N edge of the field. To the W (W of Polling, shown in Fig. 77, upper left), Late Glacial erosional terraces in the PLT of the Inn adjoin the Neukirchener Feld. According to Troll (1924), these are not connected with the last end moraines from the Ebersberg and Ölkof stages, but formed after the ice had retreated back into the mountains. By this time the permafrost in the forelands must have largely or completely disappeared. The ice rind effect and other influences favoring periglacial valley-cutting no longer existed. Under these circumstances, what was the result of vigorous undercutting along the N edge of the Neukirchener Feld? The result was that very short, narrow, and steep ravines (*Tobel*) were produced, which tended to slit open the dell floors debouching onto the PLT, or sliced back beside them into the N edge of the gravel platform, cutting at most one km back into the gravels. They are shown in Fig. 73 at Polling, Grünbach, Fraundorf, and Guttenberg, and at four other spots between. I have deliberately added nothing to this figure, which I sketched in 1944, when this particular problem had not been investigated.

This convenient natural test site elucidates another, much more far-reaching set of problems. Even under uniform periglacial conditions, one can clearly distinguish between the very gradually formed, long, shallow dells, and the V-shaped ravines or gullies formed mainly by headward erosion at an area of energetic local undercutting. The dells are largely *climatically dependent*, while the ravines (or *Klingen*, knife-blades, as they are called in the Franconian dialect, due to their blade-like sharpness) are *bound to rivers*, particularly to areas of *local undercutting*. Figure 62 shows both types. The fact that two quite different kinds of valley-cutting are involved is shown by the fact that the ravines, cutting back from below, often (but by no means always) join up with the dells, which cut from above downwards. Many but not all of the dells empty via ravines into the main rivers of the Main, Tauber, Jagst, Kocher, and Neckar.

This difference can also be found in other climatic zones. Figure 10 and Photo. 29 show how, on the segmented plateau of Semién (in the Amhara Highlands of N Ethiopia), the shallow swales or wash depressions of the gently rolling relict relief are cut by the working rim, and run out into air. The wild gorges climbing up from the Blue Nile and segmenting the plateau have created this working rim by headward erosion of their upper course. Most of these steep ravines do not join swales at the working rim. Instead the rim completely interrupts erosion, despite the generally uniform tilt of slope; two genetically extremely different valley systems are here in conflict. Wirthmann has refined our understanding of this concept by his observations on Edge Island in SE Spitzbergen (1964), while it is Bremer (1967a) who first stressed the role of vertical erosion in discussing the dissection of the MacDonnells (subsequently uplifted) in Central Australia by older rivers such as the Finke.

According to these examples, purely mechanical headward erosion produced by undercutting is an exceptional type of valley-cutting. Most valleys are extremely old, and have a very complex developmental history involving several changes of climate and several relief generations. Valleys formed by headward erosion are very simple, easily visualized exceptions, and are hence the easiest to study in geomorphological laboratories. What can be observed here, however, is only one rare factor in valley-cutting, which is not representative of the majority of terrestrial valleys. This is shown particularly well by the fact that the very long and complicated developmental history of geomorphic phenomena, tied to highly complex mechanisms that change with each generation, cannot be reproduced in a corresponding period of time under the limited conditions of even the best-equipped laboratory (see Section 1.3 and Fig. 5).

3.3.2 THE RELIEF GENERATIONS OF THE ALPS

3.3.2.1 Basic Problems of the Pleistocene Glacial Effects in the Alps and in Scandinavia

Traces of the first and second major pre-glacial relief generations have been transmitted differently in the Alps than in non-Alpine Europe, for three important reasons. Firstly, the mountains have been greatly uplifted since around Helvetian times, so that the substages of the first major generation are vertically separated from one another. Secondly, the etchplains have been uplifted to different degrees, due to differential uplift in the various Alpine groups and to relative downwarping along the major longitudinal valleys (see Fig. 78).

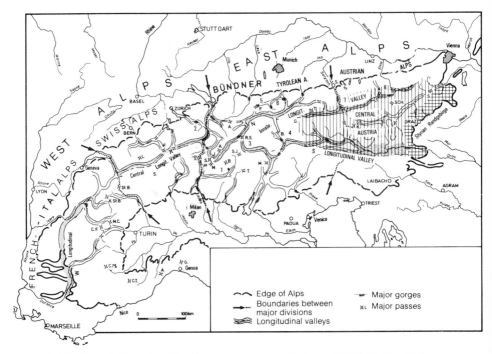

Fig. 78. Major divisions of the Alps. Valley passes mentioned in the text: (1) Maloja Pass. (2) Sargans Divide. (3) Reschen-Scheideck Pass. (4) Brenner Pass. (5) Fernpass. (6) Seefelder Sattel. (7) Zell am See. (8) Radstadt Divide. (Büdel, 1969a.)

Surfaces of the same age may therefore lie at different altitudes, while those of different ages may approximate each other in height. Thirdly, in the glaciated areas, the Ice Age and particularly the Würm glaciation were characterized by ice fields producing both thick, swift valley glaciers which scooped out valleys, and thin, slow, externally supercooled ice sheets covering the relict surfaces. These relict surfaces suffered relatively little attack from such ice sheets, and may even have been protected by them from the effects of the severe climate. The plateaus of the high limestone Alps (the Hochschwab, the Schneealpe, the Rax, and the Schneeberg) certainly had a few small cirque glaciers along their walls, but the lower areas (such as the Hohe Wand and the lower relict surfaces in the area of Vienna) did not have even these (Büdel, 1933).

The glacial history of the Alpine valleys varies greatly. The entire E half of the Austrian Alps as shown in Fig. 78 (i.e., the majority of the Lower Austrian and Styrian Alpine groups E of a line running from the Enns Valley S to the Lavant Valley in Carinthia) was not covered by any continuous Pleistocene ice cap. Here the Central Austrian Alpine groups below 2000 m morphologically resemble the Mittelgebirge, and have wholly or largely periglacial valleys (Büdel, 1933, 1944) which reach with stepless broad floors right into the heart of the mountain groups and form favorable areas for settlements and communication routes. Valley steps and lakes are lacking, but lacking also are the convenient ice-worn divides of the High Alps, such as the Moloja Pass, the low divide near Sargans, the Reschen-Scheideck Pass, the Brenner Pass, the Toblacher Feld, the Seefelder Sattel, the Fernpass, the low divide by Zell am See, and the low divide at Radstadt, all of which have served from time immemorial as the great trade routes across the Alps.

The glaciated valleys of the higher Alps developed completely differently. Besides grinding out glacial breaches between valleys, the glaciers also scooped out broad, straight, and locally deep troughs. Sometimes the deepest parts of these valleys still contain long lakes segmented by rocky barriers, but usually the bottoms of these great trough valleys were filled in after the Late Glacial ice retreat and became broad floors, onto which pour big alluvial fans from the tributary valleys. The Wallisian Rhone Valley, the Veltlin, the Alpine Rhine Valley from the Domleschg Mountains to Lake Constance, the Vinschgau, the Burggrafenamt and the Etsch Valley between Bolzano and Salerno in S Tyrol, the Inn between Landeck and Kufstein, and even the Loisach Valley between Garmisch and the Murnauer Moos are all examples of this. They are the longitudinal and transverse valleys that form the heart of the Inner Alpine passes, and it is here that the main states guarding these passes have been situated (e.g., Wallis, Graubünden or Grisons, Tyrol, Central Austria, and the tiny Werdenfelser Land).

Climbing into the higher Central Alpine groups, such as the Bern, Wallisian, and Tessino Alps, the Ötztal, or the Hohe Tauern, the landscape takes on the typical glacial morphology which characterizes the Scandinavian mountains in general, and the valleys plunging W into the Norwegian fjords in particular. True, even in this part of the Alps we may find valley sections with broad, sedimented trough floors. But these sections are divided by rocky barriers, by steps with narrow sawed-through gorges (*Klammen*), through many of which plunge great waterfalls. The tributary valley must in many cases be reached by clambering laboriously over steep discordant junctions. Proceeding up the valley one toils over one step after another, right up to the cirque staircases, often still occupied by small paternoster lakes.

The first Pleistocene glaciers certainly formed in a Late Tertiary valley relief of narrow, stepped, tropical mountain valleys. While in the periglacial regions permafrost and ice rinds destroyed these steps, broadened the valley floors, and evened out the longitudinal profiles, glacial overprinting did the opposite. As Louis (1952) showed through keen analysis of the physical principles involved, the thick, rapidly flowing ice streams were forced not only to broaden the cross-profiles of the narrow valleys, but also to accentuate the steps by scooping out the tread below each riser. The well-known glacial valley relief with its discordant junctions, confluent, diffluent, and other steps in the longitudinal profile, is not purely a glacial product. Glacial effects often only accentuated a pre-existing stepped structure. It will be shown in the third edition of this book that the pre-glacial parent form of intramontane basins still shows through in many Alpine basins.

In the Alps, glacial erosion ceased abruptly at the edge of the mountains. Only the largest piedmont glaciers scooped out their terminal basins some-what. Beyond this, deposition of glacial debris began, producing thick moraine sheets whose many shapes will be discussed in the next edition. In Scandinavia the great continental ice sheet did not begin to deposit along the E edge of the mountains in Central Sweden, but only in the extreme S, in Scania, Denmark, and on the German Baltic Sea coast. Between these lies a wide expanse of Late Tertiary etchplain, which was crossed and attacked in planar fashion by the ice, accentuating the inherited forms. The well-known roches moutonnées (*Rundhöcker*) where the bedrock frameworks of Sweden, Finland, and Canada were stripped and polished, derive from the basal knobs (*Grundhöcker*) of the basal weathering surface of the pre-Pleistocene etchplain (see Wilhelmy, 1971-72, Vol. III, p. 85, and Brochu, 1959). The ice removed the old soil blanket quickly and thoroughly, at the same time altering the basal knob relief and polishing it in the direction of the ice flow. Due to general chemical weathering, the original basal knobs were adapted to the bedrock joints to the same degree in all directions. The streaming ice, however, scraped unidirectionally. This is why the roches moutonnées and glacial lakes of Sweden are so highly oriented, and it is for this reason that all morphologically visible jointing structures correspond so strikingly with the much younger direction of ice flow.

3.3.2.2 Climate and Glaciation during the Würm.
The Position of the Glacial Maximum

We have previously emphasized (Büdel, 1960) that the discrepancy between the climatic fluctuations and the Alpine and Scandinavian glaciations grew with time. We also showed that, as they expanded, the great ice sheets became increasingly independent of the smaller climatic fluctuations. While the major glaciations may characterize the length and intensity of an Ice Age as a major climatic event, the smaller second- and third-rank fluctuations subordinate to this major event are best studied not by means of the glacier itself, but by the climatic effects around the glaciers, e.g., by loess stratification based on paleosols, by generations of ice wedge casts, and the like. The second-rank fluctuations at the end of an Ice Age and in the Holocene are easily read in the pollen stratigraphy. The onset and the maximum of an Ice Age must surely have been similarly subdivided.

In the last decades, investigations of periglacial phenomena, and especially of the Würm loess stratification (studied by horizons of wet soils, brown forest soils, reworked loesses, and the like), have led to a minute breakdown of the Würm climatic history. This has tended to outweigh considerations of the major glaciations and the first order climatic events which they represent. Several authors have interpreted the loess deposits to show 3 to 17 Würm interstadials, leaving the Würm no coherent unity at all. Woldstedt (1958, 1960, 1962) set up 16 (later 7) minor Würm glaciations, based on the loess stratigraphy. From this he then drew the conclusion (stated expressly in 1958, and implied later) that each of these minor climatic fluctuations recorded in the loess corresponded to an advance or a retreat in the Scandinavian and Alpine ice shields. In fact, however, these great ice sheets responded only to first order climatic fluctuations and were governed by their own rules of development, independent of the smaller fluctuations recorded in the loess. All attempts to correlate so-called "overrun moraines," "mid-Würm soils," and "retreats"[60] with the loess stratigraphy are easily disproved (Büdel, 1957e, 1962; Kaiser, 1963). It has been shown repeatedly that the attempt to correlate the development of the great ice caps with second or third order climatic events is misguided and useless.

Our interpretation arose out of discussions with Louis, which led to the realization that an ice mass the size of the Scandinavian Würm glacier takes an enormously long time to build up to its greatest size (Büdel, 1960). The latest estimates suggest that this might have taken 35,000 to 40,000 years. If one sets the beginning of the Würm Ice Age at 70,000 years B.P., then the

---

[60] This term is particularly misleading, for the ice cannot contract like a rubber band, but *always flows forward*, even flowing uphill if the ice sheet is thick enough. Ice flows forward even when its edge is melting back. At most the movement may come to a halt, turning the glacier into dead ice.

Scandinavian ice can only have achieved its maximum height at around 35,000 to 30,000 B.P. If we assume, as does Büdel (1949), that the snow line at the beginning of the Würm lay at around 500-600 m in the W part of S Scandinavia, and at around 700-800 m on the E side, then we must assume that at first only very small glaciers formed along both sides of the Scandinavian ridge, which at that time formed the divide between the ice masses (Stage I in Fig. 79). The western glaciers sped down through the steep U-shaped valleys into the fjords along the coast, and must have calved by the outer edge of the Norwegian "strandflat." The sluggish eastern glaciers must soon have filled the Baltic Sea basin (Stage II in Fig. 79), deepening it by their weight. They must then have streamed as an enormous Baltic Sea glacier

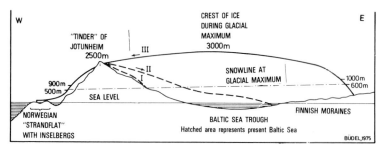

I, II & III = STAGES IN THE DEVELOPMENT OF
THE WÜRM SCANDINAVIAN ICE SHEET

Fig. 79. The Scandinavian continental ice sheet during the Würm. The ice sheet only reached its maximum development (with a crest at around 3000 m) late in the Würm Ice Age, forming its own gigantic accumulation zone high above snow line. Vertical displacement of snow line between 500 and 1000 m only played along the steep outer sides of this ice cap, and had little effect on the size of its catchment area. While second and third order climatic fluctuations had marked effects on the periglacial regions (resembling interstadials in some loess profiles), they had little effect on the area of the large continental ice caps.

down into Denmark and N Germany. By this time, the rest of the ice mass and the ice divide must have overlain the Gulf of Bothnia. From there the ice flowed between the "*tinder*" (the Norwegian term for nunataks) of the Scandian Mountains toward the W (Stage III in Fig. 79). We cannot rule out the possibility that the Scandinavian ice sheet may also have been fed (from about Stage II on) by another ice sheet which had formed independently over the Kola Peninsula, whose highest point lies at 1240 m.

As Meinardus (1926) has shown, such continental ice sheets present a standard appearance (shown in Stage III in Fig. 79), having steep sides and gently arching flat tops. By Stage III the snow line had presumably fallen lower, to 400-500 m in the W, and 600-700 m in the E. But much more important was the fact that the glacier was no longer dependent on scattered firn fields above snow line, for it had itself become an enormous catchment area, receiving hundreds of times more snow than before. It was this which must have first permitted the glacier to advance southwards, and even to push uphill across N Germany. But it is clear that this stage of the Würm Maximum can only have been reached quite late, and certainly not before a period between 40,000 and 30,000 B.P.

The most important consequence of this is that from the time when the ice shield grew above snow level, any fluctuations of 50, 100, 200, or even 400 m in the snow line (fluctuations which would be clearly recorded in the loess stratigraphy and other periglacial deposits) would scarcely affect the body of the continental ice; the snow line would only wander up or down the steep flanks of the ice body, barely affecting its firn field, and hardly limiting the size of its accumulation surface at all. The fluctuations could scarcely have affected the overall size of the glacier, for even major changes in the firn region affect the peripheries only after hundreds or even thousands of years. Such climatic fluctuations, however, would have led to a marked interstadial, recorded in the loess and plant cover. Correlation of minor climatic fluctuations with the size of the continental ice sheets is therefore completely uncalled-for. On the contrary, the ice sheets provide a very good overall curve for the climatic history of the Würm Glacial.

There can be no doubt that the Alpine ice reached its full development in the Glacial Maximum, after a long growth period in the Early Glacial. By the time it formed a continuous sheet over valleys and passes, the ice dome had reached a height of nearly 2600 m (at the Reschen-Scheideck and at the Brenner, according to A. Penck and Brückner, 1909), and was 1300-1400 m high even at the edge of the Alps near Murnau (Büdel, 1957e). Nearly the entire ice sheet was above snow line, which at that time lay at around 1300 m along the N edge of the Alps. During the entire Early Glacial, the firn fields of the individual Alpine groups cannot have been much larger than those of today. By the Glacial Maximum, when the glacier had grown to its greatest extent, its accumulation area had expanded tremendously. This is what permitted the great piedmont glaciers to spread out upon emerging from the mountains.

This model has been brilliantly verified by the investigations of Fliri and his colleagues (1970, 1971, 1972) on the Inn Valley terraces at Gnadenwald, downstream of Hall in Tyrol. Near the village of Baumkirchen, the Inn Valley lies at 580 m, flanked by the Gnadenwald terrace, which runs for nearly 10 km at a width of 1-1.5 m and a height of 830-880 m along the foot of the Karwendel Mountains. The surface of this terrace is covered by Würm Maximum moraine, overlain by scree. When climbing up from Baumkirchen one can see the base of the terrace exposed between 630 and 750 m (these are minimum values, the actual extent probably being greater both above and below), revealing a complete sequence of banded lacustrine clays. The rapid changes in these bands (about four layers per year) are probably not due to annual cycles, as would be expected if the lake were glacially fed, but to irregular storms during deposition. The enormous horizontal and vertical expanse of these clays, which are undisturbed and contain no shore facies, suggests that they belong to a very large Inn Valley Lake, to which other banded clays in the Inn Valley between Innsbruck and Jenbach may also belong. This lake did not lie in front of a glacier, nor was it dammed up by one, for in that case we would find the regularly alternating annual cycle of varves here. The ponding was probably produced by an enormous alluvial fan from the Zillertal, which was pushed into the Inn Valley near Jenbach by a local glacier ending in the inner Zillertal. The continuous deposition of at least 120 cm of clays shows that the situation remained stable over an extremely long period of time. This is further supported by two other pieces of evidence. For one thing, the Inn Valley Lake contained a rich fauna of large fish (up to one m in size) which must have fed on plants or small bottom-dwelling fauna. Fin imprints of these fish have been found in countless strata by Fliri. This excludes the possibility, suggested by some authors, that they could have been pikes or sculpins, which are not bottom feeders. It seems most likely to me that they were large coregonids (relatives of the salmon), which feed on benthic fauna. Coregonids (*Renken*) of about 80 cm in size still live more or less as faunal relics in the cold depths of the Alpine foreland lakes today. The clays also contain remnants of wood and pollen, mostly of pine (largely the dwarf pine, *Pinus montana*), but also of juniper, birch, seabuckthorn (*Hypophae rhamnoides*), and willow, as well as the herbaceous mountain avens (*Dryas octopetala*), grasses, and sedges. In short, the plant community was a wooded steppe and elfin forest, such as is found today at an altitude of 1800-1900 m. Radiocarbon dating undertaken by Fliri has yielded dates of 31,000 B.P. for the lowest exposed strata (at 648 m), and of 26,800 B.P. for the highest datable strata (at 681 m). Since these include neither the lowest nor the highest levels, one may assume that the lake existed between 32,000 and 25,000 B.P., that is, that conditions remained stable over a period of around 7,000 years. This lasted until the copious silt in the glacier milk from the Inn Valley glacier (which may have ended just above Innsbruck) caused the shallow lake to silt up.

Based on the standard conception of the Würm interstadials, derived partly from Milankovitch's curve and partly from the minor climatic fluctuations recorded in the loess, these clays are commonly regarded as the product of a Würm interstadial. But the glacial outwash which A. Penck thought he had found (1909) at the base of these clays has not been rediscovered, despite intensive search. Should it exist, it is far more likely to be a fluvially reworked Riss moraine. The idea of an interstadial equal in dimensions to a glacial period is contradicted, for one thing, by the accompanying wooded tundra flora (in the sheltered Inn Valley, of all places), and for another, by the stability of the glaciers, which were large enough to have advanced simultaneously into the Inn Valley and the Zillertal. Finally, it is highly unlikely that the Alpine glaciers could already have achieved full size before such a long period of melting (32,000 to 25,000 B.P.). *Certainly no evidence of such an occurrence has been found.* It is much more probable that the Central Alpine glaciers (the Engadine, Ötztal, and Zillertal glaciers) very gradually filled their valleys during the Early Glacial (the Inn Valley up to Finstermünz and Landeck, the Ötztal and Zillertal up their entrances into the Inn Valley). A slight warm phase at the end of the Early Glacial may then have interrupted glacial growth from 32,000 to 25,000, as suggested by the Inn Valley Lake clay deposits at Baumkirchen. Fliri himself pointed out that by 15,000 B.P. (when deposition began in the former Rosenheim Lake), the entire triple advance of the Würm Maximum into the outermost end moraines of the Ammersee, Isar, and Inn piedmont glaciers must have ended, and the ice must already have melted back to deep within the Alps. The main advances of the glacial maximum must have taken place within at most five to seven thousand years (around 25,000 to 18,000 B.P.). The suddenness with which this occurred is probably due far less to any particularly severe cold spell with sudden lowering of snow lines, than to the fact that the glaciers, having finally expanded to their fullest (for which only slight renewed cooling was necessary), now had an enormously enlarged accumulation area above snow line. With the full development of the glacier, this catchment area must have grown by twenty or thirty times within a very brief period. This permitted the dramatic advances of the ice tongues into the Alpine forelands. The extent to which the cross-valleys in the N Alps were used for this ice inundation is shown by the distribution of Julier granites in the young moraines, a feature which we have already discussed in footnote 59.

Should this theory be confirmed, then any parallels between general climatic history (as documented by periglacial features) and glacial stands can be assumed only for the first phases of the early glacials, for which no glacial traces can be seen in the field anyway. During the Glacial Maxima, the two phenomena (the major climatic event causing the growth of the ice and the smaller fluctuations during this event, recorded periglacially) become fully independent. At any rate, it is certain today that the three sets of end moraines found in the Alpine forelands (the Schaffhausen, Diesenhofen, and Stein-Singen stages of the Rhine glacier, and the Kirchseon, Ebersberg, and Ölkofen stages of the Inn glaciers, first separated by means of their outwash gravel trains by Troll, 1924), as well as the marginal stands of the Scandinavian ice, can in no way be correlated with any periglacial remains, and particularly not with those before around 20,000 B.P. The differing orders of magnitude between the major Ice Age event and the subordinate climatic fluctuations will be seen in Fig. 79. Finally, the larger the ice cap, the longer the delay between climatic influences on its accumulation zone and fluctuations at the distant margins.

3.3.2.3 The Relief Generations of
the Hohe Tauern and of the N Limestone Alps

There are two portions of the Alps in which both the relief inheritance from the first major relief generation and the glacial overprinting from the third have been well investigated, or, as in the Limestone Alps (the *Kalkalpen*), are clearly visible of themselves. In discussing the *Hohe Tauern* we will largely follow the very detailed work by Späth, "Die Grossformen des Glockner-gebietes" (1969), which I have followed before (Büdel, 1969a), and which contains very thorough discussions of the earlier literature, including Mo-rawetz (1930), Klimpt (1943), Seefeldner (1964, 1973), Pippan (1957, 1965), and Slupetzky (1968).

It was Späth who divided the first major relief generation here into finer subgenerations. This was facilitated by nature, for while the relict surface on the Franconian Alb was lowered only 15 m between the Eocene and the Early Pliocene, each older surface being inherited and incorporated into the next, around 20 to 30 times as much material was removed from the Glockner group during the approximately 15 million years between the Late Oligocene and the Helvetian. It is from the end of this period that the highest and oldest surface fragments in the Hohe Tauern derive, and these were inherited from earlier surfaces. We therefore speak of the "Augenstein surface sequence" (Fig. 80), and Späth referred to the oldest surface remains which he found in the Glockner group as inherited from the Augenstein surface.

At that time the region of the E Alps geomorphologically presented more or less the following picture: "broad etchplains covered the entire area, broken at most by shallow stairways toward their center. The core area, particularly in the region of today's Central Alps, was topped by scattered inselbergs and groups of inselbergs" (Büdel, 1969a, p. 25). The surfaces were covered by a thick latosol sheet which maintained planation with little regard to petro-variance, and which for the most part was able to counteract the slow uplift of the Alpine body. The low, elongated dome, later to become the Alps, was laved on all sides by lakes and seas. The groove of the N longitudinal valleys did not exist at that time. The area in which they would later appear and the body of the Limestone Alps (which at that time barely protruded above water) were showered with large quantities of fine matter from the Central Alpine ridge of the dome. During the Oligocene and the Early and Mid-Miocene, the seas and lakes filling the Alpine forelands received and deposited fine-grained schlier (sandy clay) and molasse. Very little coarse material came out of the mountain valleys of the Central Alps, though the uplift here was gathering speed. Pebbles from there are often found now as foreign "*Au-*

gensteine" on the plateaus of the N Limestone Alps. They were uplifted with the latter, and were often redeposited. It is after these stones that we have named the oldest surface fragments on the Central Alpine peaks the "Augenstein surface sequence."

Below the highest remains of the Augenstein relief (the first subgeneration), Späth found a second and third subgeneration of related character. The higher of these was named the "Firn Field Level" (*Firnfeldniveau*) by Creutzburg (1921). This includes, e.g., the rock surface of the upper Pasterzen soil. The lower of the two was named the "Shallow Cirque Level" (*Flachkarniveau*) by Seefeldner and Pippan (see Fig. 80). This includes the Piffkar N of the Edelweisspitze and the Fuschertörl. Both groups of forms are found in many other parts of the Glockner group and in the rest of the Hohe Tauern in the form of broad surface fragments. They cut smoothly across rock units of many different kinds, and are divided up by marked scarps of a former tropical mountain relief. Späth justifiably emphasized that even the lower portion of the Shallow Cirque Level in some places cut across present-day watershed divides, showing that they are *independent of the later valley network*. The higher Central Alps were thus basically an etchplain stairway. The scarps between the three subgenerations of planation surface were probably slashed by narrow stepped ravines just like those gashing scarps in the seasonal tropics today. But such steep relief elements are far less durable than flat ones.

During this development, the Glockner group was subject to uniform uplift. Though uplift was much less uniform in the Limestone Alps to the N and S, karstification and reduced glaciation there (the glaciation decreases as one goes E) helped preserve the relict surface far better than in the Central Alps. Of the S Alpine groups, the Julian Alps and the broad plains of the Ternovaner

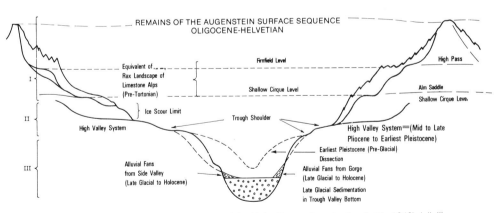

Fig. 80. Relief generations in a typical valley of the Hohe Tauern (largely after Späth, 1969). I, II, III, and IV are the areas of the major relief generations (see Figs. 61 and 62). I and II were glacially reworked during the Pleistocene Ice Age.

and Birnbaumer Wald leading down to the Adriatic are impressive examples of this kind, as is also the Triest Karst. In the NE Limestone Alps, a chain of high limestone massifs crowned with very characteristic flat relict surfaces leads from the Salzburg Alps to the edge of the Vienna Basin in the E. These massifs have steep outer precipices with marked working rims, at which the interruption of all erosion is particularly evident. They are crowned with an undulating, but in general very flat hilly relief (*Kuppenrelief*), reminiscent both of the tropical ridge relief and of certain types of tropical karst. This was given the well-known name of *Raxlandschaft* ("Rax landscape") by Lichtenecker (1926). The type remains quite standard throughout, and must have developed on all the limestone massifs of the N Alps uniformly and at relative proximity to the Early Pliocene seas and lakes (freshwater lakes still bathed the E feet of the mountains in the Early Pliocene). Subsequent uplift, however, has proceeded at varying rates. In the Steinernes Meer and on the Dachstein, the Rax relief reaches its greatest heights on peaks over 2500 m high. The Totes Gebirge and the Hochschwab reach barely over 2000 m, while the Rax and the Wiener Schneeberg are just under 2000 m. The easternmost limestone massifs have been further broken up near the recently downwarped Vienna Basin. On the Hohe Wand, the flat relief corresponding to the Rax is only at 1200 m, while in the flysch of the Wiener Wald near Vienna, the last, inconspicuous planation surfaces are barely 500 m high.

The relict surfaces of the Rax landscape, forming summit plains cutting smoothly across disturbed sedimentary sequences, are directly derived from the Augenstein surface sequence, and simply form a later and lower subgeneration. Without prolonged continued development of the very wide-flung Augenstein surfaces, the expansive surfaces of the Rax relief could never have formed. This is an important point, which is supported by the general distribution of the Augensteine on the limestone massifs, and even on some of their peaks.

Since the easternmost foot of the Rax plateau along the edge of the Vienna Basin was washed by the Tortonian sea and encircled by the latter's limestone reefs (the Leitha Limestones), the Rax landscape must have received its final formative touches in pre-Tortonian times, between the Helvetian and the Tortonian. This does not mean that in the more uplifted W portions of the Limestone Alps, flat forms of later age may not have been raised to heights of even 1000 m or more. Less uplifted portions of the Rax landscape may furthermore have been subjected to later traditional reworking. On the whole, the Rax landscape may correspond to the Firn Field Level of the Central Alps, while lower, reworked portions of the E Rax may even correspond to the Central Alpine Shallow Cirque Level.

In the following period, uplift was not only generally greater, but also much more differentiated. We have already mentioned the chopping up of the Rax relief. With the downfaulting of the Vienna Basin at the beginning of the

Tortonian, the N and S longitudinal valleys also became visible as less uplifted troughs, heralding the pleating of the ranges into the N Limestone Alps, the Central Alps, and the S Limestone Alps of today (Büdel, 1969a, pp. 27-29).

In the Glockner group Späth (again in partial agreement with earlier authors) distinguished a fourth subgeneration, which he called the "High Valley System" (*Hochtalsystem*; see Fig. 80, II). This subgeneration differs markedly from the previous subgenerations. It includes the high cirques (*Hochkare*) and high troughs (*Hochtröge*) in the interior of the Glockner group, and consists of the narrow flat surfaces of an ancient phase of dissection which seems to have been of tropical mountain relief type. "In contrast to the older relief generations, the High Valley System is completely tied to the Tauern valleys. Nowhere does it cut across watershed divides" (Späth, 1969, p. 132). We thus have in the glaciated mountains a relief feature corresponding to the Broad Valleys and Broad Terraces of the periglacial region. Similarly, the Hochtalsystem also forms most of those flat surfaces exetending out of the valleys, called trough shoulders in the glacial relief. While the remains of the Augenstein surface sequence cross the peaks, and while the Firn Field and the Shallow Cirque Level form the high passes and alm saddles cutting across today's watershed divides, the trough shoulders, part of the High Valley System, are invariably associated with the fluvial lines of today (see Fig. 80).

Further from the heart of the Hohe Tauern, the High Valley System opens out via the "Corner Surfaces" (*Eckfluren*) into wide relict plains resembling the triangular reentrants of the seasonal tropics. These cut into the higher surface relics of the older relief generations, eating into them until the law of diverging erosion came into effect. The stepped valleys which were the hinterland to these relict triangular reentrants have now been turned into cirque staircases. Glacial overprinting was of course particularly heavy in these relatively narrow and steep headwater valleys of the High Valley System.

It seems quite clear that the High Valley System is Pliocene, but whether it belongs to a middle or late stage of this epoch remains as yet unclear. The resemblance between the features along its upper edge and triangular reentrants suggests that the erosional processes were still seasonally tropic in character during its development. On the other hand, its adaptation to the later valley system, as in the Mittelgebirge, suggests a more recent age. But, as mentioned before, it seems quite possible that the narrowing down of the erosion lines could have set in fairly early in the greatly uplifted, central portions of the Alps, at a time when their less uplifted E margins were still subject to seasonally tropic or subtropic planation with deep chemical weathering. This is suggested where the E Central Alps fall off into central Burgenland. Here Fink (1961b, 1966) has found evidence of an etchplain stairway leading from the relict surface of the Hochwechsel (at around 1700 m) via surfaces at around 1000 and 800 m to a surface at around 600 m in central Burgenland (see Fig. 81). If one equates the relict surface of the Wechsel with the Rax landscape, then the younger surfaces must be of Sarmatian to early Late Pliocene age, making them chronologically equivalent to the Shallow Cirque Level and the High Valley System of the Hohe Tauern. Even the lowest of these surfaces, formed on old crystallines and basalts, are covered with red loam relicts containing montmorillonite and kaolinite, or with the corresponding grus zones. Similar red loam sheets have been described by Winkler-Hermaden (1957) on the Late Pliocene basalt sheets of E Styria. Even the lower of these surfaces (at 600 m) may therefore be regarded at least as an inherited etchplain. It is clear that this must have preceded the later planation of the second major relief generation, which we will discuss in the next section.

In the interior Alpine groups (e.g., in the area of the Hohe Tauern), the Hochtalsystem with its trough shoulders formed the parent surface for the pre-glacial dissection (probably of Earliest Pleistocene age), recognized as early as 1909 by A. Penck and Brückner as the fluvial predecessor of the great U-shaped valleys. These first valleys were probably narrow and steeply stepped (although they were widened out by the ice streams, their stepped character became greatly accentuated), and were probably formed along the lines of tropical to subtropical mountain relief. In the third relief generation, the great glaciers of the Pleistocene ice fields then completed the work of shaping this very variegated glacial relief.

3.3.2.4 The Relief Generations of the E Alpine Margin

The second major relief generation in the development of the mountains, that of the Broad Valleys, can be studied far better along the E margin of the Alps than along their N margin, for the latter underwent considerable tectonic disturbance in Late Tertiary times. In the Vienna Basin particularly, the Tortonian, Sarmatian, and Pannonian shore facies are still connected with the limestone and flysch Alps, which have been downwarped into the Vienna Basin along the line of thermal springs since Helvetian times (Hassinger, 1905; Büdel, 1933).

The Pleistocene terraces running from the NE Alps northwards along the Traisen River to the Danube, and E along the Triesting and Piesting Rivers into the Vienna Basin, are purely periglacial in character (Büdel, 1944). The mountain valleys from which these foreland terraces spring are correspondingly lacking in steps, and reach with broad gravel floors deep into the mountains, just like the periglacial valleys which, with the aid of the ice rind effect, developed in the Mittelgebirge. The well-known sequence of Pleistocene terraces on the Danube near Vienna is also characterized largely by periglacial features, for in the Pleistocene its Alpine tributaries (the Iller, Lech, Isar, Inn, Salzach, and Enns) were far removed from their glacial sources (Büdel, 1944). Yet the resemblances between this terrace sequence and the purely glacial terraces (e.g., of the Iller-Lech or the Traun-Enns rivers) are so great, that clearly it is the great ice bodies themselves which provide the best evidence regarding the major curve of the Ice Age climate.

The terrace sequence along the Danube from the Wachau Gorge to the Vienna Basin has been studied in great detail by Fink in the last two decades (1961, 1966, 1973). It has become quite clear that broad depositional and planation surfaces lie *above the Ice Age Pleistocene terrace system*. Fink attributes only five of the terraces to the Ice Age sequence (see Fink, 1973, Fig. 3, p. 101), ranging from the "Wienerberg Terrace" (at 240 m, probably of Günz age) down to the "Prater Terrace" (at 160 m, Würm in age). Another, slightly subdivided, terrace above this sequence, the "Laaerberg Terrace" (at 250-260 m) he dates for good reasons to an older phase of relief development, one that we have discussed under the second major relief generation. Fink calls this the Late Pliocene "time of pedimentation." The gist of Fink's results (though not expressly stated in his own work) is that this set of overlapping piedmont surfaces on the E margin of the Alps constitutes a broad glacis system of Latest Pliocene to Earliest Pleistocene age.

According to Fink, these foreland surfaces include not only the Laaerberg gravels (covering a strip running across the entire basin S of the Danube and containing, among other material, disk-shaped quartz pebbles weathered yellow-red and a hand's breadth in size), but also smaller gravel patches, including those built by the small Wiener Wald streams in the S part of the city. Lying at the same level as the Laaerberg Terrace, these patches are covered with coarse flysch gravel or "Plattel gravel." The large surface of the former parade grounds, the "Schmelz," is an example; it is around 250 m high and grades into yellow Pannonian sand.

A far more impressive glacis surface of the same age was shown to me by Fink. This runs E from the Alpine edge for 12 km between Rabniz and Stoberbach in central Burgenland, and can also be seen in a strip running S from there as far as Güns and the Güns mountain spur (see Fig. 81). This

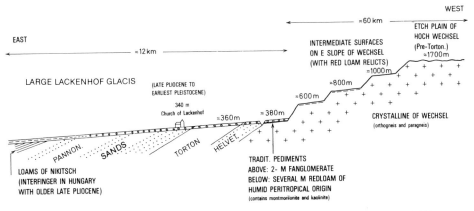

Fig. 81. Etchplain stairway and glacis at the E edge of the Central Alps in central Burgenland. (After Fink, 1961a and b.)

surface can be reached from above by way of the previously described older etchplain stairway, leading from the Hochwechsel (at around 1700 m) E across steps at 1000, 800 and 600 m E to the glacis surface of Lackenhof. The upper edge of this glacis surface consists of a relief feature at almost the same level as the glacis itself, which I would today call a traditional pediment, carved into the orthogneiss and paragneiss of the Wechsel massif. It is covered with two to three m of fanglomerates, below which are thick sheets of red weathered loam containing montmorillonite and kaolinite. These or the analagous grus zones can also be found on the etchplains above. Without doubt, the loams derive from an older wet period of semitropical to monsoonal-subtropical character. Thus we have here a typical example of a traditional pediment, i.e., a case where the younger glacis surface could expand laterally somewhat across the crystallines because an older etchplain had prepared the way for it by decomposition. From here the glacis surface drops gently eastwards from 380 to 360 and finally 340 m (near the church of Lackenhof), cutting across coarse-grained Helvetian sediments, then fine-grained Tortonian, next Pannonian sands, and finally the loams of Nikitsch. These interfinger toward the E with early Late Pliocene sediments in Hungary. Fink has therefore dated the whole glacis to the late Late Pliocene (Latest Pliocene/Oldest Pleistocene). This agrees fully with Székely's results (1969, 1970) on the S foot of the Matra, and with Stäblein's conclusions (1968) on the edge of the Pfälzer Wald in the Vorderpfalz. The second major relief generation is thus excellently documented on the E Alpine margin. A valuable future task of climato-genetic geomorphology will consist of explaining it and elucidating its relationship to the inner Alpine High Valley System and trough shoulder surfaces, which relate in so many ways to the Broad Terraces of the periglacial regions.

Another question to be studied is whether and to what extent the sharp and very straight piedmont angle (see Photo. 54) between the piedmont surface and the mountain hinterland has a traditional or basal pediment beneath its fanglomerate sheet or raña (see Fig. 53). It is of no use to determine that some of these piedmont surfaces run without a sharp piedmont angle in a gently concave curve into the arid mountain front. Such a transition to the normal, gently concave footslope is found in many different morphoclimatic zones. The question about the climatically specific development of piedmont surfaces has to do with the common, strikingly sharp piedmont angles which separate even very large piedmont surfaces from very tiny remnant mountains (see Fig. 57).

Brief mention should be made in this context of the preliminary results of some work currently in progress on the E Styrian hill country. Earlier workers (Winkler-Hermaden, 1957; Morawetz, 1967; Zötl, 1964) established that the "Hochstraden Level" forms a single surface running from about 790 m in height in the NW (at the foot of the Schöckel) across the entire Tertiary hill country for over 50 km SE to an area around 530 m high at the worn-down

54. Sharp, straight piedmont angle at upper edge of a broad glacis surface. East foot of the foothill zone of the Front Range (up to 4350 m high) at the Great Plains of Colorado, four km N of Boulder, 50 km NW of Denver; taken from 1650 m high. (Photo: Büdel, September 4, 1965.)

volcanic neck of the Stradener Kogel, from which the surface got its name. This surface was heavily dissected during the Pleistocene. It is generally agreed that it is of Latest Pliocene age (Dacian to Levantinian), and bears the character of a broad glacis surface, much like the Lackenhof Surface in central Burgenland. From the Alpine edge (its peak is where the Mur River emerges from the mountains) it slopes fanwise toward the SSE, SE, and ESE. The Pleistocene valleys of the "Grabenland" in the broad S portion of this former platform now run directly S to the Mur River between Leibnitz and Rad-kersburg. This means that they deviate from the original slope of the higher glacis surface at an angle of 40-60°. This is a problem which we have already discussed with regard to the Neukirchener Feld. The result is that all the southward-running valleys of the Grabenland (the Schwarzau, the Sulzeck-bach, the Sassbach, the Ollersbach, and the Gnasbach) receive their most vigorous tributaries from the W, according to the former slope of the land, and thus have a marked W asymmetry (great steepness of the W-facing E slopes). East of the strip running from Waasen am Berg, this asymmetry suddenly reverses itself. From there to the Stradener Kogel (at 609 m) the valley of the Poppendorf stream and of the Sulzbach show an equally marked E asymmetry (steepness of the E-facing W slopes). This has usually been attributed to later uplift in the region of the volcanic line running N-S from Straden to Gleichenberg. Such an uplift, however, has not been proved, and would hardly have caused the valleys to become so markedly asymmetrical. I therefore assume that the old direction of slope of the Hochstraden level reversed itself in the region of the Waasen strip, and that the glacis surface then climbed gently eastwards toward the volcanic ridge. This ridge may have formed a secondary radiation point for the Hochstraden glacis. At least a slight depression in the surface of the fragment near the Waasen strip suggests this. At any rate, it is a further example of the important role played by the second major relief generation in E Central Europe in the marginal area between the Hungarian Basin and its surrounding mountains.

Work in progress on the extraordinary parallel N-S-running valleys of the Pannonian Plain in Somogy, referred to by A. Penck (1906, p. 29, and in later discussions) as the "most remarkable valleys in Europe," will be discussed in the next (German) edition of this work.

3.4. The Subtropical Zone of Mixed Relief Development

Although two subtropic zones are shown in Fig. 13, we will discuss only that of the etesian subtropics. This zone, characterized by winter rains, encircles the Old World Mediterranean, covering S Europe, N Africa, and the Near East. The monsoonal zone on the E sides of the continents, characterized by hot summers and summer rainy seasons, has already been discussed (see Section 2.3.2.2 [11] as the transition from the seasonal (monsoonal) tropics to the higher latitudes.

3.4.1. GENERAL FEATURES OF THE ETESIAN REGION

This broad zone forms the transition between the trade-wind deserts and the humid ectropics, and the relief features of the two belts interfinger and overlap both horizontally in the lowlands, and vertically in altitudinal zones. Not only do both recent and relict features abound at all elevations, but also anthropogenic effects on the soil and relief play a greater role here than in nearly any other climatic zone. The plow was introduced exceptionally early to this zone (around 7000-5000 B.C.), and disastrous phases of deforestation accompanied the rise of each of the great centers of the ancient world (ancient Iran, Mesopotamia, Syria and Palestine, Egypt, Crete, Hellas, Rome, and Byzantium), as well as of the medieval city-states (Rome, the Lombardic cities, Genoa, and Venice) and medieval empires (the Arab, Norman, Staufen, Spanish, Hapsburg, and Turkish empires). Even without this relatively late human interference, the overlapping of two natural types of relief genesis in this mountainous region justifies calling it a "zone of mixed relief development," a name that has already found its way into the literature (see, e.g., Wiegand, 1970, 1972).

Our discussion will follow the altitudinal zones, starting at the top and working down.

As in Central Europe, the upper elevations are often crowned by ancient etchplains, which are also widespread at lower altitudes. These have received less attention than the more striking glacis surfaces associated with them (Panzer, 1926; Mensching, 1973; K. Fischer, 1974). We will discuss the etchplains later in context.

The upper elevations of many mountain ranges have been reworked into long chains and ridges. Glaciers today are found only along the N edge of the Mediterranean (in the Pyrenees, the S Alps, and the Caucasus), and on a few high peaks in E Anatolia and Armenia, but glaciation was greater during the Pleistocene. Work on the Pleistocene snow lines has been summarized by Messerli (1967). In places, particularly along the N of the Mediterranean, where the mountains were fully exposed to W wind low-pressure cells, the snow line fell very low, down to 1200 m in the Ligurian Appennines, and to 1300 m in the mountains of the Hercegovina and in Crna Gora (Montenegro). This is lower than in most of the Alps, which suggests that the N Mediterranean was dominated by powerful W winds during the cool and wet Early Glacial, instead of by the etesian winds of today, so that the mountains here received large quantities of snow in winter, but in summer were blanketed by clouds. The snow line rose toward the S, to 1700-1800 m on Corsica and in the Central Appennines, over 2000 m on Etna and in the Arcadian Mountains, 2500 m on the lee side of SE Spain, 2600 m on the E side of the Lycian Taurus, and 3000-3400 m in the High Atlas. Clearly the area above snow line was limited; at the most, small cirque glaciers may have sent tongues down into the valleys. The glacial influence on the relief today is therefore unimportant.

As in the Alps, the Ice Age snow line ran about where the upper forest line would lie today were it not for human interference. Anthropogenic influence, however, has been very great (see below), with the result that in this climate of dry summers, the alpine turf zone above the forest line, such a well-known feature in the Alps, is but poorly developed. This is also a reason why nomadic shepherding is practiced here, instead of alpine transhumance based on cattle.

How far the Pleistocene forest line lay below today's line (that is, below today's natural, not actual, forest limit) differs greatly according to maritime or continental situation. Where the Ice Age snow line fell very low, it may have come within 600-700 m of the forest limit of that time. Usually the distance probably equaled 800-1000 m. In the Central Appennines (70 km E of Rome), where Messerli sets the Ice Age snow line at 1800 m height, the Ice Age forest line lay slightly under 1000 m (Büdel, 1951).

But what sort of a forest was this? According to Blanc's investigations (1936) in the Agro Pontino, deciduous mixed forests reached sea level between Rome and Naples. This presumably was especially true during the wet cold Early Glacial. During the dry Glacial Maximum, when the seas retreated considerably, Near Eastern steppe vegetation pushed westwards, as it did in Central Europe. According to Frenzel (1964) and Beug (1968), steppe communities dominated the SE of Spain and other lee areas, particularly of the S Mediterranean countries. This corresponds to the increase of loess finds in the Mediterranean region, as for instance on the N Dalmatian islands in the NE Adriatic, which at that time was dry land (see Fig. 17).

Below the forest line, or at about 1000 m in the Central Appennines, lay solifluction sheets, probably of Würmian age. Solifluction-like rubble sheets are found on many slopes of the Mediterranean region at lower altitudes as well, but these are difficult to distinguish from Holocene anthropogenic rubble sheets.

Only under two conditions do true periglacial traces clearly extend into lower altitudes. One of these is common on exposed coasts where heavy winds (particularly W winds in the N Mediterranean area) caused a low maritime forest line during the Ice Age. I have found traces of such an occurrence on the W coasts of Elba and Corsica, where Tricart (oral communication) agrees with my evaluation.

The other condition consists of periglacial river terraces. The periglacial zone above about 1000 m was indeed quite extensive in many Mediterranean mountains. As Mensching has shown (1955), much of the High Atlas extended into this zone, while the Anti-Atlas was completely below it. Rivers with fair-sized drainage areas tapping such regions are accompanied by Ice Age gravel fields into the lowlands and even as far as the coasts. Where the rivers empty directly onto mountainous coastlines without alluvial plains, the terrace may grade from a periglacial (stadial) terrace to a glacio-eustatic (interstadial) one, the terrace then being adjusted to the higher interstadial sea levels (Lautensach, 1941; Büdel, 1952). Terraces of clearly cold stage character may still have been adjusted to higher sea levels. I became acquainted with an impressive example of this on the W coast of Corsica near Calvi. Here three sets of terraces consisting of very coarse but extremely well-rounded fluvial gravels run at heights of 20 and 8 m to the sea. During the deposition of these Ice Age terraces, the sea level must have been higher than it is today.[61]

Large, periglacially influenced rivers from the greater heights are the ex-

[61] A hypothetical solution to this problem may be (as concluded in Section 3.3.2.2) that the continental ice sheets of the N hemisphere, the prime cause for the eustatic drop in sea level during the Würm, only reached their maximum at the very end of this period, so that sea level only sank the very end of the Würm. The Antarctic ice cap, on the other hand, was larger during the Eem than today, and shrank slightly at the beginning of the Würm (Hoinkes, 1961). The sea may therefore have been somewhat higher during the first part of the Early Würm than it is today. Moreover, Recent eustatic fluctuations in sea level were by no means caused purely by the growth and recession of the Pleistocene ice masses. Downwarping of the sea floor (shown, for instance, in the N Atlantic by sunken guyots) may also have lowered sea level to its Holocene level. Exact data on this are not available. But even without sea-floor subsidence, the sea may have been higher during the Early Würm Glacial for purely glacio-isostatic reasons. This is precisely the period when sudden depression of the forest lines enormously expanded the periglacial zones of the Mediterranean region, providing the rivers with vastly increased amounts of rubble.

ception in the Mediterranean region. The lower tributaries and smaller rivers, draining the more common altitudinal zone between 1000 and 1200 m in height, have a quite different, subtropic-etesian character. They closely resemble tropical mountain valleys, having narrow, steeply stepped ravines, the smaller and steeper examples nearly always having narrow rock beds with little gravel. Broad gravel beds, indeed, can only form through permafrost and the ice-rind effect. The inactivated broad beds or floodplains so familiar to the inhabitants of the Central European Mittelgebirge are also lacking. Only very large Mediterranean valleys have gravel beds across the entire valley floor. These are dry in summer, but are flooded in winter by raging torrents.

This basically different valley type, which has persisted here ever since Tertiary times, gives the Mediterranean mountains an entirely different character from the Mittelgebirge. The broad, stepless floodplains (Ice Age Lower Terraces) are lacking here, and since the torrential beds of the larger rivers are uninhabitable, particularly when anthropogenically rendered swampy and malarial, the settlements of many mountainous regions are Acropolis-type settlements, crowded fortress-like on the slopes and heights. This type of settlement was also preferred due to fear of pirates, invaders, tax collectors, and army recruiters. The communication routes connecting these settlements run along the slopes and ridges, avoiding the dangerous valleys. Not for nothing have the Italians become the masters of building mountain roads and railways.

Early introduction of the plow and repeated cultural florescence and collapse have led to many periods of severe and even total deforestation. In this climate, in which summer droughts seriously damage the ground vegetation, and in which the rains of winter come in torrential onslaughts, deforestation caused extensive washing away of the soil. This turned into rill erosion, favored by the valley-type described above. The network of steep valleys continued developing in the Holocene, particularly in soft bedrock such as molasse or flysch. The sharp gorges (called *calanche* in Italian) of anthropogenic origin have not only completely stripped many slopes of their natural soil blanket (usually a form of red earth or terra rossa), but have created an agriculturally useless badlands cut into crumbling bedrock. In many parts of Sicily, once the breadbasket of the Roman and Staufen empires, the soil has been so thoroughly stripped from the fields that one may see peasants pulling their plows through the very stone itself, wherever it is brittle enough to permit this. The tragedies which must have unfolded here through the centuries, with shrinking farmland, overpopulation, and unfavorable social conditions, can well be imagined.[62] The consequences were even worse in lime-

---

[62] The terrible conditions of the past can be gathered from the report of a Spanish Benedictine monk at the beginning of the seventeenth century. ''The peasant condition in Spain today is the most miserable and impoverished of all conditions. When one says 'peasant,' one thinks of

stone regions, where soil erosion intensified karstification, and where goat pasturage prevented secondary succession by the natural communities of macchie (maquis) and phrygana (a limestone heath typical of Phrygia in Asia Minor). It is very difficult to deforest such areas, and it is often impossible to coax the dried-up springs to flow again. A small mountain range SW of Tunis, which formerly provided water for one of the two major Roman aqueducts supplying the great city of Carthage, today barely produces a tricklet sufficient for two small Arab villages.

The eroded soil, when not carried out to sea, has usually been dumped in basins such as large poljes, or more often in lowland areas along the coast. In the absence of a strong and active local government, this led to increased flooding and malarial hazards, but where deliberately exploited, such as on the mouth of the Vardar W of Thessaloniki, in the area of the Po delta, or in the Agro Pontino S of Rome, these areas became the most fertile regions in the entire Mediterranean. Continuing deposition has pushed the coastlines seawards since ancient times. This, of course, has at times cut off thriving harbor cities from the sea, such as Ephesus, Miletus, Adria, Ravenna, and Pisa. In Thessaloniki every effort is being made to avoid a similar fate.

3.4.2 THE RELIEF GENERATIONS OF THE ZONE OF MIXED RELIEF DEVELOPMENT

The interpenetrating influences on the etesian zone of mixed relief development, and the differences between this zone and the ectropic zone of retarded valley-cutting, can best be seen by studying the sequence of relief generations in an area such as the Mediterranean.

The oldest relief generation of Tertiary relict surfaces resembles that of the rest of the ectropics, since the seasonally tropic climate necessary for their creation spread into the polar regions in Early Tertiary times, and was found in the Late Tertiary at latitudes as high as those of Central Europe. Broad, ancient etchplains are found on the mesetas of Morocco and Spain, on the Algerian High Plateaus, and in Central Anatolia. In many of these places they are commonly taken for granted. Smaller examples are also found in Corsica and in the Sila of Calabria, and at lower altitudes in the Monte Gargano and in the Murge of the Apulian Plateau. We will study the broad etchplain stairways in the Dinaric karst, using an example from Bosnia and Dalmatia,

---

coarse food and a ragged tunic, of delapidated clay huts and a plot of badly farmed land, a pair of bony cows and the burden of interest, taxes, and feudal dues. When the peasant comes into the city, he finds only ridicule and boundless disillusionment. His martyrdom is only completed, however, when judicial authorities, tax collectors, and army recruiters seek out his lowly hovel'' (quoted after Rühl, 1928). Small wonder, then, that in Sicily, the Hohenstaufen domination (when Frederick II checked the power of the barons and struggled to develop the overall economy) still, after 800 years, lives in the memory of the peasants as the golden age.

where such relict surfaces exist near sea level in the form of marginal karst plains. Drowned inselberg-studded etchplains can be found in the Tuscan Archipelago and in some of the Aegean Islands. These may be considered modern counterparts to the buried inselberg landscape of the Leipzig Embayment (see Section 3.3.1.1).

The geomorphic effects of the third Central European relief generation are found only in the highest mountains of the Mediterranean region, but affected the middle and lower altitudinal zones via the major rivers.

One element of active tropical relief development, namely valley-cutting, reaches its highest elevation here. In the middle and lower altitudes, the V-shaped ravines, particularly of the smaller rivers, have steep steps and little gravel, closely resembling the mountain valleys of the humid and seasonal tropics. This form of valley-cutting has been intensified during the Holocene by calanches resulting from human interference. Another type of traditionally continued tropical relief development is found in the great karst poljes, which we will discuss in the next section. The effects of calanche development and of soil erosion on sedimentation in the lowlands will be discussed in the section after that.

The mountains are often skirted by broad, relict glacis surfaces built out into the foreland basins. These ancient relief forms, so widespread in the Mediterranean area, probably underwent their last phase of active development during the arid phases of the Earliest Pleistocene, but as recent work by K. Fischer (1974) and Wenzens (1976) shows, they are often superimposed on surfaces created by an earlier humid warm climate. We have already discussed the problems involved here (in Sections 2.6 and 3.3.2.4) and will not take them up again now.

We have thus made clear the manner in which the etesian subtropics form a zone of mixed relief development, a meeting and mixing ground of relict and current relief-forming influences from several different morphoclimatic zones. We will now focus on two less well-known features from this rich palette of influences, namely, the ancient etchplains and Recent lowland deposition.

3.4.3 Etchplain Stairways and Poljes in Central Dalmatia

The central area of the Dalmatian Bosnian karst country consists of a very well-preserved etchplain system comprised of five steps, each of which cuts smoothly across the complicated tectonic substructure. This set of relief forms has until now been ignored in favor of karst studies. Yet the etchplain stairway is distinguished by the fact that its surfaces are pocked with the largest and deepest poljes of the entire Dinaric karst region.

Poljes are the most striking and problematic of all ectropic karst features.

330   **3. CLIMATO-GENETIC GEOMORPHOLOGY**

Till now they have been studied purely with regard to current processes, with no attempt to integrate them into any general climato-genetic scheme. A thorough discussion of them is therefore all the more necessary here.

All the poljes of the S Slavic karst lands derive from pre-Pleistocene ("Upper Tertiary") times (see, among others, Roglić, 1940, 1960; Rathjens, 1960; Ridanović, 1967), when completely different climatic and erosional conditions prevailed in the Dinaric region. The well-preserved crowning etch-plains are even older, yet their development has not been correlated with that of the poljes, a striking omission which further justifies our discussion here.

Hitherto the problem has been regarded in terms of how the polje floors sunk in the karst surface could have *expanded*. A number of theories exist on this, the most recent such works, varying only slightly in emphasis, being by Kayser (1955), Louis (1956), Rathjens (1960), and Roglić (1940, 1960). The following discussion differs from these in viewing the poljes as islands of *gradual restriction* of the formerly much broader relict surface, setting them in direct relation to former planation. The poljes with their strikingly flat floors thus occupy an unusual position in the climato-genetic view of the earth, being the highest latitude form of intramontane plain still undergoing active or traditional development. This development is permitted at such latitudes by the exceptional solubility of the karstified limestone substrate.[63] The difference between the poljes and the tropical intramontane plains is that in the latter the entire removal of chemically decomposed material occurs above ground through one or more gorges. In the karst poljes, drainage and removal of material are nearly always subterranean, through ponors (swallets) and their cave systems. For this reason, neighboring poljes often have floors at strikingly different levels (see Fig. 82).

Let us begin by looking at the etchplain system of this region, as shown

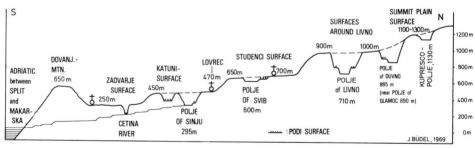

Fig. 82. Etchplain stairway (vertically exaggerated), pocked with polje and podi surfaces. Central Dalmatia and the neighboring region of Bosnia. Cross-section 80 km long, 1300 m high. (Büdel, 1973.)

[63] Karst regions further N lack poljes. It is as yet unclear to what extent the famous "Malm-theta bowls" of the Swabian Alb may be considered poljes or relict intramontane basins.

in Fig. 82. The highest surface, around the poljes of Duvno and Glamoc, is at about 1100-1300 m altitude. Widespread surfaces of similar appearance can be reconstructed from large remnants found near Livno Polje at around 900-1000 m altitude. The very large and famous poljes of this region, Kupresko Polje (the highest of all the Dinaric poljes, lying at around 1130 m), Glamoc Polje (around 890 m high), Duvno Polje (around 865 m high) and Livno Polje (around 710 m high) all have steep walls and are sunk a hundred to a few hundred m into this relict surface. The largest polje floor is that of Livno Polje, 65 km long and 14 km wide at its broadest point (see Photos. 55 and 56).

55. N part of Livno Polje in Bosnia, here over seven km wide; total length, 64 km. View looking E, 55 km NE of Split. Foreground: level polje floor at 710 m; midground: intermediate podi surface at 760 m; background: relict surface at around 1000 m. See Fig. 79. (Photo: Büdel, September 9, 1970.)

From the surface at 900-1000 m S of the Livno Polje, a magnificent etchplain stairway leads SW down to the Adriatic Coast. Descending to a marked surface at Studenci (650-700 m), we find Svib Polje (just over 600 m), containing an intermediate podi surface. At Lovrec, the Studenci surface drops sharply to the next surface of Katuni at 450-470 m (see Photo. 57). A further escarpment, particularly well-marked N and NE of Sertanovac, leads to the lowest, extremely flat relict surface of Zadvarje, at 250 m (Photo. 58). Slightly W of this, a similarly well-formed surface contains Sinj Polje, whose floor is at barely 300 m. Unlike the other poljes today, this is drained above ground through the narrow cataract-filled gorge of Cetina, which cuts razorsharp into the Zadvarje surface (see Photo 58). The narrow coastal range of the Dovanj Mountains (at 650 m), running NW into the higher Mosor Mountain Range and SE into the Biokovo Mountain Range, forms a wall cutting

56. W side of Livno Polje. Above the polje floor are podi surfaces at around 760 m height. (Photo: Büdel, September 9, 1970.)

the well-formed Zadvarje surface (250 m high) off from the sea (see Büdel, 1973, for the following discussion).

It must be emphasized that the major steps, consisting of the surface at

> 900-1000 m S of Livno Polje, the
> 650-700 m Studenci surface, the
> 450-470 m Katuni surface, and the
> 250 m Zadvarje surface

show little disturbance. Although general epeirogenic uplift has taken place here, this has not caused much fracturing. Only along the coast SW of the Dovanj, Mosor, and Biokovo mountain ranges has the land sunk below sea level. This has produced the Dalmatian island strings of Braĉ, Hvar, Korcula-Peljesac, and Lastovo-Mljet, which are clearly parallel rows of drowned mountain ridges. The recent nature of this downwarping is shown by the fact that these round-topped limestone islands have hardly any cliffs or marine abrasional terraces. So far as sounding methods can determine, these are even lacking at the lowest Pleistocene eustatic sea levels.

The three lower surfaces of the stairway are the most even and best preserved, and the two lowest are studded with particularly sharp inselbergs (see Photo. 57). According to Roglić (oral communication), they are also the youngest. Presumably they were still being traditionally formed in the Pliocene. They can be characterized as being on the whole very flat and soil-

57. Karstified etchplain of Katuni-Lovrec, 470 m. Background: etchplain escarpment rising to Studenci surface, with inselberg in front. The surface cuts smoothly across limestones block-faulted in the Saxonian phase. Bosnia, Central Dalmatia, 50 km E of Split; see Fig. 79. (Photo: Büdel, September 10, 1970.)

free, but on a fine scale they are broken up with karren and sinkholes, some uvulas, and a few valley heads. Figure 57 gives a good impression of the type of broad, finely karst-riddled expanses.

Descending from these surfaces to the almost perfectly flat polje floors, we find an even richer series of surfaces. Above the fine alluvium of the polje floor, crossed by streams running from karst spring to swallet, rise the podi surfaces. These are often but little higher, and look much the same, yet they lack the covering of fine material, just like the higher karst surfaces. The podi lie below the higher surfaces skirting the deeper polje surfaces. They may be considered former polje floors, no longer reached by the high groundwater levels of winter. They are also found as complete surfaces in the large closed depressions of Dicmo and Dugo Polje somewhat S of Sinj Polje; here the podi surfaces extending across the whole basin floor are bare, for being no longer flooded, they no longer receive sediment. They are heavily riddled with karren, but still form an even karst floor.

High podi surfaces often form narrow terraces around "active" polje floors, e.g., in Duvno Polje (Roglić, 1940), or along several sides of Livno Polje (Fig. 82, Photos. 55 and 56). The heights of these terraces do not always correspond, showing that down-cutting proceeded in stages, as can be observed in intramontane basins of the tropics. Relict polje surfaces close to the active floor can be very wide, as in the region of Svib and Sinj Poljes.

Looking at the general relief structure, there is little difference between the crowning relict surface and the broad, shallow, podi or relict polje surfaces. Given overall uplift and increasing karstification, planation gradually became

58. Cetina Gorge with high waterfall, cut into the karstified Zadvarje etchplain (see Fig. 79). This extremely narrow gorge was formed partly by roof solution and collapse of an old cavern system. 38 km E of Split. (Photo: Büdel, September 10, 1970.)

restricted to favored spots. The extremely flat, highly karstified podi erosional surfaces differ from the higher karst etchplain only in their somewhat lower position, not in morphologic appearance. We therefore see no reason to draw any basic distinction between the two flat features. We consider the striking podi surfaces of this region to be simply *an intermediate form between the relict surface above and the polje surface below*.

Zötl (oral communication), much like earlier workers (e.g., Grund), has stressed that many poljes are of tectonic origin. This applies particularly to karst basins in the broader sense, including not only true poljes with flat floors, but also highly sculptured and subterraneously drained karst basins such as the Semriach basin in Styria.

Zötl is certainly right in saying that many of the ''classic'' Dinaric polje are ''endogenously predetermined.'' But this endogenous predetermination grades through many stages, starting from tectonic grabens creating sharply defined, flat-floored basins even in non-calcareous areas. Usually these are filled with thick Neogene sediments and are drained above ground (e.g., the basins of Thessaly in Greece). Certainly, Neogene endogenous subsidence may have played a role in some true poljes, especially as the Dinaric poljes all follow the NW-SE strike of the old nappe structures and of the younger block faulting. Furthermore, many Dinaric poljes (such as Duvno Polje) contain ancient lake sediments.

But Duvno Polje shows, as Roglić (1940) has convincingly demonstrated, that these lake sediments, which are probably Pannonian, are much older and covered a much larger area than the polje which developed in them. In its present form, Duvno Polje is probably of Middle to Late Pliocene age, and corresponds only roughly to an area of previous subsidence, having either renewed or continued this downworking.

One can continue listing stages between overwhelmingly endogenous origin and overwhelmingly exogenous origin. The endogenous incitement to planation may be limited to a shallow trough or depression in an Early Tertiary relict surface system, particularly when this depression is associated with greater break-up or mylonitization of the limestone, as we have already described in Fig. 48 for intramontane basins in the tropical mountains. Heavy jointing will favor karst solution, even in ectropic latitudes, which will promote the development of a closely meshed cave system in the area of the depression, leading to subterranean drainage of dissolved matter. In this manner exogenous processes may lead to the development of large, closed karst concavities.

Kayser has shown (1934, 1955) that there are in fact poljes in the Dinaric karst whose thick soil sediments (usually 4-10 m of light yellow to terra rossa-colored, dense alluvial clays under mature relief-covering soils) overlie a surface of heavily corroded limestone, not older sediments. This karst surface, though roughened by karren and widened joints, is parallel to and as flat as the upper soil surface. Where all or part of a polje cuts directly into the bedrock in this manner, it forms an astonishing example of the actual processes by which these karst concavities are formed. Even given various types of endogenic predetermination, the active process is exogenically caused corrosion, a purely chemical and solutional erosion. The actual fashion in which the poljes are formed should be studied in terms of how this type of polje floor is formed, especially when such floors are very broad and wide.

Obviously, in order to form and maintain such a corrosive basal polje surface, the water exits must be sealed off below, so that the soil sediment keeps the basal weathering surface permanently moist. Nearly all the karst springs lie along the edge of the sealing surface. From here the polje streams flow across the upper polje surface, often for long distances, without cutting in at all. Rather they deposit further alluvial loam or clay, sometimes even sands and gravels. Since the entire steep-walled and sharply defined surface is sunk over a hundred to several hundred m into the surrounding countryside, it is clear that the basin as a whole is an erosional form. Yet the streams flowing across the polje floor have not cut down erosively, or they would have dissected the floor long ago. Erosion here has been even and planar, occurring chemically through solution at the basal surface of weathering. Otherwise the floor could never have become so smooth. Certainly streams and annual floods could never have kept it so smooth. The rivers which enter the polje (sometimes above ground through impermeable strata along the sides of the polje) aid erosion only by bringing material, above all sealing material consisting of alluvial loam (see Louis, 1956). Such rivers may at times build alluvial fans out into the polje. Cases in which fans, through limited lateral erosion, help enlarge the polje floor are considered here as highly localized exceptions which fall under the rubric of traditional development of the entire pre-existing polje floor. We will discuss this in more detail below.

Other areas along the edge of the polje floor show local traces of recent traditional development, e.g., where recently collapsed sinkholes adjoin the polje, as at the W end of Imotsko Polje. The major swallets or ponors of the larger rivers are also nearly always found along the polje edges. Collapsing limestone around ponors may in places contribute somewhat to the enlargement of the polje, but this is really not a significant occurrence. We therefore fully agree with Rathjens (1960, p. 143), when he says that "the polje floors of the Dinaric karst . . . are relict features, which owe their development to an earlier, warmer [Late Tertiary] and probably more markedly seasonal climate." On the other hand, we cannot agree with his next sentence, "I would add today: a climate in which lateral corrosion at the level of the polje floor was favored more than it is in the present climate."

This brings us to the problem of the corrosively planed polje floors of the present. It is clear that these existed in similar form in pre-glacial times, but their origin may date back into Late Tertiary times. If one descends from the crowning etchplains of this region across the broad, barren, intermediate podi surfaces and the lower polje terraces to the "active" polje floor, two things become clear.

For one thing, the *intermediate podi surfaces*, like the higher etchplains, *are broad erosional surfaces* cutting evenly across the tectonically complex substructure of Cretaceous limestone. Even though they no longer bear a thick soil blanket, they are still very flat, and look exactly as would as "active" polje if stripped of its soil sediment. The fact that these relict polje floors, these basal surfaces of corrosion, have been preserved even after the removal of their fine-grained covering, and even in a karstified area, shows that, like the higher etchplains, *they are extremely durable features*.

Secondly, looking down into the polje from above, one begins to see the active corrosion of the polje floor in a new light. There is much to suggest that these surfaces originated in a warmer (probably wetter) Late Tertiary climate, in the manner of tropical intramontane basins, through chemical decomposition and solution of a basal weathering surface under a thick sheet of soil or soil sediment externally subject to surface wash. This produced the ponors, which became karst springs during the rainy season. Springs and slope wash provided the growing polje floor with a constant supply of fine-grained material, which maintained planation in this now increasingly favored area. The agent at work here cannot have differed much from the mechanism of double planation surfaces, which according to the most likely theory was also responsible for the development of the higher relict surface. The only difference was that planation on the higher surface was much more wide-spread, for at that time (at around the Mio-Pliocene boundary) the climate was more favorable, there had been far less uplift, and karstification was not yet so advanced.

Looking from the crowning etchplains across the lower podi surfaces to the polje terraces and finally to the "active" polje floors, we see that since the Mio-Pliocene transition, planation has not expanded, but has narrowed down. During this entire time, particularly in the Late Pliocene, planation has been restricted to increasingly favored ribbons of land. This closely parallels the first narrowing down of planation to the "Trough Surface" and "Broad Terraces" of Central Europe at the Latest Pliocene-Earliest Pleistocene transition.

The gradually cooling climate with increasing uplift and karstification of the region meant that eventually down-cutting was restricted to isolated strips which provided the favorable endogenic conditions, including fairly pure limestone, downwarping, fragmentation, and mylonitization. By the end of the Pliocene, active planation from tropicoid paleo-earth times had become limited to favorable ribbons following the strike of the mountains. The mechanism whereby planation occurred will not basically have changed, but of course local lateral enlargement may have occurred, just as it does today. The products of this narrowing down of planation to favored ribbons are the "active" polje floors with their soil sediment blanket today. These must already have existed in their present form in pre-Würmian times.

But they are not fully active to the extent of being able to form themselves anew today, even given the necessary endogenous conditions. The poljes that we see today are remnant islands of humid-tropic Tertiary planation which have survived into the present. Their very existence and their great durability have permitted them to continue functioning locally wherever a fine-grained sheet still covers them. But where they are uplifted too far, or where the karst water in the cave systems of the surrounding limestones sinks too far, they lose their soil blanket and become inactive; like Duvno and Dicmo Polje, they turn into relict poljes made up of podi surfaces. The alluvium-covered polje floors are thus an example of traditionally continued development of flat surfaces.

The similarity of poljes to intramontane plains of the seasonal tropics is striking. Like them, the poljes have developed independently of a river network, which in both cases fulfills only the passive function of removing material (nearly exclusively in dissolved form in the case of the polje). Like intramontane plains, the poljes often have very gentle watershed divides resembling etchplain passes; the rivers flow to their ponors in different directions without affecting the smooth down-cutting of the polje floor. This is true even over long periods of development.

One final conclusion seems important. In discussing the development of intramontane plains, we stressed that the rivers flowing through the plains or exiting through narrow gorges fulfilled only the passive role of transporting material, without helping to carve out the basin through lateral erosion. The basins originated primarily through local planation caused by chemical decomposition and solution. It is evident that the poljes also developed in this fashion. For it is clear that the streams, disappearing into subterranean courses through the ponors, could not have any function in the development of the overall polje landform, other than that of transporting dissolved and suspended matter.

3.4.4 FLUVIAL ACTIVITY IN THE MEDITERRANEAN SINCE EARLY ANTIQUITY: THE GROWTH AND BURIAL OF OLYMPIA

Human interference in the balance of nature has been particularly marked in the etesian subtropics. As we have already stressed, this is primarily due to the early introduction of the plow here (as early as 7000 B.C. in the East) and to the accompanying deforestation. That these activities had such drastic morphological consequences was due in great part to the mountainous nature of these countries and to the climate; for the dry summers, in which much of the ground vegetation withered away even without man's aid, followed by the sudden onslaught of the winter rains, greatly accelerated soil wash and slope dissection. Even more important for man, both positively and negatively, was the accompanying deposition and flooding in the lowlands, and the resulting silting-in along the coasts.

The area around the mouth of the Alpheios River on the W side of the Peloponnese forms an exceptionally favorable "natural test site" for studying the causes, main periods, and results of such lowland events as they evolved in interaction with man.

The rare quantitative investigations of these events have mainly studied the erosional processes alone. Their effect on morphogenetic fluvial behavior, and above all, on lowland deposition has largely been ignored, so that the connecting link between the erosional processes and the seaward growth of the coastline has been lacking. Many of the coastlines of antiquity have changed position (e.g., the Po delta, the Vardar-Axios mouth near Thessaloniki, the Menderes-Meandros near ancient Miletus, the Spercheos near Thermopylae, etc.): this has usually been attributed to eustatic or tectonic changes in sea level, without attempting to correlate this with Late Holocene delta development.

The Alpheios system is particularly suited to the study of the morphologic activity of Mediterranean rivers for several reasons (see Büdel, 1963a). Running W to the Ionian Sea, its three branches, each 100-125 km long, drain the entire center of the Peloponnese, or the Arcadian Highlands (one of the main karst regions of Greece). Despite its position in the S Mediterranean (at 37° 38' lat., Olympia lies on the same latitude as Granada and Mount Etna, and just slightly N of the N tip of Africa which is Cap Blanc in Tunisia), the river is fed by copious, winter convective rains and permanent karst springs. It therefore flows all year round. At floodwater stage, its waters may rise to three or four m above its average spring water levels, measured in early May. The biggest advantage of the Alpheios system is that German archeologists excavating Olympia since 1874 (to investigate the central role and history of the sanctuary) have provided geomorphology with a unique opportunity to follow the activity of such a river continuously and with precision over the last 3000 years. The results showed unexpected extremes of erosion, deposition, and renewed erosion.

On the S side of the ancient region of Elis on the W side of the Peloponnese (which, as of old, is one of the largest agricultural plains of Greece), the lower course of the Alpheios, around 25 km long, forms the ancient region of Pisatis, which is up to eight km wide. From the coast (now at most 4.5 km further seaward than the ancient coast, as indicated by remains in the old harbor site of Olympia), the plain rises more steeply than the river's floodplain, which, around one km wide, cuts around ten m into the plain. Upstream the plain quickly shrinks, until it is about one km wide where it cuts into the higher molasse mountain country. At Olympia, around 25 km from the coast, the plain dwindles to a marked ten m terrace running along both sides of the river, whose floodplain is still 800 m wide.

The floodplain looks like a barren gravel field. The ten m terrace, on the other hand, is well built up, with many new settlements and communication routes. The visitor from Central Europe will at first be struck by the terrace's resemblance to a Würmian Lower Terrace. A small stream entering near Olympia from the N, the Kladeos, forms a small ravine which also cuts ten m into the terrace. The sanctuaries, the Altis of Olympia, lie just upstream of the Kladeos mouth, *within the terrace body, not on top of it*. The foundations of the lower structures are covered by over five m of fine terrace sediment (see Figs. 83-85, and Photos. 59 and 60).

The Olympic Games have been celebrated since around 1000 B.C. When the first winner's lists were published in 766 B.C., the Games had already acquired supraregional significance, and this year was used as the basis for Greek time reckoning. The last Games took place in 393 A.D., after the christianization of the Roman Empire. This was followed by a period of

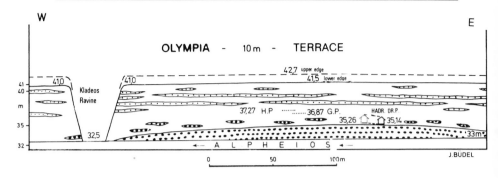

Fig. 83. Longitudinal profile through the terrace of Olympia. Bluff ten m high along the Alpheios River. (Elis region, W side of the Peloponnese). Spotted areas: basal conglomerate from the first deforestation period, 5000-4000 B.C. White area with finely stippled lenses: fine-grained medieval deposits, 550-1500 A.D. G.P.: Greek pavement of sanctuary (behind plane of drawing), ca. 500 B.C. H.P. and HADR. DR. P.: Hadrianic pavement (behind plane of drawing) and Hadrianic drain pipe, ca. 130 A.D.

decay, brought about by human and natural influences (particularly by the two earthquakes of 522 and 551 A.D.). The site was last occupied by a Byzantine fortress, and then by a Byzantine village, which was abandoned around 680 A.D. Precise dating is thus possible for roughly one-and-a-half thousand years (from around 1000 B.C. to 550 A.D.). The written records then fall silent until, after around 1200 years, the first Western European revisited the ancient cult centers in 1766. Since the liberation of Greece and the accession of the Bavarian King Otto in 1832, settlement and construction on the

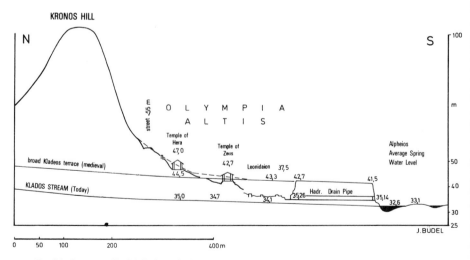

Fig. 84. Cross-profile (N-S) through the terrace of Olympia. (Equals left side of Fig. 85.)

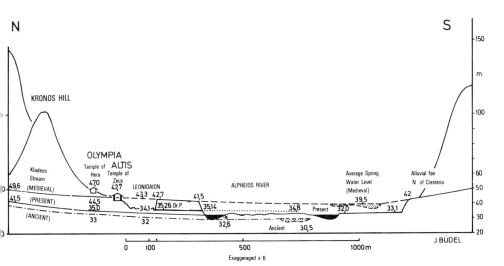

Fig. 85. Cross-profile (N-S) through the Alpheios Valley near Olympia. Depositional levels in antiquity, the Middle Ages, and today.

ten m terrace has flourished. In 1874 the German excavations began, which together with the first exact survey of the Alpheios Valley in 1880 revealed the history of the river and its valley. Renewed measurements undertaken in 1961 by myself and my colleagues revealed that astounding changes have taken place since 1880, and that even more astonishing changes have taken place over the last 3000 years.

The ten m terrace has been exposed for 300 m where the Alpheios undercut in the post-1880 period. The sequence in this exposure is as follows (see Fig. 83). The base consists of coarse, partly conglomeratically indurated river gravels, containing well-rounded cobbles of up to double-fist size. This size fully corresponds to that of the river gravels of the Alpheios and Kladeos today, while the composition (71% limestone, 22% chert, 7% sandstone) is somewhat closer to that of the Kladeos than of the Alpheios. The shape of the basal conglomerate also shows that it must be an old lateral gravel fan of the Kladeos, for its upper margin in the center of the exposed bank is a full three m above the level of today's Alpheios, while dropping off to river level at both sides. The ten m terrace today therefore covers an old gravel fan of the Kladeos, which at that time deposited material largely from the Arcadian Highlands.

Above this are thin intermediate layers, at most 0.8-1.2 m thick, containing a few very narrow bands of coarser gravel. The upper seven m of the terrace are completely different in character, consisting of fine, at times highly calcareous alluvial marls resembling reworked loess, interrupted only by thin lenses of fine sand. The composition clearly shows that we are dealing here

59. View out over the Alpheios Valley at Olympia (Elis, W Peloponnese), looking SE. In the midground, the sacred precinct, excavated from beneath more than five m of medieval alluvial loam (see Figs. 83-85). (Photo: Büdel, April 17, 1962.)

60. View from the interior of the Altis at the lowest point of the excavation in the Leonidaion. The base of the pillar (center of picture) served as the base point for the survey. View looking NW toward the profile wall (over four m high) of medieval fine-grained deposits (alluvial loam). Below center: start of Hadrianic drain, which originally led above ground toward the Alpheios (left); subsequently buried deep in medieval alluvial loam. See Figs. 84 and 85. (Photo: Büdel, April 18, 1962.)

with material washed off the adjacent molasse slopes: it is purely local in character, and contains no coarse gravel from the Arcadian Highlands. This conclusion is also supported by the quantities of marine shells found secondarily deposited in the sands, along with potsherds and domestic animal bones.

The surprising feature in this exposure, as we have already mentioned, is that the sacred precinct was not built on top of the terrace, on the alluvial marls, but rather on the surface of the basal conglomerate, within the gravel bands just above this. According to its composition, the basal conglomerate must represent a period of vigorous erosion in the Arcadian Highlands, much like that of today. It was therefore probably not a product of deposition under natural, early Holocene conditions, but derives from the period of the first deforestation in this region (5000-4000 B.C.). For it is always the transitional phases in climate or vegetation that cause the greatest extremes in fluvial morphologic behavior.

The position of the sanctuary relative to the basal conglomerate and the former river level is shown best by the position of a drainage pipe, part of a sewer system built under the Emperor Hadrian in 130 A.D. to drain the athletes' baths, the kitchens, and the sanitation facilities. This pipe was uncovered along its entire length of 200 m from the lowest part of the sanctuary to the edge of the ten m terrace, at which point it lies within the floodwater region of the present Alpheios. It is quite impossible that the Greek and Roman engineers would have planned the sacred cult site and the Hadrianic drainage pipe (which had a Greek predecessor built at the same level around 350 B.C.) in such a way that flooding could have occurred there or that sewage could have backed up. There can be no doubt that the average level of the Alpheios and the Kladeos must have been at least two m lower during antiquity than today, and that the average spring water levels must have been at about 30.5 m instead of at 32.5 m as today. The many literary sources of the period between around 1000 to 500 B.C. make no mention of floodwater damage to the Altis.

During the 1500 years in which the Games flourished, conditions in the neighborhood of Olympia clearly remained stable. The entire Pisatis was protected by a sacred truce, and the peasants there were freer than in any other part of Greece at that time. The slopes were well cultivated, and were assuredly protected by artificial terracing. For not only the entire Grecian world from Massilia to Colchis, but later the Roman Emperor as well, were all interested in ensuring that the thousands of pilgrims who flocked every four years to the site and camped on the bed of the Alpheios, were well supplied with running water, food, and firewood. At no time did famines or plagues break out during the Games. Even during the thirty fratricidal years of the Peloponnesian War (432-404 B.C.), athletes and spectators were allowed free passage through enemy territory, for they were protected by the sacred truce.

All this changed within a few decades. With the introduction of Christianity, the ancient gods were dethroned, and their power no longer protected the region. The Games, once a sort of United Nations, world athletic contest, and church council wrapped into one, shrank into a sort of local fair. In 393 A.D. the last Games were forbidden by Byzantium. The stormy Migration Period swept over the land. From the point of view of Byzantium, the Pisatis dwindled into a petty marginal province, barely sufficing for the despoiling of the forests and the exploitation of the people through tithes, taxes, and recruiting. In the face of all these threats, the population fled into the remotest corners of the mountains. The terraces were abandoned, the roads became disused; earthquakes destroyed the temple, and it was not rebuilt. Finally, the river, no longer tamely meandering, but wildly braided, mercifully covered the once holy sanctuary with over five m of local fine matter (Photos. 59 and 60).

Deposition can only have begun around 500 A.D., for it is at this time that the first reports of flood damage to the sanctuary appear in the sources. The subsequent flooding and deposition must be largely medieval in age, and may fall into two phases. The first of these would have coincided with the decay of the Games and the fall of the Roman Empire. After this a brief lull may have set in. But the peasants did not return from their high, remote settlements to the lowlands. The strife between the four powers of Byzantium, the Crusaders, Venice, and Turkey, lasting from 1150 till after 1500 A.D., brought renewed danger to the peasants, and caused a second phase of forest devastation, slope erosion, and fluvial aggradation. A graveyard peace then set in under Turkish rule. The abandoned fields in the lowlands were taken over by macchie, and slope wash was reduced. But the transportation routes were not rebuilt in the valleys, although by 1776 the Alpheios had cut into its deposits down to about its present level. When the Bavarian prince entered in 1832, there was not a single through road left in all of Greece. Even the invention of the wagon had been forgotten, and only narrow mule tracks wound up to the hidden villages in the mountain heights. In the lowlands, the once flourishing cities were forgotten. Of ancient Corinth, only a few lonely pillars remained. Around the ruins of the Acropolis of Athens clustered a village of 2500 inhabitants, which had not even kept the ancient name, but only lived ''near Athens.'' Where ancient Sparta lay is not known to this day. As a place to establish a first residence, only Nauplia, the Venetian fortification, came into question.

The vast sums of money which Ludwig of Bavaria pumped into Greece between 1832 and 1848 brought about a rapid recovery. At the beginning of the first German mapping and excavation of Olympia in 1880, the average spring water level of the Alpheios was about the same as today. Our survey of 1960 and 1961 showed, however, that the river has not only completely changed its course over its gravel bed, but also that it has changed character, having turned in the last eighty years from a braided into a meandering stream.

To summarize, we found that the river level in antiquity must have been at least two m lower than today, lying at 30.5 m. During medieval deposition, on the other hand, the level of the Alpheios must have been at least seven m higher than today (nine m higher than in antiquity), lying at at least 39.5 m, perhaps even somewhat over 40 m above sea level. The deposits can be followed as a widening ten m terrace down to the coast, which has been displaced by 4.5 km beyond the remains of the ancient harbor. The Late Holocene two m rise in sea level found by Hafemann (1960, 1965) cannot have been the cause of this deposition, for then the coast would have retreated, instead of advancing.

These frequent changes in fluvial activity are far more likely to have been anthropogenically induced. The deposition of the basal conglomerate was probably due largely to the first period of deforestation between 5000 and 4000 B.C. Then agriculture became better adapted to the demands of nature. Fairly uniform conditions in the Alpheios bed from Early to Late Antiquity (from around 1000 B.C. to 500 A.D.) reflect political and agricultural stability during the height of the Games. The cause of the medieval deposition is to be found less in renewed deforestation than in destruction of the well-ordered peasant agriculture, lapse of the sacred truce, and withdrawal of the populace to safer heights. The hillside terraces became ruinous in the upheavals of the Migration Period and the Turkish wars. The Alpheios became braided and covered over the ruins of Olympia. By the time the interest of Europe awakened again, the landscape had stabilized once more, and the river had returned to its regular meandering: it cut down, the groundwater sank, and the excavation of Olympia was made possible. Thus nature herself preserved the ancient ruins until such time as the civilized world should look for and find its sources.

A series of further conclusions may be drawn from this for geomorphology. It has been shown that the etesian subtropics are so sensitive to human interference that the lower river reaches may react as much to disturbance as do rivers of the mid-latitudes to a full Ice Age. That this is not simply a local situation pertaining to the Alpheios River, but rather an evolutionary sequence found throughout the etesian climatic zone (following much the same sequence of cultural periods) is shown by the analogous (in some places even more striking) deposits at Thermopylai, the mouth of the Vardar-Axios, the Menderes-Meandros, the Tiber, the Arno, and the Po rivers.

Secondly, the observations at Olympia show that such effects follow directly upon the heels of an anthropogenic change in the landscape. After a time, the morphodynamic processes achieve a new equilibrium with the changed features. This is completely analogous to the way in which the morphodynamics adjust to natural changes in climate, vegetation, and soil cover.

At the same time it must be stated that such sensitivity to human interference is not a general phenomenon, but specifically limited to the etesian subtropics. Two other morphoclimatic regions are similarly sensitive. The first of these consists of loess areas (or areas blanketed by loose soils of fine material) where arid periods alternate regularly and sharply with periods of heavy rains. When such areas (forested steppe, dry savannas, and steppes of all kinds) are stripped of their plant cover by fire and plow, extreme soil erosion will set in. Climate, the type of dissected substrate, and the intensity of cultivation will all determine if the damage can be healed by being left alone or by the introduction of appropriate agricultural methods. In some places, as in much of the United States, thoroughgoing and expensive protection measures have helped to curb the damage. In some of the loess gorges of China, on the other hand, the damage is final and irreparable.

The second anthropogenically endangered region consists of the forested mountains of the interior perhumid tropics. Here deforestation can produce an extraordinary increase in open landslides and soil slips, leading to the proliferation of red scars in the forest cover, as described in Section 2.4.1.

Afterword and Conclusions

On geomorphological field trips in the twenties and thirties (and even later) attention was directed to isolated features in the landscape, picked out as typical examples of young meanders, undercut slopes, terraces, sinkholes, dunes, structural ledges, loess or gravel exposures. The intervening areas were apparently devoid of interest. In the mid-latitudes only two regions formed any significant exception to this, these being the Würm moraine regions, and the structural escarpment reliefs. These were once referred to by Troll as "*the* morphological landscapes."

It is clear enough why this was the case. Where petrovariance was well revealed, as in the scarps of the classic structural scarp landscape, it was easy to find a correlation (but only one of many) between structure and landform. The structure is a purely endogenous characteristic to which the relief adapts, according to exogenic conditions of the successive relief generations. The adaptation may be greater or lesser, and sometimes may not occur at all. The Würm moraines, on the other hand, formed a depositional landscape whose glacially produced dynamics were explained by A. Penck and Brückner (1909) and later in greater detail by Troll (1924) and others. Here the correlation between structure and form was not only clearly evident, but the physics of its development could be concretely established.

The other examples mentioned above, such as young meanders or dunes, could be putatively associated with present (as in the case of terraces or undercut slopes) or past processes (as in the case of sinkholes). But what lay between these exceptional cases, i.e., the majority of the relief sphere, was ignored or explained only in the vaguest of terms.

In 1945 the French geomorphologists Cailleux and Tricart (1959, etc.) introduced a method which they called "dynamic geomorphology." This attempted to bring greater precision into the study of the relief by observing and measuring current processes which appear to be involved with relief development. This meant applying physical and mathematical quantitative methods from basal integration spheres (see Fig. 5 and Section 1.3). Such procedures yielded some important results, and they have since been taken up and refined by English and American geomorphologists, leading to increasingly specialized analyses.

Such quantitative methods, based on actual or theoretical breakdown of the very complex processes of relief development, may help clarify specific details of the processes involved, but on their own they rarely contribute much to the overall understanding of the entire relief. This is shown particularly well by the results achieved with quantitative methods extrapolated from mathematical models. Certainly one can measure some of the basic components of the whole process with all-too-delusive precision, but one can hardly use this method to formulate a balanced, comprehensive picture of the highly intricate relief-forming mechanisms developing out of the specific interaction of many already complex elements (see Sections 1.6-1.9). This is true for a number of reasons.

(1) Individual analyses at the basal level, no matter how precise, can say nothing as to the role and effect of these components at the highly complex level at which *all the elements interact* to produce the relief-forming processes. This is instead best studied by means of the product, the relief appearance itself. The methods to be used here will be discussed in the third edition under the name of "active geomorphology."

(2) The present processes at work in the mid-latitudes are exceptionally *weak*. This is shown quite simply by the fact that the unconsolidated Würm moraines have hardly been altered during the 10,000 years of the Holocene. If this is true, what effect can the Holocene have had on the hard rock underlying the total relief? In fact, 95% of the mid-latitude relief consists of relict forms. Even were it possible for us to depict the balanced interaction of all the current mechanisms, simple projection of this picture millions of years into the future would not produce a relief corresponding to that around us, for most of the latter was largely produced by totally different relief-forming mechanisms of the past.

(3) Current processes are always dependent on *local conditions* which cannot be reproduced in the laboratory. These conditions include microclimate, pedology, hydrology, the presence or absence of paleosols, the type of underlying bedrock (see point 5 below), the gradient, and the microrelief (that is, whether the site lies on an upper or lower slope, how far away it is from the next nickpoint upslope, and what the slope above the nickpoint is like in terms both of height and of the above-listed conditions). Average values produced by repeated parallel measurements may be representative, but they can also obscure exceptional but decisive factors.

(4) Measurements of any kind usually apply only to the normal working of present processes, instead of to the *minor catastrophes* which are often primarily responsible for relief development (see the next edition for relativity of "geological actualism"). Minor catastrophes are especially important directly after major climatic changes (e.g., at the beginning and end of a glacial). Once conditions have achieved a new balance, the processes weaken and play a much smaller role in shaping the total relief. In cultural landscapes this "normal condition" has been anthropogenically affected, to some extent intensified (through deforestation, soil wash, etc.; see Section 3.3.4), and to

some extent weakened (through protective measures against snow and mud-slides, flood control measures, and the building of dikes).

(5) All present processes in all climatic zones of the world *act upon a stage set by ancient processes.* The current processes may in some cases have behaved in the same or similar fashion for long periods of time, and may continue the activities of the ancient mechanisms with little or no change (as in the seasonal tropics). They may be traditionally shaping the inherited relief, rapidly destroying it, or having no effect on it at all. (These variations are all dealt with by synactive geomorphology, to be discussed in the next edition.) Flat features (relict planation surfaces) are particularly resistant to change, especially in the humid intermediate latitudes under Holocene conditions. The cumulative processes predominating here today, and erosion in particular, have nothing to do with their inherited setting. In many cases, as on nearly all humid mid-latitude slopes of less than 27°, the Holocene erosional processes have not even been able to remove the frost debris sheets from the last Ice Age. The basal skeletons of the podzols and brown forest soils contain cryoturbation and solifluction structures with traces of the ice rind and of ice wedge casts. Here the processes at work have no relation to the setting on which they operate, and can attack it only rarely and insignificantly. This is shown particularly well in our example of the "reinterpretation" of the trough shoulders (see Section 2.3.9).

In order, therefore, to fulfill the goal of geomorphology and explain the total relief, one must reconstruct those processes which led up to this relief. This is done by climato-genetic geomorphology, which is therefore the very core of our field.

The stages by which a total relief was produced are called here relief generations. Each generation corresponds to a specific period of an ancient climate, and was produced by the specific relief-forming mechanisms governed by that climate. But the mechanisms invariably unfolded upon the stage inherited from earlier relief generations, a stage upon which they acted in various ways, transforming or traditionally working it. Before a mature product can be created by the relief-forming mechanisms of any generation, climatic change often sets in, bringing a new type of relief development with it. This has happened several times in the ectropics since the Late Pliocene. A particular relief generation is therefore often represented today only by scattered traces in the total relief. Collecting all the remains or overprinted traces of a particular relief generation, along with the traces of its companion spheres, and weeding out the remains of the previous and succeeding relief generations is the most important task of climato-genetic geomorphology.

Every relief generation is integrated with the preceding generations which it transforms, and with the succeeding generations, which in turn transform it. This is even true of relief generations like the Ice Ages, which differ radically from their predecessors. The Ice Age Alps, described so exhaustively by A. Penck and Brückner (1909) in terms of the features relating directly to the glaciations, are fully integrated into the sequence of quite different pre-glacial generations and the succeeding Holocene generation. Nearly all the glacially scoured features in the Alps and in Scandinavia are actually over-printed relict forms (though this overprinting was extremely thorough in the valley bottoms), and the relief here cannot be really satisfactorily explained without taking this inherited relief into account. Our example of the trough shoulder, for instance, is not Holocene, and is also not purely glacial. The trough shoulders in the Alpine valleys are gentle slopes slightly modified by glacial activity, and their basic shape was created by a completely different Middle to Late Pliocene warm climate, which in turn was working upon the stage inherited from a previous generation of etchplains (see Section 3.3.2.3 above).

Only when all generations of the pre-Holocene relief have been thoroughly analyzed can one then recognize what has been changed during the Holocene (the present relief generation) by the cumulative processes adapted to the present climate. Not even the most precise study of the basic elements of the current relief-forming mechanisms (even given that a balanced study of dynamic geomorphology were possible) could produce a picture of the accumulated effects of these processes on the bedrock.

Dynamic geomorphology must therefore be supplemented in two ways in its study of the Holocene. The first supplement regards the features which would mature under constant Holocene conditions, that is, which would result from these conditions were the older features to be entirely transformed. Determining what these features would be is the goal of *active geomorphology*. The second supplement concerns the effects produced when Holocene processes only lightly overprint the total relief. This is *synactive geomorphology*. Both of these will be taken up in more detail in the next edition. They are difficult to distinguish, and include much of what has hitherto gone under the name of slope research. Naturally this trinity of dynamic and active processes and their synactive effects applies to all older relief generations as well.

The best evidence available on which to interpret the features of a relief generation and to distinguish it from older and younger generations consists of an exact analysis of the inherited features themselves. Although the four companion spheres (see Section 1.7) may provide additional clues in the way of soil or sedimentary remains, the relict relief features themselves are evidence of typical cumulative effects (the active effects, as we have defined them above) produced over long periods of time. From here it is still a long step to recognize the dynamic mechanisms which created these relict features. We will explain this by means of the following example.

The mid-latitude periglacial region has become increasingly well investigated during the last forty years. We know pretty well *what* the Ice Age (especially the Würm) climate created in the way of soils and relief in the non-glaciated areas. But this does not tell us *how*, by what processes, the climate of that time created this plenitude of features, including frost-patterned ground, ice wedges, loess sheets, dells, and the ectropic deep, stepless, flat-floored valleys first created in the ectropics by this climate. For the processes then at work have long vanished and can no longer be observed and measured in the mid-latitudes.

Only one method can be applied here, that which we call *chrono-geographic comparison*. Keeping in mind all recognizable primary and secondary clues of the relict generation, we search for a morphoclimatic zone today in which active processes are working toward similar effects. Naturally an area with identical climate or chain of processes will rarely be found; it is enough that it resemble the ancient relief in question as closely as possible. Having discovered such a zone, one must then locate a special area within it providing a "natural test site," where the dynamics of the Holocene processes can clearly be recognized, and their part in the total Holocene relief-building rigorously identified.

As we have shown, the periglacial climate of Central Europe and the relief effects produced there correspond well to the conditions found in SE Spitzbergen today. The differences, namely, that the latter has no loess and no deeply involuted soils, are minor and show that although the temperatures were similar in both areas, Ice Age Central Europe had no arctic night, far more radiation in summer, and a more continental climate. Such differences, when recognized and explained, often provide additional insights.

The same is true, *mutatis mutandis*, for other relict relief generations. The chrono-geographic comparisons which we use today are by no means self-evident. The etchplain development which dominated the entire earth throughout the long Tertiary was thought by many authors to have its current equivalent in the arid regions. But only in the seasonal tropics can the processes which created this striking relief type be followed in fine detail.

In order to make chrono-geographic comparisons, one must naturally be familiar with the processes and morphologic effects in all the climatic zones of today. *Climatic geomorphology is a prerequisite for climato-genetic geomorphology*. Both our research and the layout of this book have followed this sequence.

Climatic geomorphologists have divided the continental surfaces into a series of zones (see Section 2.1 and Fig. 13 on the endpaper) of fairly uniform current relief-forming mechanisms. This division was based not only on analysis of the processes, but also of their product: each zone is characterized by a fairly uniform type of actively developing relief.

Three of these zones are particularly marked in character. The first of these to be recognized was the zone of sub- and circumglacial effects, for this was so strikingly different as to be obvious nearly from the beginning. The mechanisms which created the typical form assemblage of this zone (sands, gravel plains, and fluvio-glacial terraces) have been well investigated during the last century, and are so familiar that we have taken the reader's knowledge of them for granted.

Submarine and coastal features, though covering a very narrow zone, are as striking as glacial features, even though their effects are not as long-lasting. Where found, submarine and perimarine forms create significant, easily recognized breaks in relief development. Old coastal features are often important, because their common association with fossiliferous marine sediments may provide valuable clues for separating relief generations.

Volcanic flows which have buried and preserved relict surfaces, paleosols, or fossiliferous sediments may play a similar role (see Sections 2.5.1.3 and 3.3.1.1). Karst forms and structural escarpment landscapes, however, we do not regard as special, independent forms, but as (often very interesting) variations of the major morphoclimatic zone in which they are found.

The major emphasis in climatic and climato-genetic geomorphology is on the subaerial relief, which includes most of the continental surface today and throughout the geomorphic era (the time span ranging from the Late Cretaceous to the Holocene). This was considered practically a unit in the older geomorphology: the region of "normal erosion," the other regions being considered the "glacial" and "marine." True, Penck and Davis distinguished a larger "humid region" and a much smaller "arid region," but the latter was defined largely on the basis of minor features such as dunes, wind polishing, rinds, and hollow blocks, but not through any characteristics of the major landform assemblage.

The great advance represented by climatic geomorphology lies in the fact that it has divided up the enormous subaerial region into a series of morphoclimatic zones, each fully as striking as the glacial zone, not only in dominant mechanisms, but also in the relief style produced. This is true despite the fact that all these zones lie in the "humid" or "fluvial" region. But the word "fluvial" says relatively little in the framework of climatic geomorphology. The characteristic features of the various zones in the subaerial relief are based on the differing kinds and intensities of their elements, and above all, on the different ways in which their elements interact to create the respective relief-forming mechanisms (see Sections 1.6-1.9).

These zones were established by means of geographical comparions. Areas of extremely similar endogenous structure (according to tectonics, petrovariance, and epeirovariance) were identified in different climatic zones. The exogenic relief types of the two regions were then determined as precisely as possible. The present features were sorted out from the earlier relief generations, and the active developmental mechanisms were then investigated as they work there today. This included not only the study of all the individual elements of the system, but also of their mutual strengths and manner of interaction. In analyzing these elements, lower integrational levels, sometimes even basal integretation levels had to be used (see Fig. 5). In the polar zone of extreme valley-cutting it was necessary to delve more into the physical aspect. The subtropical zone of extreme planation, required more attention to the chemical aspect of things (see Sections 2.2. and 2.3).

These two zones, along with the subglacial relief zone, form the three main templates of the continental relief, the subglacial being the least of these both in time and space. All three zones were active in former relief generations in areas widely removed from their present zones. The zone of extreme valley-cutting was a major agent in shaping the periglacial region of the entire mid-latitudes from around 45° to as far S as 40° latitude. This occurred during the very brief period of the Ice Age Pleistocene, probably largely in the early phases of each Glacial. The zone of extreme planation, on the other hand, dominated the entire area of the continents from the equator into the polar regions through the Paleocene. The oldest relief generations that can be found today derive from this time, and traces can even be found from as far back as the Late Cretaceous (see Section 3.3.1.5). Since plains formed under these conditions are particularly durable and lend themselves to traditional development, they are the oldest members in the relief today. Tracing the history of the relief generations from there on, we find that the zone of excessive planation shrank during the course of the Neogene down to lower and lower latitudes. Active planation continued in Central Europe into the Early and Mid-Pliocene, while traditional planation lasted even into the Late Pliocene.

The major features of recent etchplain development, discussed in Section 2.3, have provided us with a series of further insights. For one thing, this morphoclimatic zone has remained stable on parts of the continents for extremely long periods of time: reckoning from the start of the Late Cretaceous on, we have a period of nearly 100 MY, or over one hundred times the entire Ice Age Pleistocene. During this time, the lowland relief of inselberg-studded etchplains produced there has become extremely accentuated.

A second insight is that the higher surfaces were important predecessors to the younger and lower surfaces. Many younger surfaces would not be as expansive as they are, had they not inherited a flat relief. We have stressed this in reference to both the Augenstein surface sequence in the Alps (Section 3.3.2.3) and the development of the Franconian Alb (3.3.1.5). Presumably, even the oldest Late Cretaceous surface remains were preceded by older, now destroyed surfaces.

This throws new light on the development of many structural scarp landscapes. Most of the surfaces above the upper scarp rims are actually etchplains. Even when fully adapted along brief stretches to harder strata, they are not endogenously created forms, but the products of exogenic, subaerial erosion. It can be shown that many hundreds of m of rock of varying hardness have been removed from above these surfaces. Yet nowhere is there any evidence that this occurred through a valley relief. Since the end product is nearly always a plain, there can be no other conclusion than that this occurred through planation under peritropical conditions. There are many indications of this in the Franconian Gäuland, and in the Franconian and Swabian Alb (see Sections 3.3.1.4 and 3.3.1.5). The fact that the surface, during the course of parallel lowering, may at times have become slightly adapted to a horizontal or gently dipping harder stratum in no way contradicts this. But it is worthy of note that such surfaces may be adapted at the same level a short distance away to a completely different stratum, where displacement along faults has aligned two different resistant strata at the surface. This has, for instance, been shown by Späth (1973) in the Hassberge.

Inselberg-studded-etchplains are the most characteristic relief assemblage of the zone of excessive planation, which takes its name from them. But they are not the only relief type of this zone. Uplift may lead to various kinds of more energetic ridge relief with wash valleys. This may resemble a hummocky *Kuppenrelief*, like that of the Highlands of Semién, the Bavarian forelands of the Böhmerwald, the Raxlandschaft, or the Kuppenalb. The latter two examples could also have developed out of tropical karst. The most extreme form of tropical karst is tower or cone karst, consisting of swarms of steep inselbergs sitting on a very soluble substrate. The "marginal karst plains" (*Karstrandebenen*) with their even-floored narrow extensions between the mogotes show the effects of the double planation mechansim, while the steep walls of the mogotes exemplify the rule of diverging erosion (see Fig. 52).

Another feature here consists of etchplain escarpments, usually very steep features with abrupt piedmont angles and working rims. These develop along monoclines. Once formed, the etchplain escarpment divides a higher, older etchplain from a lower one, whose inner edge will continue to expand actively (see Fig. 49). Such escarpments are usually broken up along their course by triangular reentrants. In cross-section escarpments lack the thicker lateritic soils of the surface, and have rock ledges where they are adapted to any harder outcropping strata. The ledges disappear wherever the escarpment runs across more consolidated rock. Etchplain escarpments and structural escarpments grade into each other in the tropics. The next etchplain, or system of etchplains, or even mountain area above an escarpment, will be pitted with shallow intramontane plains. These can be regarded as outlying "islands" of planation. They are formed not by lateral fluvial erosion, but rather by the narrowing down of planation to endogenously favored areas (see Section 2.3.3.3. and Figs. 47 and 48). The highest-latitude cases which resemble this are the poljes, found in highly soluble limestone (Section 3.4.3 and Fig. 82).

The tropical mountain relief type is found on etchplain escarpments, on the outer rises of inselberg ranges, along the walls of intramontane plains, and, of course, in the tropical mountains. In places, as in the highest parts of the tropical Andes, or in the deep river furrows of the mountain masses at the base of Indochina, it may produce the most stupendous valleys on earth.

This type of valley system is completely different from that familiar to the inhabitants of the ectropics. All upper courses and tributary valleys are narrow and V-shaped, and lead with steep and heavily stepped longitudinal profiles quickly up to the heights of the uplifted area without cutting horizontally far back into the uplifted mass. The base level of the relief never deviates far from the highest level of relief. The margins of the mountain mass are heavily scored, but its interior remains undissected. Since these upper courses and tributary valleys are largely controlled by chemical decomposition of the substrate, they are often joint-controlled (see Fig. 50d), have narrow floors, and carry little or no gravel. The main valleys and water gaps may be similar in character, but often have relatively low gradients compared to the stepped tributary gorges, and may widen out at times. This is particularly true near the edge of the mountain, where elongated triangular reentrants may reach far into the mountains along such major rivers.

An important feature is that this tropical mountain relief contributed to the stepped structure of the Alpine glacial valleys. Glacial scour did not create the valley steps, but only emphasized them, in some places overdeepening them (see Sections 3.3.2.1 and 3.3.2.3).

The same pre-glacial valley network was transformed differently in the periglacial region through the ice rind effect. Late Miocene and Early Pliocene valleys, filled in and later uncovered through mining or through later uplift and dissection (e.g., the old valleys along the Alpine margin S of the Vienna Basin and the S Franconian Alb N of Regensburg), often display such a steeply stepped tropical mountain valley type. As we have mentioned, the polar zone of excessive valley-cutting reached far into the mid-latitudes during the Early Glacial period, uniformly dissecting the entire relief from the high mountains (where unglaciated) to the most recent lowland platforms (the old moraines of N Germany, Central Poland, and Central Russia, the Tertiary hill country of the Alpine forelands, and the Late Tertiary sedimentary platforms of the Ukraine and of the N American Midwest). Broad, gravel-filled, flat-floored valleys were created, reaching far back into the heart of the mountains. Here the base level and the highest level of the relief were forced far apart. Solifluction and minor catastrophes of sudden wash, rill wash, and slope dissection with sharp ravines, all furthered slope dissection. On the relict surface, cryoturbation and frost rubble pavements permitted a very minor amount of traditional planation, and in some cases even protected the inherited surface. At most, small blockgipfels and felsklippen could be broken down and flattened off, but new creation of etchplains through "cryoplanation" cannot occur in this climate. The developmental character is instead directed far more toward destruction of relict surfaces through excessive valley-cutting, wherever a working rim above a bipartite or tripartite frost slope does not interrupt all erosional processes. Above such a working rim, flat dell systems can expand, while steep gullies develop below.

On the relict surfaces of the loess belt, dells from older Ice Ages were sometimes completely filled in with loess. Along the edges of the large valleys, gullies from older Ice Ages were at times later reformed into dells.

In addition to the two main subaerial relief templates, the zones of excessive valley-cutting and excessive planation, there are four weaker template zones. These include the morphoclimatic zone of partial planation in the perhumid tropics, whose relief assemblage is very similar to that of the seasonal tropics. In the lowlands of the trade-wind deserts, which have little or no frost, we have the "warm arid zone of plains preservation and traditional development" (Section 2.5) in which the etchplain relief from more humid Late Tertiary times is well preserved, although its ancient latosol sheets have been replaced by fluvio-eolian sandplains. The upper, more frost-prone mountain edges of the Sahara are part of the zone of glacis development, and have narrow pediments along their upper edges. At higher latitudes, this morphoclimatic/altitudinal zone reaches down into the lowlands in continental deserts like that of Iran, forming the "winter-cold arid zone of surface overprinting," which occurs primarily through glacis development (see Section 2.6). Here the surfaces are obscured by broad glacis fans, whose upper edges join the mountain via pediments. We have distinguished four types of pediment-cutting (Fig. 53a-d). Finally, the taiga valley-cutting zone (Section 2.2.21) resembles that of excessive valley-cutting in linear erosion, but in areal erosion is more like the zone of retarded valley-cutting.

In all six zones, the older relief generations are still more or less visible in the overall relief. This visibility is far greater in the two remaining zones, those of retarded valley-cutting in the mid-latitudes and of mixed relief development in the etesian subtropics; the older relief generations, in fact, are of predominant importance here. These two zones are therefore the most important study areas for climato-genetic geomorphology. It is here that influences from the humid tropics, from the glacial and periglacial polar areas, and from the cold arid regions, all come into play. Unraveling this picture, classifying all the traces of the different relief generations, and explaining their developmental conditions through chrono-geographic comparison is therefore particularly difficult here, but also particularly fruitful. Since the geomorphologists in N America and in Europe dealt first and foremost with the regions in which they grew up, it was at first thought that these areas presented the norm for terrestrial relief development and could clearly be interpreted on that basis. Actually, however, these two regions are the very ones which are particularly intricately formed. In order to explain them, one must gather experience throughout the entire world. Moreover, it was necessary to lay the methodological groundwork for climato-genetic geomorphology in order to distinguish the relief generations of other zones.

We will here summarize once again the core of this method.

(1) The first step is to analyze the features of a clearly defined relief

generation, for the relief assemblage itself is the prime evidence for the longevity of a specific relief-forming mechanism.

(2) At the same time one must investigate the petrovariance and all other endogenous influences.

(3) The next step is to search for further clues from the former companion spheres of the relief generation concerned. These should help indicate the type of developmental process going on at that time. They consist of remains from the pedosphere and decomposition sphere, including paleosols, crusts and rinds, traces of deep weathering, crevice fillings, traces of microweathering, remnant or lag gravels, pisolites and other hard or durable concretions, heavy and clay mineral types, sands, specific kinds of rounding, grain-sizes, windkanters, karst relicts, striated gravels, peat remnants, lignites, fossiliferous sediments, and all other kinds of sediments which are correlatable or whose age and origin can be determined.

(4) Along with (1) and (3), one must at all times endeavor meaningfully to associate all evidence found. Field research should not be conducted at random, but should rather aim at finding the exposures and areas of the relief most likely to prove productive. This "selective field research," constantly combining induction and deduction, leads to the locating of "natural test sites," where clear answers may be obtained as to which of the various possible solutions to a problem is correct. We have already provided various examples of this (see Sections 2.2.4, 3.3.1.1, 3.3.1.7, 3.3.1.8, and 3.4.4).

(5) Once these traces have been properly assembled, the next step is that of chrono-geographic comparison with a current morphoclimatic zone in which the active processes and the resulting relief assemblages closely resemble the relict relief generation under consideration.

(6) A prerequisite for the investigation of older relief generations, of course, is the exact description of the present relief generation. One of the most important means of doing this is through geographical comparison of its processes and features with those of an endogenously similar area in another morphoclimatic zone. It was through such comparisons that the nature and extent of the zones were first established. Once this has been done, it is possible to begin climato-genetic studies through chrono-geographic comparisons. At all times, a systematic, logical chain of evidence is what produces results. Such a chain of evidence may help identify specific questions which can be solved by investigation of appropriate natural test sites. And it is such a logical chain of evidence that can lead one to seek the help of neighboring branches of science working at less complex integrational levels.

(7) Finally, one must investigate how and to what extent the relief assemblage of the older relief generation was later preserved, traditionally developed, transformed, or destroyed. This can be called "synactive geomorphology."

In contrast to the situation described at the beginning of this epilogue, climato-genetic geomorphology is now in a position to interpret all of the relief. All elements, be they surface, slope, or rim, all discontinuities, and the incredible wealth of forms in the relief sphere can be accounted for by a well-founded genetic explanation. In this manner, the study of relief history through the geomorphic era takes an equal place beside geology and paleontology, which in the last century and a half have explored the history of the earth's crust. The latter two sciences have described the result of all the endogenously caused processes in the crust. The relief history describes the sum of all the exogenous processes which have been at work upon the crust since the Late Cretaceous and which have created its surface structure. Only both of these together can truly provide us with a history of the earth.

Considering its role as the main energy-conversion surface of the earth, as the sunlit setting of all life and of our future, the relief sphere and its developmental history are of great practical significance. The soils and waters, the rivers and coasts, the relief types and geographic settings are of inestimable value to us, while the natural plant and animal world must depend on us for their protection. Yet the world population has begun proliferating beyond all bounds; the relief sphere and its companion spheres, the sole basis of our existence, are being subjected to stress and even to attack by man. In order to find out what this, the basis of our existence, can tolerate in the way of exploitation, a deeper look into the laboratory of nature which created our environment may be appropriate and useful in many ways. Applied geomorphology (see Section 3.3.4) deals with this in a purely practical, technical sense. Yet such studies also provide us with deepened, more theoretical insight, appeasing our hunger for knowledge and teaching us how to behave in accordance with the great act of creation which surrounds us.

Glossary

This glossary is not intended to explain all or even most of the geomorphological terms used in the translation of this book. It includes only those terms for which English equivalents could not be found or had to be coined; terms that are not widely used in the English geomorphological language, but are important in Büdel's work; or terms that are used with a meaning differing from standard usage. For the explanation of words not found in this list the reader is referred to the subject index.

Areal erosion. *Breitenabtragung, Breitendenudation.*
  Erosion on interfluves. (See Section 1.8.1.)
Arid front slope. *Trockenfronthang.* Slope rising above a pediment in an arid environment. (See Fig. 53.)
Augmented solifluction. *Zufuhrsolifluktion.* Accelerated solifluction at the foot of a tripartite frost slope, due to increased supply of lubricating water and fine material from upslope, as well as to more intense chemical weathering in the moss tundra often found at the foot of such slopes. (See Section 2.2.13.)
Basal knob. *Grundhöcker.* Joint-controlled convexity of the basal surface of weathering, forming a shield inselberg when cropping out at the surface of the weathering mantle. (See Section 2.3.2.1 and Figs. 44 and 45.)
Basal surface of weathering. *Verwitterungsbasisfläche.* Irregular surface separating fresh unweathered bedrock from deeply weathered red loam blanket at the weathering front. (See Section 2.3.2.1.)
Break in slope, concave (Cv B/S). *Konkave Arbeitskante (Akv). See* piedmont angle.
Break in slope, convex (Cx B/S). *Konvexe Arbeitskante (Akx). See* working rim.
Broad Terrace. *Breitterrasse.* Late Pliocene to Early Pleistocene wide, rock-cut river terrace. Broad Terraces make up Büdel's second relief generation, separating the Central European relict etchplain surfaces (first generation) from the deeply incised Pleistocene valleys (third relief generation). (See Section 3.2.)
Cold stage. *Kaltzeit, kaltzeitlich.* Equivalent to glacial stage, but also applied to cold periods of the Early (non-glacial) Pleistocene and to regions

which were never glaciated, e.g., the subtropical lowlands during the time of Pleistocene glaciation elsewhere.

Companion spheres. *Begleitsphären.* The terrestrial hydrosphere and cryo-sphere, the biosphere, pedosphere, and decomposition sphere. These coexist with and determine the nature of the relief sphere. (See Section 1.7.)

Decomposition sphere. *Dekompositionssphäre.* The transitional zone of weathering between the pedosphere and the unweathered bedrock. (See Section 1.3.)

Deep weathering. *Tiefenverwitterung.* Deep-reaching and intense tropical weathering eating downwards along the basal surface of weathering. Typically produces red loam.

Dell. *Delle.* Shallow valley having a saucer-shaped profile and poorly defined rim, usually dry. Formed under periglacial conditions. *Cf.* swale for a similar feature of non-periglacial origin. (See Section 3.3.1.7.)

Divergent erosion. *Divergierende Abtragung.* Term originally introduced by Bremer (1971, 1973b) to describe the difference between erosion in the subtropics, where intense chemical weathering is most effective on the subhorizontal surfaces and weakest on steeper slopes, and the mid-latitudes or ectropics, where erosion is weakest on horizontal surfaces and most effective on slopes. (See Section 2.3.3.2.)

Double planation. *Doppelte Einebnung.* Areal erosion process lowering the subaerial upper denudation surface (*obere Einebnungsfläche*) at the top of the red loam blanket by surface wash, at the same time as it attacks the basal weathering surface (*basale Verwitterungsfläche*) by chemical disintegration and removal in solution. Neither surface is really level, the upper having a low relief of wash divides and wash depressions, and the basal surface (tens of meters below) a parallel relief of etchsurface divides and etchsurface depressions. In addition, the basal surface is more minutely sculptured by joint-controlled basal knobs. (See Sections 2.3.2ff.)

Ectropics, ectropical. *Ektropen, ektropisch.* Non-tropics or extra-tropic.

Epeirovariance. *Epirovarianz.* Variations in epeirogenic conditions.

Etchplain. *Rumpffläche.* A plain created by the double planation process. Active forms today are found only in the seasonal tropics. *Rumpffläche* is often translated as "peneplain," a term which we have avoided in order to prevent confusion with the very different Davisian approach to geomorphology. (See footnote 39.)

Etchplain escarpment. *Rumpfstufe.* Escarpment formed under tropicoid con-ditions by divergent erosion where uplift has warped an etchplain. Fre-quently contrasted in the German literature with *Schichtstufe* (strati-graphic scarp), said to be entirely lithologically controlled. (See Section 2.3.4.)

Etchplain pass. *Flächenpass*. Usually a fairly broad, level pass linking two etchplains across a divide between inselbergs or cutting through an inselberg mountain area. Relict etchplain passes in arid regions have been described as "pediment passes." (See Section 2.3.3.1.)

Etchsurface depression. *Rumpfmulde*. Concavity in the basal weathering surface of an etchplain, running roughly parallel to a wash depression on the external surface above the soil blanket. (See Section 2.3.2.3 [7].)

Etchsurface divide. *Rumpfschwelle*. Gentle rise in the basal surface of an etchplain, running roughly parallel to a wash divide on the external surface above the soil blanket. (See Section 2.3.2.3 [7].)

Fluvio-eolian sandplain. *Sandschwemmebene*. Very sandy plain in an arid environment, created by traditionally continued development of an etchplain. Slow sanding of exposed rock contributes sand which, transported by wind and water, attacks the tropical loam and converts it into a very mobile sandy sheet. (See Section 2.5.1.2.)

Frost slope. *Fosthang*. Slope formed by periglacial processes. Two types are recognized: the simple convex-concave frost slope (see Fig. 28) and the tripartite frost slope (see Fig. 30).

Geomorphic era. *Geomorphologische Ära*. The period which produced most of the exposed, non-exhumed relief of today. Begins with the Paleocene, following the Late Cretaceous marine transgressions. (See Sections 1.1 and 2.3.2.2 [12].)

Glacis. *Glacis*. Veneer of increasingly fine-grained arid debris extending out from a pediment into a basin and cutting discordantly across the basin fill beneath. (See footnotes 44 and 50.)

Ice rind. *Eisrinde*. Top layer of permafrost below the active layer, characterized by much ice which fragments the bedrock. Facilitates slope and fluvial erosion upon thawing by supplying readily transported debris. (See Section 2.2.11.)

Intramontane plain (intramontane basin). *Intramontane Ebene (intramontane Becken)*. Island of planation within a mountain or plateau area, created where double planation and divergent erosion continue in favored localities. Characterized by a single continuous surface containing several rivers which exit in different directions through separate gorges. (See Section 2.3.3.3.)

Moat (marginal depression). *Randsenke (Bergfussniederung)*. Depression in the basal weathering surface around an inselberg or in front of an etchplain escarpment. Created where deep weathering is favored by extra runoff from higher slopes. Visible as marked rings or trenches when the weathering mantle is deflated under arid conditions. (See Section 2.3.3.1.)

Pediment. *Pediment*. Debris-covered rock ledge in an arid mountain front at the head of a glacis fan. (See footnote 45 and Section 2.6.)

Basal pediment. *Basispediment*. Debris-covered rock ledge at the head of a glacis fan, formed by slight backweathering of the arid front slope (See Fig. 53d and Section 2.6).

Lateral pediment. *Lateralpediment*. Debris-covered rock ledge formed by lateral erosion of an aggrading desert stream or alluvial fan (see Fig. 53c and Section 2.6).

Traditional pediment. *Traditionspediment*. Relict etchplain margin at the foot of an arid mountain front, striped of much of its original red loam blanket and subsequently covered by fanglomerate in a winter cold desert environment. (See Section 2.6.)

Peritropical. *Randtropisch*. The regions bordering the seasonally humid tropics.

Piedmont angle (Cv B/S). *Konkave Arbeitskante (Akv), Fussknick*. Relatively sharp angle at the foot of a slope.

Relief-covering soil. *Ortsboden*. Any ectropic soil. The German term literally means "*in situ* soil," i.e., soil which has no effect on the relief and which does not change once it has reached maturity. (See Sections 1.6.6 and 2.3.2.2.)

Relief-forming soil. *Arbeitsboden*. Tropicoid soil, typically red loam. The German term literally means "working soil," i.e., a soil which is the agent of double planation, constantly eating away at the basal surface of weathering below, while being removed by surface wash above (see "double planation"). (See Sections 1.6.6 and 2.3.2.2.)

Relief generation. *Reliefgeneration*. The relief assemblage produced by one cycle of relief development under a given set of climatic conditions.

Shield inselberg. *Schildinselberg*. Protrusion of the basal surface of weathering barely visible at the surface of the soil blanket. (See Section 2.3.2.1.)

Subsurface basal sapping. *Subkutane Rückwärtsdenudation*. Chemical attack under the red loam blanket at the foot of an inselberg is intensified due to added water supply from inselberg slopes; this steepens inselberg flanks and forms moats. (See Section 2.3.3.1.)

Surface. *Fläche*. Following French geomorphological usage, refers to a more or less horizontal or gently sloping plain.

Swale. *Mulde*. Similar to dell, a spoon-shaped valley with a saucer-shaped cross-profile and no well-defined rim, but without periglacial connotations of dell. Used for any very gentle valley without genetic implications.

Traditional. *Traditional*. Refers to continued existence of a landform after climatic change, despite the new set of relief-forming processes working on it.

Traditionally continued development. *Traditionale Weiterbildung*. Relief development governed by the paleo-relief. New processes acting on an inherited feature may weakly carry on its development along the same lines as the former relief-forming mechanisms, but are incapable of creating it from scratch. Examples are: cryoplanation on a Tertiary relict surface, or fluvio-eolian sandplains on a former etchplain. (See Section 2.3.9.)

Triangular reentrant. *Dreiecksbucht.* Growth apex by which an etchplain extends into an adjacent upland. Ideally resembles an equilateral triangle in plan. (See Section 2.3.3.3.)

Triangular slope facet. *Dreieckshang.* Mid-slope area of a tripartite periglacial frost slope. Seen from above, resembles a triangle, separated from neighboring facets by ravines extending from the plateau rim down to the footslope. Augmented solifluction on the footslope and slight erosion on the face of the facet cause the base of the facet to have a straight working edge. The facet is actually the remnant of the original valley wall (the glacial trough wall). (See Section 2.2.15.)

Tropicoid. *Tropoid.* Tropic or tropic-like in nature. Includes both the tropic and the subtropic climates.

Tropicoid paleo-earth. *Tropoide Alterde.* Phase of worldwide tropic to subtropic climates, lasting from the Cretaceous/Tertiary transition into mid-Pliocene times. (See Sections 1.1 and 2.3.2.2 [12].)

Valley. *Tal.* A concavity which does not merely contain a river, but was actively created by it, by linear fluvial erosion working ahead of and therefore controlling slope erosion. Wash depressions in the tropics and drainage lines formed between coalescing alluvial fans are therefore not valleys. (See Sections 1.8.3, 2.3.2.3 [5], 2.3.2.3 [6], and footnote 37.)

Wash depression. *Spülmulde.* Active etchplains have an undulating relief at right angles to their overall slope, the undulations being 200-500 m from crest to crest. The depressions are not valleys in Büdel's genetic definition (see ''valley''): since no fluvial down-cutting occurs in them, they cannot have been created by the streams which flow in them. (See Section 2.3.2.3 [5] and Figs. 40 and 41.)

Wash divide. *Spülscheide.* The broad crest between two wash depressions. Both wash divides and wash depressions are created and lowered by surface wash. (See Section 2.3.2.3 [5] and Figs. 40 and 41.)

Wash ramp. *Spülsockel.* The gentle rise sometimes observed in the basal surface of weathering toward the foot of an inselberg. Covered by several meters of red loam on active etchplains (See Figs. 40 and 42d and e.)

Wash slope. *Spülmuldenflanke.* The flank of a wash depression, usually having a slope of less than 20‰.

Wash valley. *Spültal.* Valley which develops out of a wash depression when an etchplain is being dissected. Double planation becomes restricted to the latosol along these drainage lines, while the wash slopes and wash divides, stripped of their red loam blanket, become inactive. (See Section 2.3.5.)

Working rim (Cx B/S). *Konvexe Arbeitskante (Akx).* Sharp convex break in slope.

Bibliography

Achenbach, H., 1963, Die Halbinsel Cap Bon: Strukturanalyse einer mediterranen Kulturlandschaft in Tunesien: *Jahrbuch der Geographischen Gesellschaft zu Hannover*, v. 19, pp. 1-176.

Ackermann, E., 1936, Dambos in Nordrhodesien: *Wissenschaftliche Veröffentlichungen des Deutschen Instituts für Länderkunde*, Neue Folge, 4, pp. 148-157.

————, 1962, Büssersteine—Zeugen vorzeitlicher Grundwasserschwankungen: *Zeitschrift für Geomorphologie*, Neue Folge, 6, pp. 148-182.

Agadzhanyan, A. K., T. O. Boyarskaya, N. T. Glusenkova, and N. G. Sudakova, 1973, Paleogeography and stratigraphy of Mamontova Gora, Central Yakutia: *Biuletyn Peryglacjalny*, v. 23, pp. 19-36.

Ahlmann, H. W. *et al.*, 1931, Scientific results of the Swedish-Norwegian Arctic Expedition in the summer of 1931: *Geografiska Annaler*, v. 15, pp. 1-312.

Ahnert, F., 1955, Die Oberflächenformen des Dahner Felsenlandes: *Mitteilungen der Pollichia*, v. III, pp. 3-105.

————, 1970a, An approach towards a descriptive classification of slopes: *Zeitschrift für Geomorphologie*, Supplement-Band 9, pp. 71-84.

————, 1970b, A comparison of theoretical slope models with slopes in the field: *Zeitschrift für Geomorphologie,* Supplement-Band 9, pp. 88-101.

Angenheister, G., 1970, Die Erforschung der tieferen Erdkruste, Untersuchungsmethoden und Ergebnisse: *Physik unserer Zeit*, v. 1, no. 2, pp. 59-66.

Atlas Baikala (Atlas of the Baikal Lake), 1969: Akademija Nauk, Irkutsk-Moscow, 30 pp.

Bachmann-Vögelin, F., 1966, Fossile Strukturböden und Eiskeile auf jungpleistozänen Schotterfluren im NE-schweizerischen Mittelland: Dissertation, Zurich, 176 pp.

Bachmann, F., and G. Furrer, 1968, Die Situmetrie (Einregelungsmessung) als morphologische Untersuchungsmethode: *Geographica Helvetica*, v. 23, pp. 1-14.

Bakker, J. P., 1957a, Zur Granitverwitterung und Methodik der Inselbergforschung in Surinam: *31. Deutscher Geographen Tag, Würzburg*, pp. 122-131.

Bakker, J. P., 1957b, Einleitung zum Schwerpunkt "Flächenbildung in den feuchten Tropen": *31. Deutscher Geographen Tag, Würzburg*, pp. 86-89.

———, 1958, Zur Entstehung von Pingen, Oriçangas und Dellen in den feuchten Tropen, unter besonderer Berücksichtigung des Voltzberggebietes (Surinam): *Maull Festschrift, Abhandlungen des Geographischen Instituts Freie Universität Berlin*, v. 5, pp. 1-20.

———, 1960, Some observations in connection with recent Dutch investigations about granite weathering and slope development in different climates and climate changes: *Zeitschrift für Geomorphologie*, Supplement-Band 1, pp. 69-92.

———, 1963, Grossregionale Verwitterungszonen und Ferntransport von Ton durch Meeresströmungen: *Tijdschrift van het koninklijk Nederlandsch Aardrijkskundig Genootschap*, v. 80, pp. 109-120.

———, 1965, A forgotten factor in the interpretation of glacial stairways: *Zeitschrift für Geomorphologie*, Neue Folge, 9, pp. 18-34.

———, 1968, Contribution to a discussion in Saarbrücken on planation, Oct. 1967: *Zeitschrift für Geomorphologie*, Neue Folge, 12, pp. 470-489.

Bakker, J. P. and H. J. Müller, 1957, Zweiphasige Flussablagerungen und Zweiphasenverwitterung in den Tropen unter besonderer Berücksichtigung von Surinam: *Stuttgarter Geographische Studien*, v. 69, pp. 365-397.

Barsch, H. D., 1970, Die Poljen im Schweizer Jura: guest lecture, Department of Geography, Würzburg, June 23, 1970.

Barth, H. K., 1970, Probleme der Schichtstufenlandschaften Westafrikas am Beispiel der Bandiagara-, Gambaga-, und Mampong-Stufenländer: *Tübinger Geographische Studien*, v. 38, 215 pp.

Beaty, E. B., 1963, Origin of alluvial fans, White Mountains, California and Nevada: *Annals of the Association of American Geographers*, v. 53, pp. 516-535.

Becker, B., 1971, Absolute Chronologie der Nacheiszeit (Holozän) mit Hilfe der Dendrochronologie: *Akademie der Wissenschaften und der Literatur*, v. 1971, pp. 139-145.

———, 1972, Möglichkeiten für den Aufbau einer absoluten Jahrringchronologie des Postglazials anhand subfossiler Eichen aus Donauschottern: *Berichte der Deutschen Botanischen Gesellschaft*, v. 85, pp. 29-45.

Behrmann, W., 1924, Das westliche Kaiser-Wilhelmsland in Neuguinea: *Zeitschrift der Gesellschaft für Erdkunde*, Ergänzungs-Heft 1, 72 pp.

———, 1927, Die Oberflächenformen im feuchtheissen Kalmenklima: *Düsseldorfer Geographische Vorträge*, v. 3, pp. 4-9.

Berckhemer, H., 1968, Erdkruste und Erdmantel, in: *Vom Erdkern bis zur Magnetosphäre*, edited by H. Murawski, Umschau-Verlag, Frankfurt, pp. 131-145.

Beug, H. J., 1968, Probleme der Vegetationsgeschichte in Südeuropa: *Berichte der Deutschen Botanischen Gesellschaft*, v. 80. pp. 682-689.

Bibus, E., 1971, Zur Morphologie des südöstlichen Taunus und seines Rand-gebietes: *Rhein-Mainische Forschungen*, v. 74, 291 pp.

―――――, 1973, Untersuchungen zur jungtertiären Flächenbildung, Verwitterung und Klimaentwicklung im südöstlichen Taunus und in der Wetterau: *Erdkunde*, v. 27, pp. 10-26.

―――――, 1975, Eigenschaften tertiärer Flächen in der Umrahmung der nördlichen Wetterau (Taunus- und Vogelsbergrand): *Zeitschrift für Geomorphologie*, Supplement-Band 23, pp. 49-61.

Bird, B., 1967, *The Physiography of Arctic Canada:* Johns Hopkins University Press, Baltimore, 336 pp.

Birkenhauer, J., 1972, Modelle der Rumpfflächenbildung und die Frage ihrer Übertragbarkeit auf die deutschen Mittelgebirge am Beispiel des Rheinischen Schiefergebirges: *Zeitschrift für Geomorphologie*, Supplement-Band 14, pp. 39-53.

Birot, P., 1951, Sur le problème de l'origine des pédiments: *Compte Rendu du 16. Congrès International de Géographie, Lisbon, 1949*, v. 2, pp. 9-18.

―――――, 1955, *Les méthodes de la morphologie:* Presses Universitaires de France, Paris, 175 pp.

―――――, 1958, *Morphologie structurale:* Presses Universitaires de France, Paris, 2 vols., 463 pp.

―――――, 1960, *Le cycle d'érosion sous les différents climats:* Universidad do Brasil, curso 1, Rio de Janeiro, 137 pp.

―――――, 1963, Réflexions sur les caractères des glacis d'érosion en roches tendres: *Bulletin de la Société Hellénique de Géographie, Athens*, pp. 133-138.

―――――, 1970, Étude quantitative des processus érosifs agissants sur les versants: *Zeitschrift für Geomorphologie,* Supplement-Band 9, pp. 10-43.

Birot, P. and J. Dresch, 1966, Pédiments et glacis dans L'Ouest des États Unis: *Annales de Géographie*, v. 75, pp. 513-552.

Birzer, F., 1939, Verwitterung und Landschaftsentwicklung in der südlichen Frankenalb: *Zeitschrift der Deutschen Geologischen Gesellschaft*, v. 91, 57 pp.

Blackwelder, E., 1928, Mudflow as a geologic agent in semiarid mountains: *Bulletin of the Geological Society of America*, v. 39, pp. 456-484.

Blake, W., Jr., 1965, The late Pleistocene chronology of Nordaustlandet, Spitzbergen. Abstract in: *Vorträge des Fridtjof-Nansen-Gedächtnis Symposions; Ergebnisse der Stauferland-Expedition*, edited by J. Büdel and A. Wirthmann, v. 3, p. 29.

Blanc, A. C., 1936, Über die Quartärstratigraphie des Agro Pontino und der Bassa Versilia: *Verhandlungen der III Internationalen Quartärkonferenz, Wien*, pp. 273-279.

Blenk, M., 1960, Ein Beitrag zur morphometrischen Schotteranalyse: *Zeitschrift für Geomorphologie*, Neue Folge, 4, pp. 202-242.

Blüthgen, J., 1942a, *Die polare Baumgrenze in Lappland*: Veröffentlichungen des Deutschen Wissenschaftlichen Instituts zu Kopenhagen, Reihe I: Arktis, v. 10, 92 pp.

———, 1942b, Die diluviale Vereisung des Barentsee-Schelfes: *Naturwissenschaft*, v. 30, pp. 674-679.

———, 1966, *Allgemeine Klimageographie*, de Gruyter, Berlin, 720 pp.

Blume, H., 1968, Mangho Pir, eine Schichtstufenlandschaft im ariden Nordwesten Vorderindiens: *Geographische Zeitschrift*, v. 56, pp. 295-306.

———, 1970, Karstmorphologische Beobachtungen auf den Inseln über dem Winde: *Tübinger Geographische Studien*, v. 34, pp. 33-42.

———, 1971, Probleme der Schichtstufenlandschaft: *Erträge der Forschung*, v. 5, 117 pp.

———, 1974, Die Schichtstufen in der Umrahmung des Murzukbeckens: unpublished lecture delivered to the first symposium of the Deutscher Arbeitskreis für Geomorphologie, April 2.

Blume, H., and H. K. Barth, 1972, Rampenstufen und Schuttrampen als Abtragungsformen in ariden Schichtstufenlandschaften: *Erdkunde*, v. 26, pp. 108-116.

Bobek, H., 1934, Reise in Nordwestpersien 1934: *Zeitschrift der Gesellschaft für Erdkunde*, v. 69, pp. 359-369.

———, 1961, Die Salzwüsten Irans als Klimazeugen: *Anzeiger der Österreichischen Akademie der Wissenschaften*, v. 3, pp. 7-19.

———, 1963, Nature and implications of Quaternary climatic changes in Iran: *Proceedings of the Rome Symposium, UNESCO, WHO, Arid Zones Research*, 20, pp. 403-413.

———, 1969, Zur Kenntnis der südlichen Lut: *Mitteilungen der Österreichischen Geographischen Gesellschaft*, v. 111, pp. 155-192.

Bodechtel, J., 1965, Die südlichen Osterseen bei Iffeldorf in Oberbayern: *Erdkunde*, v. 19, pp. 150-155.

Bögli, A., 1956, Der Chemismus der Lösungsprozesse und der Einfluss der Gesteinsbeschaffenheit auf die Entwicklung des Karstes: *XVIII International Geographical Union, Rio de Janeiro, Report of the Commission on Karst Phenomena*, pp. 7-71.

———, 1960, Kalklösung und Karrenbildung: *Zeitschrift für Geomorphologie*, Supplement-Band 2, pp. 4-21.

———, 1969, Neue Anschauungen über die Rolle von Schichtfugen und Klüften in der karsthydrographischen Entwicklung: *Geologische Rundschau*, v. 58, pp. 395-408.

Bornhardt W., 1900, *Zur Oberflächengestaltung und Geologie Deutsch-Afrikas:* Berlin.

Brandt, B., 1917, Die tallosen Berge an der Bucht von Rio de Janeiro: *Mitteilungen der Geographischen Gesellschaft Hamburg*, v. 30, pp. 1-68.

Bremer, H., 1959, Flusserosion an der oberen Weser. Ein Beitrag zu den Problemen des Erosionsvorganges, der Mäander und der Gefälls-Kurve: *Göttinger Geographische Abhandlungen*, v. 22, 192 pp.

———, 1965a, Musterböden in tropisch-subtropischen Gebieten und Frostmusterböden: *Zeitschrift für Geomorphologie*, Neue Folge, 9, pp. 222-236.

———, 1965b, Ayers Rock, ein Beispiel für klimagenetische Morphologie: *Zeitschrift für Geomorphologie*, Neue Folge, 9, pp. 249-284.

———, 1965c, Der Einfluss von Vorzeitformen auf die rezente Formung in einem Trockengebiet—Zentralaustralien: *Deutscher Geographentag*, Heidelberg, pp. 184-196.

———, 1967a, Zur Morphologie von Zentralaustralien: *Heidelberger Geographische Arbeiten*, v. 17, 224 pp.

———, 1967b, Ein Beitrag zur Deutung der süddeutschen Schichtstufenlandschaft: die "Geologie des Schilfsandstein" von P. Wurster: *Zeitschrift für Geomorphologie*, Neue Folge, 11, pp. 352-355.

———. 1968, Der Fluss als Gestalter der Landschaft: *Geographische Rundschau*, v. 20, pp. 372-381.

———, 1971, Flüsse, Flächen- und Stufenbildung in den feuchten Tropen: *Würzburger Geographische Arbeiten*, v. 35, 194 pp.

———, 1972, Flussarbeit, Flächen- und Stufenbildung in den feuchten Tropen: *Zeitschrift für Geomorphologie*, Supplement-Band 14, pp. 21-38.

———, 1973a, Der Formungsmechanismus im tropischen Regenwald Amazoniens: *Zeitschrift für Geomorphologie*, Supplement-Band 17, pp. 195-222.

———, 1973b, Grundsatzfragen der tropischen Morphologie, insbesondere der Flächenbildung: *Geographische Zeitschrift, Beiheft: Geographie heute, Einheit und Vielfalt*, pp. 114-130.

———, 1975, Intramontane Ebenen, Prozesse der Flächenbildung: *Zeitschrift für Geomorphologie*, Supplement-Band 23, pp. 26-48.

Brochu, M., 1959, Genèse des moraines des boucliers cristallines (exemple du Bouclier Canadien): *Zeitschrift für Geomorphologie*, Neue Folge, 3, pp. 105-113.

Brosche, K.-U., 1969, Über die Beziehungen von Rumpfflächen zu Schichtkämmen und Schichtstufen sowie Beobachtungen an einigen wichtigen Strukturformtypen, erläutert an Beispielen aus dem nördlichen und nordwestlichen Harzvorland: *Zeitschrift für Geomorphologie*, Neue Folge, 13, pp. 207-216.

Brückner, W., 1955, The mantle rock ("laterite") of the Gold Coast and its origin: *Geologische Rundschau*, v. 43, pp. 307-327.

Brunnacker, K., 1957, Die Geschichte der Böden im jungeren Pleistozän in Bayern: *Geologica Bavarica*, v. 34, pp. 1-95.

Brunnacker, K., 1958, Über junge Bodenverlagerungen: *Geologische Blätter für Nordost Bayern*, v. 8, pp. 13-24.

———, 1959, Jungtertiäre Böden in Nordbayern: *Geologische Blätter für Nordost Bayern*, v. 9, pp. 55-63.

———, 1964, Grundzüge einer quartären Bodenstratigraphie in Süddeutschland: *Eiszeitalter und Gegenwart*, v. 15, pp. 224-228.

———, 1967, Die regionale Stellung der niederrheinischen Lössprovinz: *Sonderveröffentlichungen des Geologischen Instituts der Universität Köln*, v. 13, pp. 55-63.

———, 1970, Reliktböden und Landschaftsgeschichte zwischen Frankenhöhe und Rednitztal: *Geologische Blätter für Nordost Bayern*, v. 20, pp. 1-17.

———, 1973a, Gesichtspunkte zur jungeren Landschaftsgeschichte und zur Flussentwicklung in Franken: *Zeitschrift für Geomorphologie*, Supplement-Band 17, pp. 72-90.

———, 1973b, Einiges über Löss-Vorkommen in Tunesien: *Eiszeitalter und Gegenwart*, v. 23-24, pp. 89-99.

———, 1975, Der stratigraphische Hintergrund von Klimaentwicklung und Morphogenese ab dem höheren Pliozän im westlichen Mitteleuropa: *Zeitschrift für Geomorphologie*, Supplement-Band 23, pp. 82-106.

Brunner, H., 1968, Geomorphologische Karte des Mysore Plateaus (Südindien)—ein Beitrag zur Methodik der morphologische Kartierung in den Tropen: *Wissenschaftliche Veröffentlichungen des Deutschen Instituts für Länderkunde*, Neue Folge, v. 25-26, pp. 5-17.

———, 1969, Verwitterungstypen auf den Granitgneisen (Peninsular Gneis) des östlichen Mysore-Plateaus (Südindien): *Petermanns Geographische Mitteilungen*, v. 113, pp. 241-248.

———, 1970, Pleistozäne Klimaschwankungen im Bereich des östlichen Mysore-Plateaus (Südindien): *Geologie*, v. 19, pp. 72-82.

Bryan, K., 1935, The formation of pediments: *Report of the 16. International Geological Congress, Washington*, part 2, pp. 765-775.

Bubnoff, S. von, 1954, *Grundprobleme der Geologie*: Akademisches Verlag, Berlin, 234 pp.

Büdel, J., 1933, Die morphologische Entwicklung des südlichen Wiener Beckens und seiner Umrandung: *Berliner geographische Arbeiten*, v. 4, 73 pp.

———, 1935, Die Rumpftreppe des westlichen Erzgebirges: *Verhandlungen und wissenschaftliche Abhandlungen des deutschen Geographen-Tags, Bad Nauheim*, pp. 138-147.

———, 1936, Die quantitative Bedeutung der periglazialen Verwitterung, Abtragung und Talbildung in Mitteleuropa: *Verhandlungen der III Internationalen Quartärkonferenz*, pp. 169-172.

———, 1937, Eiszeitliche und rezente Verwitterung und Abtragung im ehemals nicht vereisten Teil Mitteleuropas: *Petermanns Geographische Mitteilungen*, Ergänzungs-Heft 229, 71 pp.

———, 1938, Das Verhältnis von Rumpftreppen und Schichtstufen in ihrer Entwicklung seit dem Alttertiär: *Petermanns Geographische Mitteilungen*, v. 7-8, pp. 229-238.

———, 1944, Die morphologischen Wirkungen des Eiszeitklimas im gletscherfreien Gebiet: *Geologische Rundschau (Klimaheft)*, v. 34, pp. 482-519.

———, 1948a, Das System der klimatischen Geomorphologie: *Deutscher Geographen-Tag, München*, pp. 65-100.

———, 1948b, Die klima-morphologischen Zonen der Polarländer: *Erdkunde*, v. 2, 1/3, pp. 22-53.

———, 1949, Die räumliche und zeitliche Gliederung des Eiszeitklimas: *Naturwissenschaften*, v. 36, pp. 105-112 and 133-139.

———, 1951, Fossiler Tropenkarst in der Schwäbischen Alb und den Ostalpen; seine Stellung in der klimatischen Schichtstufen und Karstentwicklung: *Erdkunde*, v. 5, pp. 168-170.

———, 1952, Bericht über klimamorphologische und Eiszeit-Forschung in Nieder Afrika: *Erdkunde*, v. 6, pp. 104-132.

———, 1954a, Sinai, "die Wüste der Gesetzes-Bildung:" *Abhandlung der Akademie für Raumforschung und Landesplannung*, v. 28, pp. 63-85.

———, 1954b, Klima-morphologische Arbeiten in Äthiopien im Frühjahr 1953: *Erkdunde*, v. 8, pp. 139-156.

———, 1955a, Das alte und das neue Äthiopien: *Deutscher Geographen-Tag, Hamburg*, pp. 97-133.

———, 1955b, Reliefgenerationen und plio-pleistozäner Klimawandel im Hoggar-Gebirge: *Erdkunde*, v. 9, pp. 100-115.

———, 1957a, Die "Doppelten Einebnungsflächen" in den feuchten Tropen: *Zeitschrift für Geomorphologie*, Neue Folge, 1, pp. 201-228.

———. 1957b, Die Eiszeit in den Tropen: *Universitas*, v. 12, no. 7, pp. 741-749.

———, 1957c, Die Flächenbildung in den feuchten Tropen und die Rolle fossiler solcher Flächen in anderen Klimazonen: *Deutscher Geographen-Tag, Würzburg*, pp. 89-121.

———, 1957d, Grundzüge der klimamorphologischen Entwicklung Frankens: *Würzburger Geographische Arbeiten*, v. 4-5, pp. 5-46.

———, 1957e, Die angebliche Zweiteilung der Würm-Eiszeit im Loisachtal bei Murnau (Südbayern): *Stuttgarter Geographische Studien*, v. 69, pp. 121-141.

———, 1959, Periodische und episodische Solifluktion im Rahmen der klimatischen Solifluktionstypen: *Erdkunde*, v. 13, pp. 297-314.

———, 1960, Die Frostschuttzone Südost-Spitzbergens: *Colloquium Geographicum*, v. 6, 105 pp.

Büdel, J., 1961a, Die Abtragungs-Vorgänge auf Spitzbergen im Umkreis der Barents-Insel: *Deutscher Geographen-Tag, Köln*, pp. 337-375.

————, 1961b, Morphogenese des Festlandes in Abhängigkeit von den Klimazonen: *Die Naturwissenschaften*, v. 48, pp. 313-318.

————, 1962, Die beiden interstadialen Würm-Böden in Südbayern. Eine Richtigstellung zu E. C. Kraus: *Eiszeitalter und Gegenwart*, v. 13, pp. 178-180.

————, 1963a, Aufbau und Verschüttung Olympias. Mediterrane Flusstätigkeit seit der Frühantike: *Deutscher Geographen-Tag, Heidelberg*, pp. 179-183.

————, 1963b, Die pliozänen und quartären Pluvialzeiten der Sahara: *Eiszeitalter und Gegenwart*, v. 14, pp. 161-187.

————, 1965, Die Relieftypen der Flächenspül-Zone Süd-Indiens am Ostabfall Dekans gegen Madras: *Colloquium Geographicum*, v. 8, 100 pp.

————, 1968a, Hang- und Talbildung in Südost-Spitzbergen: *Eiszeitalter und Gegenwart*, v. 19, pp. 240-243.

————, 1968b, Die junge Landhebung Spitzbergens im Umkreis des Freeman-Sundes und der Olga-Strasse: *Würzburger Geographische Arbeiten*, v. 22, pp. 1-21.

————, 1969a, Der Werdegang der Alpen, Europa und die Wissenschaft, in: *Neue Forschungen im Umkreis der Glocknergruppe*, edited by J. Büdel and U. Glaser, Wissenschaftliche Alpenvereinshefte, v. 21, pp. 13-45.

————, 1969b, Der Eisrinden-Effekt als Motor der Tiefenerosion in der exzessiven Talbildungszone: *Würzburger Geographische Arbeiten*, v. 25, 41 pp.

————, 1969c, Geomorphologie von Zentral-Australien nach H. Bremer: *Zeitschrift für Geomorphologie*, Neue Folge, 13, pp. 217-230.

————, 1970a, Pedimente, Rumpfflächen und Rückland-Steilhänge; deren aktive und passive Rückverlegung in verschiedenen Klimaten: *Zeitschrift für Geomorphologie*, Neue Folge, 14, pp. 1-57.

————, 1970b, Der Begriff: Tal: *Tübinger Geographische Studien*, v. 34, pp. 21-34.

————, 1971, Das natürliche System der Geomorphologie mit kritischen Gängen zum Formenschatz der Tropen: *Würzburger Geographische Arbeiten*, v. 34, 152 pp.

————, 1972a, Die Eisanschwellung im Würm-Hochglazial—Ursachen und Dauer: lecture delivered on Sept. 25 at the Deutsche Quartärgesellschaft, Stuttgart-Mannheim.

————, 1972b, Typen der Talbildung in verschiedenen klimamorphologischen Zonen: *Zeitschrift für Geomorphologie*, Supplement-Band 14, 20 pp.

————, 1973, Relief generationen der Poljenbildung im Dinarischen Raum:

in: *Neue Ergebnisse der Karstforschung in den Tropen und im Mittel-meerraum*, edited by A. Semmel, *Erdkundliches Wissen*, v. 32, pp. 134-142.

————, 1975a, Das Relief der Sahara zwischen Anti-Atlas und Hoggar-Gebirge: *Würzburger Geographische Arbeiten*, v. 43, pp. 66-78.

————, 1975b, Die Stellung der Geomorphologie im System der Naturwis-senschaften: *Zeitschrift für Geomorphologie*, Supplement-Band 23, pp. 1-11.

Busche, D., 1972a, Untersuchungen zur Pedimententwicklung im Tibesti-Gebirge: *Zeitschrift für Geomorphologie*, Supplement-Band 15, pp. 21-38.

————, 1972b, Vorläufiger Bericht über Untersuchungen an Schwemmfächern auf der Nordabdachung des Tibesti-Gebirges (République du Tschad): *Berliner Geographische Abhandlungen*, v. 16, pp. 95-104.

————, 1974, Die Entstehung von Pedimenten und ihre Überformung, un-tersucht an Beispielen aus dem Tibesti-Gebirge, République du Tschad: *Berliner Geographische Abhandlungen*, v. 18, 110 pp.

————, 1980, On the origin of the Msāk Mallat and the Hamādat Mānghīnī Escarpment: *Proceedings, Second Symposium on the Geology of Libya, Tripoli* (in press).

Butzer, K. W., 1957, The recent climatic fluctuation in lower latitudes and the general circulation of the Pleistocene: *Geografiska Annaler*, v. 39, pp. 105-113.

————, 1958, Das ökologische Problem der neolithischen Felsbilder in der östlichen Sahara, in: *Studien zum vor- und frühgeschichtlichen Land-schaftswandel in der Sahara*, Akademie der Wissenschaften und der Literatur Mainz, Jahrbuch 1958, pp. 197-198.

————, 1961, Climatic change in arid regions since the Pleistocene: *Arid Zones Research*, v. 17, pp. 31-56.

Cailleux, A., 1951, Morphoskopische Analyse der Geschiebe und Sandkörner und ihre Bedeutung für die Paläoklimatologie: *Geologische Rundschau*, v. 40, pp. 11-19.

Cailleux, A., and J. Tricart, 1959, *Initiation à l'étude des sables et galets*, Centre de Documentation Université Paris, 3 vols., 765 pp.

Capot-Rey, R., 1953, *Le Sahara Français*: Presses Universitaires de France, Paris, 202 pp.

Carner, H. F., 1974, *The Origin of Landscapes, A Synthesis of Geomorphology*: Oxford University Press, 734 pp.

Carson, M. A., and M. F. Kirby, 1972, *Hillslope, Form, and Process*: Cambridge University Press, Cambridge, 475 pp.

Castell, W.-D., Graf zu, 1935, *Chinaflug*: Atlantis Verlag, Berlin-Zürich.

Chorley, R. J., 1956, The relationship between angle of landscape and soil profile characteristics in the USA: *Premier rapport de la Commission pour l'étude des versants*, Union Géographique Internationale, pp. 42-44.

Choubert, G., 1957, Essai sur la corrélation des formations continentales et marines du pleistocène au Maroc: *Notes et Mémoires du Service Géologique du Maroc*, v. 25.

Clayton, R. W., 1956, Linear depressions (*Bergfussniederungen*) in savanna landscapes: *Geographische Studien*, v. 3, pp. 102-126.

Cooke, R. V., 1970, Stone pavements in deserts: *Bulletin of the Association of American Geographers*, v. 60, pp. 560-577.

Corbel, J., 1954, Les phénomènes karstiques en climat froid, abstract: *Erdkunde*, v. 8, p. 119.

————, 1957, Les Karsts du Nord-Ouest de l'Europe: *Institute des études Rhodaniennes de L'Université de Lyon, Mémoires et Documents*, v. 12, 547 pp.

————, 1959, Vitesse de l'érosion: *Zeitschrift für Geomorphologie*, Neue Folge, 3, pp. 1-28.

————, 1965, Soulèvement isostatique et englacement ancien (Spitzberg et mer de Barentz): in: *Vorträge des Fridtjof-Nansen-Gedächtnis Symposions, Ergebnisse der Stauferland-Expedition*, edited by J. Büdel and A. Wirthmann, v. 3, pp. 59-67.

Cotton, C. A., 1962, Planes and inselbergs of the humid tropics: *Transactions of the Royal Society of New Zealand, Geology*, v. 1, pp. 269-278.

Cramer, P., 1937, Die Störungszonen im Osten der Rhön: *Jahrbuch des Preussischen Geologischen Landesanstalts zu Berlin*, v. 58, pp. 746-777.

Credner, W., 1931, Das Kräfteverhältnis morphogenetischer Faktoren und ihr Ausdruck im Formenbild Südost-Asiens: *Bulletin of the Geological Society of China*, v. 11, pp. 13-34.

————, 1933, Zur Problematik einiger Durchbruchstäler in Kwantung (Süd-China): *Geologische Rundschau (Salomon-Calvi Festschrift)*, v. 23a, pp. 155-167.

————, 1935, *Siam, das Land der Thai*: Engelhorn, Stuttgart, 422 pp.

Creutzburg, N., 1921, *Formen der Eiszeit in Ankogelgebiet*: Ostalpine Formenstudien, Berlin.

Czudek, T., 1964, Development of the Surface of Levelling in the Bohemian Mass with special reference to the Nízký Jeseník Mountains (Gesenke): *Journal of the Czechoslovakian Geographical Society* (Congress supplement), v. 20, pp. 47-53.

————, 1971, The geomorphology of the eastern Nizky Jesnik Basin: *Rospravy Ceskoslovenske Akademie ved rada matematickych a prirodnich ved*, v. 81, pp. 1-90 (in Czechoslovakian with a German summary).

————, 1973a, Die Talasymmetrie im Nordteil der Morovská Bròna (Mährische Pforte): *Acta scient. nat. Academiae Sc Bohemoslovacae*, v. 7, new series 3, pp. 1-50.

————, 1973b, Zur klimatischen Talasymmetrie des Westteiles der Tschechoslovakei: *Zeitschrift für Geomorphologie*, Supplement-Band 17, pp. 49-57.

Czudek, T., and J. Demek, 1960, The importance of Pleistocene cryoplanation at the landscape development of the Bohemian Highland: *Anthropos*, v. 14, new series 6, pp. 57-69.

Czudek, T., and J. Demek, 1972, Present day cryogenic processes in the mountains of Eastern Siberia: *Geographia Polonica*, v. 23, pp. 5-20.

Czudek, T., and J. Demek, 1973, Die Reliefentwicklung während der Dauerfrostboden-Degradation: *Rozpravy Československé Akademie ved rada matematickych a prirodnich ved*, v. 83, pp. 1-83.

Dahlke, J., 1970, Beobachtungen zum Phänomen der Hangversteilungen in Südwestaustralien: *Erdkunde*, v. 24, pp. 285-290.

Davis, W. M., 1899, The geographical cycle: *Geographical Journal*, v. 2, pp. 481-504.

————, 1912, *Die erklärende Beschreibung der Landformen* (German translation by A. Rühl): B. G. Teubner, Leipzig, 565 pp.

————, 1938, Sheetfloods and streamfloods: *Bulletin of the Geological Society of America*, v. 49, pp. 1337-1416.

Dege, W., 1941, Landformende Vorgänge im eisnahen Gebiet Spitzbergens: *Petermanns Geographische Mitteilungen*, v. 87, pp. 81-97 and 113-122.

————, 1960, Wissenschaftliche Beobachtungen auf dem Nordostland von Spitzbergen 1944-1945: *Berichte des Deutschen Wetterdienstes*, v. 10, no. 72, 99 pp.

De Geer, G., 1912a, A Geochronology of the last 12,000 years: *Congrès Géologique International, Compte Rendu de la XI. Session, Stockholm*, pp. 241-257.

————, 1912b, Geochronologie der letzten 12,000 Jahre: *Geologische Rundschau*, v. 3, pp. 457-471.

Dehm, R., 1961a, Spaltenfüllungen als Lagerstätten fossiler Landwirbeltiere: *Mitteilungen der Bayerischen Staatssammlung für Paleontologie und historische Geologie*, v. 1, pp. 57-72.

————, 1961b, Über neue tertiäre Spaltenfüllungen des Süddeutschen Jura- und Muschelkalk-Gebietes: *Mitteilungen der Bayerischen Staatssammlung für Paleontologie und historische Geologie*, v. 1, pp. 27-56.

Demangeot, J., 1973, *Les milieux naturels désertiques*: Centre de Documentation Universitaires, Paris, 300 pp.

————, 1975, Recherches géomorphologiques en Indie du Sud: *Zeitschrift für Geomorphologie*, Neue Folge, 19, pp. 229-272.

Demek, J., 1964, Altiplanation terraces in Czechoslovakia and their origin: *Journal of the Czechoslovakian Geographical Society* (Congress supplement), v. 20, pp. 55-65.

―――, 1969, Cryoplanation terraces in Yacutia: *Biuletyn Periglacjalny*, v. 17, pp. 91-116.

―――, 1972, Die Pedimentation im subnivalen Bereich: *Göttinger Geographische Abhandlungen*, v. 60, pp. 145-153.

―――, 1973, Die Klimamorphologie des vulkanischen Gebietes von Chanbaisan in der Koreanischen Volksdemokratischen Republik: *Zeitschrift für Geomorphologie*, Supplement-Band 17, pp. 58-71.

Demek, J., and J. Kukla, 1969, *Periglazialzone, Löss und Paläolithikum der Tschechoslovakei*: Czechoslovakian Academy of Science, Geographical Institute, Brno, 155 pp.

De Ploey, J., 1963, Quelques indices sur l'évolution morphologique et palaeoclimatique des environs du Stanley-Pool (Congo): *Studia Universitatis "Lovanium,"* v. 17, pp. 1-16.

Dietrich, G., 1964, *Ozeanographie, Physische Geographie des Weltmeeres*: Westermann, Braunschweig, 96 pp.

Do Amaral, I., 1973, Formas des "Inselberge" (on montes-ilhas) e de meteorizacão superficial e profunda em rochas graniticas do Deserto de Macâmedes (Angola) na margem direita do rio Coroca: Júnta de investigaciones do ultramar, *Ser. Geogr. Lisboa*, v. 1, pp. 1-34.

Dongus, J., 1970, Über die Beobachtungen an Schichtstufen in Trockengebieten: *Tübinger Geographische Studien*, v. 34, pp. 43-55.

―――, 1972, Schichtflächenalb, Kuppenalb, Flächenalb (Schwäbische Alb): *Zeitschrift für Geomorphologie*, Neue Folge, 16, pp. 374-392.

Douglas, J., 1967, Man, vegetation and the sediment yields of rivers: *Nature*, v. 215, pp. 925-928.

―――, 1973, Rates of denudation in selected small catchments in Eastern Australia: *University of Hull, Occasional Papers in Geography*, v. 21, 127 pp.

Dresch, J., 1941, *Recherches sur l'évolution du relief dans le Massif Central du Grand Atlas, le Haouz et le Sous*: Arrault et Cie, Tours, 708 pp.

―――, 1957, Pédiments et glacis d'érosion, pédiplaines et inselbergs: *L'Information géographique*, v. 21, pp. 183-196.

―――, 1959, Notes sur la géomorphologie de l'Air: *Bulletin de l'Association de Géographes Français*, v. 280, pp. 2-20.

Dresch, J., and G. Rougerie, 1960, Observations morphologiques dans le Sahel du Niger: *Revue de Géomorphologie dynamique*, v. 11, pp. 49-58.

Dresch, J., M. Gigont, F. July, J. Le Coz, and R. Raynal, 1952, Aspects de la Géomorphologie du Maroc: *Notes et Mémoires du Service Géologique du Maroc*, v. 96, 182 pp.

Elsasser, H., 1968, *Untersuchungen an Strukturböden im Kanton Graubünden*: Juros-Verlag, Zürich, 63 pp.

Emiliani, C., 1966, Paleotemperature analysis of the Caribbean cores and a generalized temperature curve for the past 425,000 years: *Journal of Geology*, v. 74, pp. 109-124.

Ergenzinger, P., 1968, Rumpfflächen, Terrassen und Seeablagerungen im Süden des Tibesti-Gebirges: *Deutscher Geographen-Tag, Bad Godesberg*, pp. 412-427.

——, 1971-1972, Das südliche Vorland des Tibesti. Beitrag zur Geomorphologie der südlichen Sahara: unpublished manuscript.

Fahn, H. J., 1975, Die Sedimentschüttung in die Bayerischen Voralpenseen seit dem schmelzen des Würmeises (zugleich als Mass der Abtragungsleistung in den Einzugsgebieten): Dissertation, Würzburg, in press.

Fairbridge, R. W., ed., 1968, *The Encyclopedia of Geomorphology*, Reinhold Book Corporation, New York, 1295 pp.

Faust, B., 1973, Morphologische Entwicklungsphasen und heutige Morphodynamik in der Thessalischen Beckenzone und ihre Umwandlung: Dissertation, Technische Universität Braunschweig.

Fermor, L. L., 1911, What is laterite?: *Geological Magazine*, v. 5, pp. 454-462, 507-516, 559-566.

Feyling-Hanssen, R. W., 1955a, Stratigraphy of the marine Late Pleistocene of Billefjorden, Vestspitsbergen: *Norsk Polarinstitut Skrifter*, v. 107.

——, 1955b, Late Pleistocene deposits at Kapp Wijk, Vestspitsbergen: *Norsk Polarinstitut Skrifter*, v. 108.

——, 1965a, A marine section from the Holocene of Talavera on Barentsöya in Spitsbergen, in: *Vorträge des Fridtjof-Nansen-Gedächtnis Symposions, Ergebnisse der Stauferland-Exkursion*, edited by J. Büdel and A. Wirthmann, vol. 3, pp. 30-58.

——, 1965b, Problème du soulèvement des Terres autour de la Mer de Barentz, in: *Vorträge des Fridtjof-Nansen-Gedächtnis Symposions, Ergebnisse der Stauferland-Expedition*, edited by J. Büdel and A. Wirthmann, v. 3, pp. 68-72.

Feyling-Hanssen, R. W., and I. Olsson, 1960, Five radiocarbon datings of postglacial shorelines in central Spitsbergen: *Norsk Geografisk Tidscrift*, v. 17, pp. 1-4.

Fezer, F., 1969, Tiefenverwitterung circumalpiner Pleistozänschotter: *Heidelberger Geographische Arbeiten*, v. 24, 144 pp.

Finck, A., 1963, *Tropische Böden*: Parey, Hamburg und Berlin, 188 pp.

Fink, J., 1956a, Zur Korrelation der Terrassen und Lösse in Österreich: *Eiszeitalter und Gegenwart*, v. 7, pp. 49-77.

——, 1956b, Zur Systematik fossiler und rezenter Lössböden in Österreich: *VI. Congrès International de la Science du Sol*, pp. 585-592.

Fink, J., 1961a, Die Gliederung des Jung-Pleistozäns in Österreich: *Mitteilungen der Österreichischen Geologischen Gesellschaft*, v. 54, pp. 1-25.

———, 1961b, Die Südostabdachung der Alpen: *Mitteilungen der Österreichischen Bodenkundlichen Gesellschaft*, v. 6, pp. 123-183.

———, 1966, Die Paläogeographie der Donau, in: *Limnologie der Donau*, edited by R. Liepolt, Schweizerbart, Stuttgart, second edition, pp. 1-50.

———, 1973, Zur Morphogenese des Wiener Raumes, *Zeitschrift für Geomorphologie*, Supplement-Band 17, pp. 91-117.

Firbas, F., 1969, *Spät- und nacheiszeitliche Waldgeschichte Mitteleuropas nördlich der Alpen*, G. Fischer, Jena, 480 pp.

Fischer, H., 1963-1964, Geomorphologie des unteren Mühlviertels im Einzugsgebiet der Naarn: *Geographischer Jahresbericht aus Österreich*, v. 30, pp. 49-130.

Fischer, K., 1974, Die Pedimente im Bereich der Montes de Toledo, Zentralspanien: *Erdkunde*, v. 28, pp. 5-13.

Fittkau, E. I., 1970, Limnological conditions in the headwater region of the Xingu River, Brazil: *Tropical Ecology*, v. 11, pp. 20-25.

Flint, R. F., 1971, *Glacial and Quaternary Geology*: J. Wiley, New York, 892 pp.

Fliri, F., and E. W. von Grüning, 1971, Beiträge zur Stratigraphie und Chronologie der Inntalterrasse im Raum von Innsbruck: *Veröffentlichungen des Museums Ferdinandeum*, v. 51, pp. 5-21.

Fliri, F. *et al.*, 1970, Die Bändertone von Baumkirchen (Inntal, Tirol): *Zeitschrift für Gletscherkunde und Glazialgeologie*, v. 6, pp. 5-35.

Fliri, F., H. Hilscher, and V. Markgraf, 1971, Weitere Untersuchungen zur Chronologie der alpinen Vereisung (Baumkirchen Inntal, Nordtirol): *Zeitschrift für Gletscherkunde und Glazialgeologie*, v. 7, pp. 5-24.

Fliri, F., H. Felber, and H. Hilscher, 1972, Weitere Ergebnisse der Forschung am Bänderton in Baumkirchen: *Zeitschrift für Gletscherkunde und Glazialgeologie*, v. 8, pp. 203-213.

Flohn, H., 1969, Ein geophysikalisches Eiszeit-Modell: *Eiszeitalter und Gegenwart*, v. 20, pp. 204-231.

Fölster, H., 1964, Morphogenese der südsudanesischen Pediplane: *Zeitschrift für Geomorphologie*, Neue Folge, 8, pp. 393-423.

———, 1969, Slope development in SW Nigeria during the Late Pleistocene and Holocene: *Göttinger bodenkundliche Berichte*, v. 10, pp. 3-56.

Fournier, F., 1960, *Climat et érosion: la relation entre l'érosion du sol par l'eau et les précipitations atmosphériques*: Presses Universitaires de France, Paris, 201 pp.

Frechen, J., 1967, Vulkane, in: *Lehrbuch der allgemeinen Geologie*, edited by R. Brinkmann, Ferdinand Enke, Stuttgart, v. 3, pp. 45-91.

Freise, F. W., 1936, Erscheinungen des Erdfliessens im Tropenurwalde (Beobachtungen aus brasilianischen Küstenwäldern): *Zeitschrift für Geomorphologie*, v. 9, pp. 88-98.

Frenzel, B., 1964, Zur Pollenanalyse von Lössen: *Eiszeitalter und Gegenwart*, v. 15, pp. 5-39.

———, 1967, Die Klimaschwankungen des Eiszeitalters: *Die Wissenschaft*, v. 129, pp. 1-291.

———, 1968a, Grundzüge der pleistozänen Vegetationsgeschichte Nord-Eurasiens: *Erdwissenschaftliche Forschung*, v. 1, pp. 1-326.

———, 1968b, The Pleistocene vegetation of northern Eurasia: *Science*, v. 161, pp. 631-649.

Fugel, A., 1975, Beiträge zur Morphologie der Muschelkalkstufe zwischen Neckar und Fränkischer Saale: Dissertation, Würzburg, 105 pp.

Furrer, G., 1954, Solifluktionsformen im Schweizerischen Nationalpark: Dissertation, Zürich.

———, 1959, Untersuchungen am subnivalen Formenschatz in Spitzbergen und in den Bündener Alpen: *Geographica Helvetica*, v. 14, pp. 277-309.

———, 1969, Vergleichende Beobachtungen am subnivalen Formenschatz in Ostspitzbergen und in den Schweizer Alpen: *Ergebnisse der Stauferland-Expedition*, v. 9, 40 pp.

Furrer, G., and F. Bachmann, 1968, Die Situmetrie (Einregelungsmessung) als morphologische Untersuchungsmethode: *Geographica Helvetica*, v. 23, pp. 1-14.

Gabriel, A., 1934, Beobachtungen im Wüstengürtel Innerpersiens 1933: *Mitteilungen der Geographischen Gesellschaft in Wien*, v. 77, pp. 53-77.

———, 1942, Die Lut und ihre Wege: *Zeitschrift für Erdkunde*, v. 10, 423-442.

———, 1964, Zum Problem des Formenschatzes in extrem ariden Räumen: *Mitteilung der Österreichischen Geographischen Gesellschaft*, v. 106, pp. 3-15.

Gabriel, B., 1972a, Neuere Ergebnisse der Vorgeschichtsforschung in der östlichen Zentral-Sahara: *Berliner Geographische Abhandlungen*, v. 16, pp. 153-156.

———, 1972b, Terrassenentwicklung und vorgeschichtliche Umweltbedingungen im Enneri Dirennao (Tibesti, östlich Zentralsahara): *Zeitschrift für Geomorphologie*, Supplement-Band 15, pp. 113-128.

Gaertner, R. von, 1968, Zur stratigraphischen und morphologische Altersbestimmung von Altflächen: *Geologische Rundschau*, v. 58, pp. 1-9.

Gams, H., 1927, Die Ergebnisse der pollenanalytischen Forschung in bezug auf die Geschichte der Vegetation und des Klimas von Europa: *Zeitschrift für Gletscherkunde*, v. 15, pp. 161-190.

Ganssen, R., 1957, *Bodengeographie*: Koehler, Stuttgart, 219 pp.

———, 1965, *Grundsätze der Bodenbildung*: B-I Hochschultaschenbücher, Mannheim, 132 pp.

Gardi, R., and J. Neukom-Tschudi, 1969, *Felsbilder in der Sahara im Tassili n'Adjer*: Orbis Pictus, Bern, v. 52, pp. 1-4.

Gardner, H. F., 1974, *The Origin of Landscapes: A Synthesis of Geomorphology*: Oxford University Press, 734 pp.

Gellert, J. F., 1931, Geomorphologie des mittelschlesischen Inselberglandes: *Zeitschrift der deutschen geologischen Gesellschaft*, v. 83, pp. 431-447.

————, 1958, *Grundzüge der physischen Geographie von Deutschland*, v. 1: *Geologische Struktur und Oberflächengestaltung*: Deutscher Verlag der Wissenschaften, Berlin, 402 pp.

————, 1965, Neue morphologische Untersuchungen und Probleme in den sächsisch-thüringischen Rumpfgebirge und in ihrem Vorland zwischen Elbe und Pleisse: *Forschung und Forschritt*, v. 39, pp. 70-76.

————, 1967, Zur Problematik der verschütteten Bergländer (Inselbergländer) im sächsischen und schlesischen Gebirgsvorland und der "fossilen Inselberge" in den Mittelgebirgen Mitteleuropas: *Wissenschaftliche Zeitschrift der Pädagogischen Hochschule Potsdam*, v. 11, pp. 281-286.

Gerasimov, I. P., and A. Velitschko, 1969, *Loess-Periglaciaire Paléolithique sur le territoire de l'Europe moyenne et orientale*: Académie des Sciences of the German Democratic Republic, Hungary, Poland, Czechoslovakia, and the Soviet Union at the VIII INQUA Congress in Paris.

Gerasimov, I. P., and A. W. Sidorenko, 1971, *Map of the Denudation Surfaces and Weathering Blankets of the USSR* (in Russian): Ministry of Geology and the Academy of Sciences, Moscow.

Gerstenhauer, A., 1960, Der tropische Kegelkarst in Tabasco (Mexiko): *Zeitschrift für Geomorphologie*, Supplement-Band 2, pp. 22-48.

————, 1967, Ein karstmorphologischer Vergleich zwischen Florida und Yucatan: *Deutscher Geographen-Tag, Bad-Godesberg*, pp. 332-344.

————, 1972, Der Einfluss des Co_2-Gehaltes der Bodenluft auf die Kalklösung: *Erdkunde*, v. 26, pp. 116-120.

Gerstenhauer, A., and K. H. Pfeffer, 1966, *Beiträge zur Frage der Lösungsfreudigkeit von Kalkgesteinen*: Abhandlungen zur Karst- und Höhlenkunde, Reihe A., v. 2, 46 pp.

Geyh, M. A., 1971, *Die Anwendung der ^{14}C-Methode und anderer radiometrischer Datierungsverfahren für das Quartär*: Claustaler tektonische Hefte, v. 11, 118 pp.

Geyh, M. A., and D. Jäkel, 1974, Spätpleistozäne und holozäne Klimageschichte der Sahara auf Grund zugänglicher ^{14}C-Daten: *Zeitschrift für Geomorphologie*, Neue Folge, 18, pp. 82-98.

Gierloff-Emden, H. G., and U. Rust, 1971, Verwertbarkeit von Satellitenbildern für geomorphologische Kartierungen in Trockenräumen (Chihuahua, New Mexico, Baja California): Bildinformation und Geländetest: *Münchener Geographische Abhandlungen*, v. 5, 97 pp.

Gitermann, R. E. *et al.*, 1968, The main development stages of the vegetation of North Asia in Anthropogen: *USSR Academy of Sciences, Geological Institute, Transactions*, v. 177, 270 pp.

Gladfelter, B. G., 1971, *Meseta and Campiña, Landforms in Central Spain, a geomorphology of the Alto Henares Basin*: University of Chicago, Department of Geography Research Paper, v. 130, 204 pp.

Glaser, G., 1969, Der Zinnstein-Abbau in Rondônia: *Geomorphologische Zeitschrift*, v. 57, pp. 241-267.

Glaser, U., 1964, Die miozäne Strandzone am Südsaum der Schwäbischen Alb: *Würzburger Geographische Arbeiten*, v. 11, 99 pp.

————, 1968, Junge Landhebung im Umkreis des Storfjord (SO-Spitzbergen): *Würzburger Geographische Arbeiten*, v. 22, pp. 1-22.

Glen, J. W., 1952, Experiments on the deformation of ice: *Journal of Glaciology*, v. 2, pp. 11-114.

Glinka, K. D., 1914, *Die Typen der Bodenbildung, ihre Klassifikation und geographische Verbreitung*: Borntraeger, Berlin, 365 pp.

Gossmann, H., 1970, Theorien zur Hangentwicklung in verschiedenen Klimazonen. Mathematische Hangmodelle und ihre Beziehungen zu den Abtragungsvorgängen: *Würzburger Geographische Arbeiten*, v. 31, 146 pp.

Goudie, A., 1973, *Duricrusts in Tropical and Subtropical Landscapes*: Clarendon Press, Oxford, 174 pp.

Gradmann, R., 1931, *Süddeutschland. Vol. 1. Allgemeiner Teil*: Engelhorn, Stuttgart, 215 pp.

Graf, K., 1973, Vergleichende Beobachtungen zur Solifluktion in verschiedenen Breitenlagen: *Zeitschrift für Geomorphologie*, Supplement-Band 16, pp. 104-154.

Graul, H., 1952, Zur Gliederung der mittelpleistozänen Ablagerungen in Oberschwaben: *Eiszeitalter und Gegenwart*, v. 2, pp. 133-146.

————, 1960, Der Verlauf des glazialeustatischen Meeresspiegel-Anstieges, berechnet an Hand von 14-Datierungen: *Deutscher Geographen-Tag, Berlin*, pp. 232-242.

Grunert, J., 1972a, Die jungpleistozänen und holozänen Flussterrassen des oberen Enneri Yebbigué im zentralen Tibesti-Gebirge und ihre klimatische Deutung: *Berliner Geographische Abhandlungen*, v. 16, pp. 105-116.

————, 1972b, Zum Problem der Schluchtbildung im Tibesti-Gebirge (République du Tschad): *Zeitschrift für Geomorphologie*, Supplement-Band 15, pp. 144-155.

————, 1975, Beiträge zum Problem der Talbildung in ariden Gebieten am Beispiel des zentralen Tibesti-Gebirges (République du Tschad): *Berliner Geographische Arbeiten*, v. 22, 96 pp.

Gurlitt, D., 1949, Das Mittelrheintal, Formen und Gestalt: *Forschungen zur Deutschen Landeskunde*, v. 46, 159 pp.

Gwinner, P., 1968, Paläogeographie und Landschaftsentwicklung im weissen (oberen) Jura der Schwäbischen Alb (Baden-Württemberg): *Geologische Rundschau*, v. 58, pp. 32-41.

Haefke, F., 1926, Karsterscheinungen im Südharz: *Mitteilungen der Geographischen Gesellschaft Hamburg*, v. 37, pp. 77-105.

Hafemann, D., 1960, Die Frage des eustatischen Meeresspiegelanstiegs in historischer Zeit: *32. Deutscher Geographen-Tag, Wiesbaden*, v. 32, pp. 218-231.

―――, 1965, Die Niveauveränderungen an den Küsten Kretas seit dem Altertum nebst einigen Bemerkungen über alte Strandablagerungen auf West Kreta: *Abhandlungen Mathematisch-Naturwissenschaftlicher Klasse, Akademie der Wissenschaften und der Literatur Mainz*, v. 12, 82 pp.

Haffner, H., 1967, Landschaft und Oberfläche des Mondes: *Bild der Wissenschaft*, v. 4, pp. 288-301.

―――, 1969, Zur Morphologie des Mondes nach Lunar Orbiter V: *Naturwissenschaft*, v. 56, pp. 49-56.

Hagedorn, H., 1966, Landforms of the Tibesti Region, in: *Southern Libya and Northern Chad: A Guide Book*, Petroleum Exploration Society of Libya, Tripoli, pp. 53-58.

―――, 1967a, Studien über den Formenschatz der Wüste an Beispielen aus der Südost Sahara: *36. Deutscher Geographen-Tag, Bad Godesberg*, pp. 401-411.

―――, 1967b, Beobachtungen an Inselbergen im westlichen Tibesti-Vorland: *Berliner Geographische Abhandlungen*, v. 5, pp. 17-22.

―――, 1968, Über äolische Abtragung und Formung in der Südost-Sahara: *Erdkunde*, v. 22, pp. 257-269.

―――, 1970, Zum Problem der Glatthänge: *Zeitschrift für Geomorphologie*, Neue Folge, 14, pp. 103-113.

―――, 1971, Untersuchungen über Relieftypen arider Räume an Beispielen aus dem Tibesti-Gebirge und seiner Umgebung: *Zeitschrift für Geomorphologie*, Supplement-Band 11, 251 pp.

―――, 1972, Commentary in discussion to Anders, Beobachtungen zur jungquartären Formungsdynamik am Südrand des Anti-Atlas (Morokko): *Zeitschrift für Geomorphologie*, Supplement-Band 14, p. 80.

―――, 1974, Sumatra, Entwicklungsprobleme eines tropischen Waldlandes: unpublished lecture to the Geographische Gesellschaft, Würzburg, July 8.

Hagedorn, H., W. Haars, D. Busche, and H. Forster, 1975, Pleistozäne Vergletscherungs-Spuren in Zentral Iran: *Zeitschrift für Geomorphologie*, Supplement-Band 23, pp. 146-156.

Hallsworth, E. G., G. K. Robertson, and F. R. Gibbons, 1955, Studies in New South Wales, VI: The "Gilgai" soils: *Journal of Soil Science*, v. 6, pp. 1-31.

Hamberg, A., 1916, Zur Kenntnis der Vorgänge im Erdboden beim Gefrieren und Auftauen usw.: *Geologische Forschung*, v. 37.

Hamelin, L.-E., and F. A. Cook, 1967, *Le périglaciare par l'image. The glossary of periglacial phenomena*: Traveaux et documents du centre d'études nordiques, Quebec, 232 pp.

Hann, J., 1910, *Handbuch der Klimatologie*, V. II/I, *Klima der Tropenzone*, Engelhorn, Stuttgart, 426 pp.

Harrossowitz, H.L.F., 1926, Laterit: *Fortschritte der Geologie und Paläontologie*, v. 4, pp. 253-566.

Haserodt, K., 1965, Untersuchungen zur Höhen- und Altersgliederung der Karstformen in den nördlichen Kalkalpen: *Münchener Geographische Hefte*, v. 27, 114 pp.

Hassinger, H., 1905, Geomorphologische Studien aus dem inneralpinen Wiener Becken und seinem Randgebirge: *A. Pencks Geographische Abhandlungen*, v. 8, 205 pp.

————, 1914, Die mährische Pforte und ihre benachbarten Landschaften: *Abhandlungen der Geographischen Gesellschaft in Wien*, v. 11, no. 2, 313 pp.

Hedin, S., 1910, *Zu Land nach Indien*: 2 vols., Brockhaus, Leipzig, 801 pp.

————, 1916-1922, *Southern Tibet*: 9 vols., Lithographic Institute of the General Staff of the Swedish Army, Stockholm.

Heine, K., 1970a, Fluss- und Talgeschichte im Raum Marburg: *Bonner Geographische Abhandlungen*, v. 42, 195 pp.

————, 1970b, Einige Bemerkungen zu den Liefergebieten und Sedimentationsräumen der Lösse im Raum Marburg/Lahn auf Grund tonmineralogischer Untersuchungen: *Erdkunde*, v. 24, pp. 180-194.

————, 1972, Die Bedeutung pedologischer Untersuchungen bei der Trennung von Reliefgenerationen: *Zeitschrift für Geomorphologie*, Supplement-Band 14, pp. 113-137.

Heinzelin, J. de, 1955, Observations sur la genèse des nappes de gravats dans les sols tropicaux: *Publications de l'Institut National pour l'étude agronomique du Congo Belge (I.N.E.A.C.), Série scientifique*, v. 64, pp. 1-37.

Helbig, K., 1965, Asymmetrische Eiszeittäler in Süddeutschland und Ostösterreich: *Würzburger Geographische Arbeiten*, v. 14, 108 pp.

Hempel, L., 1951, Struktur- und Skulpturformen im Raum zwischen Leine und Harz: *Göttinger Geographische Abhandlungen*, v. 7, 66 pp.

Herold, A., 1957, Die geographischen Grundlagen des Obstbaus im Bereich zwischen Maindreieck und Steigerwaldstufe: *Würzburger Geographische Arbeiten*, v. 4-5, pp. 241-384.

Herrmann, R., 1968, Methoden und Ergebnisse einer hydrogeographischen Landsaufnahme von Ost-Nigerien: *Erdkunde*, v. 22, pp. 89-100.

Herrmann, R., 1970, Die vertikale Differenzierung des Wasserhaushaltes in einem tropischen Hochgebirge (Sierra Nevada de Santa Marta/Kolumbien): unpublished lecture delivered to the Würzburger Geographische Kolloquium, April 28.

———, 1971, Die zeitliche Änderung der Wasserbindung im Boden unter verschiedenen Vegetationsformationen der Höhenstufen eines tropischen Hochgebirges (Sierra Nevada de Santa Marta, Kolumbien): *Erdkunde*, v. 25, pp. 90-102.

Hettner, A., 1921, Die Oberflächenformen des Festlandes: *Geographische Schriften*, v. 4, 178 pp.

Hills, E. S., 1969, *Arid Lands*, Methuen, London, 461 pp.

Hirano, M., 1966, A study of mathematical models of slope development: *Geographical Review of Japan*, v. 39, pp. 324-336.

Hjulström, F., 1932, Das Transportvermögen der Flüsse und die Bestimmung des Erosionsbetrages: *Geografiska Annaler*, v. 14, pp. 244-258.

———, 1935, Studies on the morphological activity of rivers (as illustrated by the river Fyris): *Bulletin of the Geological Institute of the University of Uppsala*, v. 25, pp. 221-452.

———, 1942, Studien über das Mäander-Problem: *Geografiska Annaler*, v. 24, pp. 233-269.

Högbom, B., 1910, Einige Illustrationen zu den geologischen Wirkungen des Frostes auf Spitzbergen: *Bulletin of the Geological Institute of the University of Uppsala*, v. 11, pp. 242-251.

———, 1914, Über die geologische Bedeutung des Frostes: *Bulletin of the Geological Institute of the University of Uppsala*, v. 12, pp. 257-390.

Höhl, G., 1963, Die Siegritz-Voigendorfer Kuppenlandschaft. Ein Beitrag zur klimatisch-morphologischen Deutung einer Reliktlandschaft des Karstes aus feuchtwarmer Zeit in der nördlichen Frankenalb: *Mitteilungen der Fränkischen Geographischen Gesellschaft*, v. 10, pp. 211-223.

Höllermann, P. W., 1967, Zur Verbreitung rezenter periglazialer Kleinformen in den Pyrenäen und Ostalpen: *Göttinger Geographische Abhandlungen*, v. 40, pp. 235-260.

———, 1972, Beiträge zur Problematik der rezenten Strukturbodengrenzen: *Göttinger Geographische Abhandlungen*, v. 60, pp. 235-260.

Hörner, N., 1933, Geomorphic processes in the continental basins of Central Asia: *Report of the XVI International Geological Congress, Washington*.

Hövermann, J., 1949, Morphologische Untersuchungen im Mittelharz: *Göttinger Geographische Abhandlungen*, v. 2, 80 pp.

———, 1950, Zur Altersdatierung der Granitvergrusung: *Neues Archiv für Niedersachsen*, v. 4, pp. 489-491.

———, 1953a, Die Periglazial-Erscheinungen im Harz: *Göttinger Geographische Abhandlungen*, v. 14, pp. 7-44.

————, 1953b, Die Oberflächenformen um Göttingen: *Göttinger Jahrbuch*, pp. 65-74.

————, 1963, Vorläufiger Bericht über eine Forschungsreise ins Tibesti-Massiv: *Erde*, v. 94, pp. 126-135.

————, 1966, Hangformen und Hangentwicklung zwischen Syrte und Tschad: *L'évolution des versants*, pp. 139-156.

————, 1967, Die wissenschaftlichen Arbeiten der Station Bardai im ersten Arbeitsjahr (1964-1965). Arbeitsberichte aus der Station Bardai (Tibesti): *Berliner Geographische Abhandlungen*, v. 5, pp. 7-10.

————, 1972, Die periglaziale Region des Tibesti und ihr Verhältnis zu angrenzenden Formungsregionen: *Göttinger Geographische Abhandlungen*, v. 60, pp. 261-283.

Hoffmann, V., 1964, Die Geologie des Blattes Würzburg-Süd 1:25,000 (Stratigraphie der Trias und Tektonik im südlichen Unterfranken): Dissertation, Würzburg, 107 pp.

Hofmann, W., 1968, Geobotanische Untersuchungen in Südost-Spitzbergen 1960: *Ergebnisse der Stauferland-Expedition*, v. 8, 83 pp.

Hoinkes, H., 1961, Die Antarktis und die geophysikalische Erforschung der Erde: *Naturwissenschaft*, v. 48, pp. 354-374.

Hormann, K., 1965, Das Längsprofil der Flüsse: *Zeitschrift für Geomorphologie*, Neue Folge, 9, pp. 437-456.

Howard, A. D., 1942, Pediment passes and the pediment problem: *Journal of Geomorphology*, v. 5, pp. 3-31 and 95-136.

Hüser, K., 1972, Geomorphologische Untersuchungen im westlichen Hintertaunus: *Tübinger Geographische Studien*, v. 50, 184 pp.

Hurault, J., 1967, L'érosion régressive dans les régions tropicales humides et la genèse des inselbergs granitiques: *Institute de Géographie National, Études de photointerprétation*, v. 3, 98 pp.

Illies, H., 1951, Die paläogeographische Auswertung der Schrägrichtung: *Geologische Rundschau*, v. 39, pp. 234-237.

————, 1952, Die morphogenetische Analyse fluvialer und fluvio-glazialer Aufschüttungs-Landschaften: *Neues Jahrbuch für Geologie, Paleontologie und Mineralogie*, v. 9, pp. 385-401.

————, 1964, Kontinentalverschiebungen und Polverschiebungen: Ursachen und Probleme: *Geologische Rundschau*, v. 54, pp. 549-579.

————, 1965a, Bauplan und Baugeschichte des Oberrheingrabens: *Oberrheine Geologische Abhandlungen*, v. 14, 54 pp.

————, 1965b, Kontinentaldrift und Polverschiebung als Triebkräfte der Erdgeschichte: *Naturwissenschaftliche Rundschau*, v. 18, pp. 267-273.

————, 1967, Randpazifische Tektonik und Vulkanismus im südlichen Chile: *Geologische Rundschau*, v. 57, pp. 81-101.

————, 1970, Die grossen Gräben: Harmonische Strukturen in einer disharmonisch struierten Erdkrust: *Geologische Rundschau*, v. 59, pp. 528-552.

———, 1972, The Rhinegraben rift system—plate tectonics and transform faulting: *Geophysical Survey*, v. 1, pp. 27-60.

Jäkel, D., 1971, Erosion und Akkumulation im Enneri Bardagué-Arayé des Tibesti-Gebirges (zentrale Sahara) während des Pleistozäns und Holozäns: *Berliner Geographische Abhandlungen*, v. 10, 55 pp.

Jäkel, D., and E. Schulz, 1972, Spezielle Untersuchungen an der Mittelterrasse im Enneri Tabi, Tibesti-Gebirge: *Zeitschrift für Geomorphologie*, Supplement-Band 15, pp. 129-143.

Jahn, A., 1960, Some remarks on the evolution of slopes on Spitsbergen: *Zeitschrift für Geomorphologie*, Supplement-Band 1, pp. 49-58.

———, 1967, Some features of mass movement on Spitsbergen slopes: *Geografiska Annaler*, v. 49, Series A, pp. 213-225.

Jeje, L. K., 1973, Inselbergs' evolution in a humid tropical environment: the example of South Western Nigeria: *Zeitschrift für Geomorphologie*, Neue Folge, 17, pp. 194-225.

Jennings, J. N., and J. A. Mabbutt, 1967, *Landform Studies from Australia and New Guinea:* Cambridge University Press, Cambridge, 434 pp.

Jessen, O., 1936, *Reisen und Forschungen in Angola*: Reimer, Berlin, 397 pp.

———, 1943, Die Randschwellen der Kontinente: *Petermanns Geographische Mitteilungen*, Ergänzungs-Heft 241, pp. 1-205.

Jessup, R. W., 1960, The stony tableland soils of the southeastern portion of the Australian arid zone and their evolutionary history: *Journal of Soil Science*, v. 11, pp. 188-196.

Johnson, D. W., 1932, Rock planes in arid regions: *Geographical Review*, v. 22, pp. 656-665.

Journaux, A., 1972, Étude des phénomènes périglaciaires en laboratoire: *Bulletin du Centre de Géomorphologie de Caen*, v. 13-15, 205 pp.

———, 1974, Des phénomènes périglaciaires du Nord de l'Amerique: unpublished lecture delivered to the Geographisches Kolloquium Würzburg, July 9.

———, 1975, Recherches géomorphologiques en Amazonie brésilienne: *Bulletin du Centre de Géomorphologie de Caen*, v. 20, 67 pp.

Jungfer, E. V., 1974, Deh-Bala am Shir-Kuh. Zur Geoökologie einer Gebirgsoase im Hochland von Iran: *Marburger Geographischer Schriften*, v. 62, pp. 49-66.

Kaiser, K. H., 1963, Zur Frage der Würm-Gliederung durch einen ''Mittelwürm-Boden'' im nördlichen Alpen-Vorland bei Murnau: *Eiszeitalter und Gegenwart*, v. 14, pp. 208-215.

———, 1972, Prozesse und Formen der ariden Verwitterung am Beispiel des

Tibesti-Gebirges und seiner Rahmenbereiche: *Berliner Geographische Abhandlungen*, v. 16, pp. 49-80.

Kalle, K., 1943, *Der Stoffhaushalt des Meeres*, Becker und Euler, Leipzig, 263 pp.

Karrasch, H., 1972, Flächenbildung unter periglazialen Klimabedingungen: *Göttinger Geographische Abhandlungen*, v. 60, pp. 155-168.

Katasonov, E. M., and P. A. Solovev, 1969, Guide to a trip round Central Yakutia (Paleogeography and periglacial phenomena): Yakutsk, International Symposium (Paleogeography and Periglacial Phenomena of the Pleistocene).

Kayser, K., 1934, Morphologische Studien in West-Montenegro. II. Die Rumpftreppe von Cetinje und der Formenschatz der Karstabtragung: *Zeitschrift Gesellschaft für Erdkunde Berlin*, v. 1-2, pp. 26-49, and v. 3-4, pp. 81-102.

——, 1949, Die morphologischen Untersuchungen an der Grossen Randstufe auf der Ostseite Südafrikas, in: *Die Grosse Randstufe auf der Ostseite Südafrikas und ihr Vorland*, edited by E. Obst, and K. Kayser, Sonderveröffentlichung der Geographischen Gesellschaft Hannover, v. 3, pp. 85-275.

——, 1955, Karstrandebene und Poljeboden: *Erdkunde*, v. 9, pp. 60-64.

——, 1957, Zur Flächenbildung, Stufen- und Inselberg-Entwicklung in den wechselfeuchten Tropen auf der Ostseite Süd-Rhodesiens: *Deutscher Geographen-Tag*, v. 31, pp. 165-172.

——, 1973, Beiträge zur Geomorphologie der Namib-Küstenwüste: *Zeitschrift für Geomorphologie*, Supplement-Band 17, pp. 156-167.

Keller, R., 1961, *Gewässer und Wasserhaushalt des Festlandes. Eine Einführung in die Hydrogeographie*: Haude und Spener, Berlin, 520 pp.

——, ed., 1968, Flussregime und Wasserhaushalt: *Freiburger Geographische Hefte*, v. 6, 240 pp.

Kellogg, C. E., 1950, Tropical soils: *Transactions of the Fourth International Congress of Soil Science*, v. 1, pp. 266-276.

——, 1960, Soil map of the world, in: *Efficient Use of Fertilizers*, Food and Agricultural Organization Studies 43.

Kieslinger, A., 1958, Restspannungen und Entspannungen im Gestein: *Geologie und Bauwesen*, v. 24, pp. 95-112.

King, L. C., 1948, A theory of bornhardts: *Geographical Journal*, v. 112, pp. 83-87.

——, 1949, The pediment landform, some current problems: *Geological Magazine*, v. 86, pp. 245-250.

——, 1956, Research on slopes in South Africa: *Premier Rapport de la Commission pour l'étude des versants*, Union Géographique Internationale, pp. 105-106.

King, L. C., 1967, *The Morphology of the Earth*: Oliver and Boyd, Edinburgh, 726 pp.

Kirkby, M. J., 1969, Erosion by water on hill slopes, in: *Water, Earth, and Man, A Synthesis of Hydrology, Geomorphology, and Socio-Economic Geography*, edited by R. J. Chorley, Methuen, London, pp. 229-238.

Klaer, W., 1956, Verwitterungsformen im Granit auf Korsika: *Petermanns Geographische Mitteilungen*, Ergänzungs-Heft 261, 146 pp.

————, 1962, Die periglaziale Höhenstufe in den Gebirgen Vorderasiens: *Zeitschrift für Geomorphologie*, Neue Folge, v. 6, pp. 17-32.

————, 1970, Formen der Granitverwitterung im ganzjährig ariden Gebiet der östlichen Sahara (Tibesti): *Tübinger Geographische Studien*, v. 34, pp. 71-82.

Klammer, G., 1971, Über plio-pleistozäne Terrassen und ihre Sedimente im unteren Amazonas-Gebiet: *Zeitschrift für Geomorphologie*, Neue Folge, 15, pp. 62-106.

————, 1975, Beobachtungen an Hängen im tropischen Regenwald des Unteren Amazonas: *Zeitschrift für Geomorphologie*, Neue Folge, 19, pp. 273-286.

Klimaszewski, M., 1960, Geomorphological studies of the Western part of Spitzbergen between Kongsfjord and Eidembukta (in Polish with an English translation by Z. Domaniewska): *Prace Geograficzne*, Zeszyt 1, 168 pp.

Klimpt, H., 1943, Morphogenese der Sonnblickgruppe: *Geographische Jahresbericht aus Österreich*, v. 21-22, 130 pp.

Klinge, H., 1969, Climatic conditions in lowland tropical podzol areas: *Tropical Ecology*, v. 10, pp. 222-239.

Klitzsch, E., 1970, Die Strukturgeschichte der Zentral-Sahara: *Geologische Rundschau*, v. 59, pp. 459-527.

Knetsch, G., 1950, Beobachtungen in der libyschen Sahara: *Geologische Rundschau*, v. 38, pp. 40-59.

————, 1966, Über Boden- und Grundwasser in der Wüste (am Beispiel westägyptischer Vorkommen): *Nova Acta Leopoldina*, v. 31, pp. 67-88.

Knothe, H., 1931, Spitzbergen, eine landeskundliche Studie: *Petermanns Geographische Mitteilungen*, Ergänzungs-Heft 211, 116 pp.

Köppen, W., 1923, *Die Klimate der Erde, Grundriss der Klimakunde*, de Gruyter, Berlin, 369 pp.

Körber, H., 1956, Morphologie von Waldeck und Ostsauerland: *Würzburger Geographische Arbeiten*, v. 3, 155 pp.

————, 1962, Die Entwicklung des Maintales: *Würzburger Geographische Arbeiten*, v. 10, 170 pp.

Köster, E., 1960, *Mechanische Gesteins- und Bodenanalyse*, Hauser, Munich, 171 pp.

Krebs, N., 1928, *Die Ostalpen und das heutige Österreich*: Wissenschaftliche Buchgesellschaft, v. 1, 330 pp.

————, 1939, *Vorderindien und Ceylon*, Engelhorn, Stuttgart, 382 pp.

————, 1942, Über Wesen und Verbreitung tropischer Inselberge: *Abhandlungen der Preussischen Akademie der Wissenschaften mathematischnaturwissenschaftlicher Klasse*, v. 6, pp. 3-41.

Kremer, E., 1954, Die Terrassenlandschaft der mittleren Mosel als Beitrag zur Quartärgeschichte: *Arbeiten zur Rheinischen Landeskunde*, v. 6, 100 pp.

Krenkel, E., 1922, *Die Bruchzonen Ostafrikas. Tektonik, Vulkanismus, Erdbeben und Schwereanomalien:* Borntraeger, Berlin, 184 pp.

————, 1926, Abessomalien: *Handbuch der Regionalen Geologie*, v. 7, 119 pp.

Kroll, A., and G. Wessely, 1967, Neue Erkenntnisse über Molasse, Flysch, und Kalkalpen auf Grund der Ergebnisse der Bohrung Urmannsau I (Nieder Österreich): *Erdöl-Erdgas Zeitschrift*, v. 10, pp. 342-353.

Krygowska, L., and B. Krygowski, 1968, The dynamics of sedimentary environments in the light of hishgramm types of grain abrasion: *Geographia Polonica*, v. 14, pp. 87-92.

Kubiena, W., 1953, *Bestimmungsbuch und systematik der Böden Europas*: Enke, Stuttgart, 392 pp.

————, 1954, Micromorphology of laterite formation in Rio Muni (Spanish Guinea): *Transactions of the Fifth International Congress of Soil Science*, v. 4, pp. 77-84.

————, 1955, Über die Braunlehmrelikte des Atakor (Hoggar-Gebirge, Zentrale Sahara): *Erdkunde*, v. 9, pp. 115-132.

————, 1957, Neue Beiträge zur Kenntnis des planetarischen und hygrometrischen Formenwandels der Böden Afrikas: *Stuttgarter Geographische Studien*, v. 69, pp. 50-64.

————, 1964, Die Genese lateritischer Profile als bodenkundliches Problem: *Schriften der Gesellschaft Deutscher Metalhütten und Bergleute*, v. 14, pp. 79-84.

Kutscher, F., 1954, Die Verwitterungsrinde der voroligozänen Landoberfläche und tertiäre Ablagerungen im östlichen Hunsrück: *Notizblatt des Hessischen Landes-Amtes für Bodenforschung zu Wiesbaden*, v. 82, pp. 202-212.

Kvitkovic, J., 1968, Die geomorphologischen Verhältnisse im NO-Teil des ostslovakischen Tieflandes: *Würzburger Geographische Arbeiten*, v. 22, 34 pp.

Langer, M., 1964, Untersuchungen zur Theorie der Wasseraufnahmefähigkeit von Tonen: *Vortrag zur Baugrundtagung 1964 in Berlin*, pp. 1-56.

Lauer, W., 1951, Hygrische Klimate und Vegetationszonen der Tropen mit besonderer Berücksichtigung Ostafrikas: *Erdkunde*, v. 5, pp. 284-293.

————, 1952, Humide und aride Jahreszeiten in Afrika und Südamerika und ihre Beziehungen zu den Vegetationsgürteln: *Bonner Geographische Abhandlungen*, v. 9, pp. 15-98.

Lautensach, H., 1934-36, *Spanien und Portugal: Handbuch der Geographischen Wissenschaft*: Akademische Verlagsgesellschaft Athenaion, Potsdam, pp. 426-557.

———, 1941, Interglaziale Terrassenbildung in Nordportugal und ihre Beziehungen zu den allgemeinen Problemen des Erdzeitalters: *Petermanns Geographische Mitteilungen*, v. 87, pp. 297-311.

———, 1945, *Korea*: Koehler, Leipzig, 542 pp.

Lawson, A. C., 1915, The Epigene profiles on the Desert: *University of California, Geological Bulletin*, v. 9, pp. 23-48.

Lee, K. E., and T. G. Wood, 1971, *Termites and Soils*: Academic Press, London, 251 pp.

Lehmann, H., 1936, Morphologische Studien auf Java: *Geographische Abhandlungen*, Reihe 3, v. 9, 114 pp.

———, 1953, Karstentwicklung in den Tropen: *Umschau in Wissenschaft und Technik*, v. 53, no. 18, pp. 559-562.

———, 1954, Das Karstphänomen in den verschiedenen Klimazonen: Der tropische Kegelkarst auf den Grossen Antillen: *Erdkunde*, v. 8, pp. 130-139.

———, 1955, Der tropische Kegelkarst in Westindien: *Deutscher Geographen-Tag, Essen*, v. 29, pp. 126-131.

———, 1957, Klimamorphologische Beobachtungen in der Serra de Mantiqueria und im Paraiba-Tal (Brasilien): *Abhandlungen des Geographischen Instituts der Freien Universität Berlin*, v. 5, pp. 67-72.

———, 1964, Glanz und Elend der morphologischen Terminologie, in: *Neue Fragen der Allgemeinen Geographie,* edited by J. Büdel, *Würzburger Geographische Arbeiten*, v. 12, pp. 11-22.

———, 1970, Kegelkarst und Tropengrenze: *Tübinger Geographische Studien*, v. 34, pp. 107-112.

Lehmann, O., 1927, Das Tote Gebirge als Hochkarst: *Mitteilungen der Geographischen Gesellschaft Wien*, v. 70, pp. 201-245.

———, 1932, Die Hydrographie des Karstes, in: *Enzyklopädie der Erdkunde*, edited by G. Brown, Deuticke, Leipzig and Vienna, 218 pp.

———, 1933, Morphologische Theorie der Verwitterung an Steinschlagwänden: *Vierteljahresschrift der naturforschenden Gesellschaft in Zürich*, v. 78, pp. 83-126.

Leidlmair, A., 1962, Klimamorphologische Probleme in Hadramaut, in: *Hermann von Wissmann Festschrift*, edited by A. Leidlmair, Selbstverlag des Geographischen Instituts der Universität Tübingen, Tübingen, pp. 162-180.

Leiter, H., 1909, Die Frage der Klimaänderungen während geschichtlicher Zeit in Nordafrika: *Abhandlungen der Geographischen Gesellschaft in Wien*, v. 8, no. 1.

Leser, H., 1967, Geomorphologische Spezialkarte des Rheinhessischen Tafel- und Hüggellandes (Südteil): *Erdkunde*, v. 21, pp. 161-168.

Libby, W. F., 1965, *Altersbestimmung mit der C¹⁴-Methode*: Hochschultaschenbuch, v. 403-403a, Bibliographisches Institut, Mannheim, 204 pp.

Lichtenecker, N., 1926, Die Rax: *Jahresbericht aus Österreich*, v. 18, pp. 150-170.

Lieberoth, I., 1969, *Bodenkunde—Bodenfruchtbarkeit:* Deutsches Landwirtschaftliches Verlag, Berlin, 336 pp.

Liedtke, H., 1962, Eisrand und Karstpoljen am Westrand der Lukavica-Hochfläche (Westmontenegro): *Erdkunde*, v. 16, pp. 289-298.

————, 1967a, Die Grundzüge der geomorphologischen Entwicklung im pfälzischen Schichtstufenland: *Zeitschrift für Geomorphologie*, Neue Folge, 11, pp. 332-351.

————, 1967b, Geomorphologische Übersichtskarte, in: *Geschichtlicher Atlas für das Land an der Saar*, edited by H. Ammann and E. Meynan, Institut für Landeskunde des Saarlandes, v. 2, Saarbrucken.

————, 1968, Die geomorphologische Entwicklung der Oberflächenformen des Pfälzer Waldes und seiner Randgebeite: *Arbeiten aus dem Geographischen Institut der Universität des Saarlandes*, Sonderband 1, 232 pp.

————, 1969, Grundzüge und Probleme der Entwicklung der Oberflächenformen des Saarlandes und seiner Umgebung: *Forschung zur deutschen Landeskunde*, v. 183, 63 pp.

Löffler, E., 1974, Piping and pseudokarst features in the tropical lowlands of New Guinea: *Erdkunde*, v. 28, pp. 13-18.

Louis, H., 1928, Die Form der norddeutschen Bogendünen: *Zeitschrift für Geomorphologie*, v. 4, pp. 7-18.

————, 1934, Glazialmorphologische Studien in den Gebirgen der Britischen Inseln: *Berliner Geographische Arbeiten*, v. 6, 39 pp.

————, 1935, Probleme der Rumpfflächen und Rumpftreppen: *25. Deutscher Geographen-Tag, Bad Nauheim.*

————, 1952, Zur Theorie der Gletschererosion in Tälern: *Eiszeitalter und Gegenwart*, v. 2, pp. 12-24.

————, 1953, Über die ältere Formenentwicklung im Rheinischen Schiefergebirge, insbesondere im Moselgebiet: *Münchener Geographische Hefte*, v. 2, pp. 1-97.

————, 1956, Die Entstehung der Poljen und ihre Stellung in der Karstabtragung auf Grund von Beobachtungen im Taurus: *Erdkunde*, v. 10, pp. 33-53.

————, 1957, Der Reliefsockel als Gestaltungsmerkmal des Abtragungsreliefs: *Stuttgarter Geographische Studien*, v. 69, pp. 65-70.

————, 1964, Über Rumpfflächen- und Talbildung in den wechselfeuchten

Tropen, besonders nach Studien in Tanganyika: *Zeitschrift für Geomorphologie*, Neue Folge, 8, pp. 43-70.

————, 1968a, *Allgemeine Geomorphologie*: de Gruyter, Berlin, 522 pp.

————, 1968b, Über die Spülmulden und benachbarte Formbegriffe: *Zeitschrift für Geomorphologie*, Neue Folge, 12, pp. 490-501.

————, 1973, Fortschritte und Fragwürdigkeiten in neueren Arbeiten zur Analyse fluvialer Landformung besonders in den Tropen: *Zeitschrift für Geomorphologie*, Neue Folge, 17, 42 pp.

Lüttig, G., 1956, Eine neue einfache geröllmorphometrische Methode: *Eiszeitalter und Gegenwart*, v. 7, pp. 13-20.

Lyell, C., 1830-1833, *Principles of Geology*: John Murray, London, 811 pp.

Maack, R., 1969, *Kontinentaldrift und geologie des südatlantischen Ozeans*: de Gruyter, Berlin, 164 pp.

Mabbutt, J. A., 1955, Pediment landforms in Little Namaqualand: *Geographical Journal*, v. 121, pp. 77-83.

————, 1961, "Basal surface" or "weathering front": *Proceedings of the Geological Association*, v. 72, pp. 357-358.

————, 1963, Wandering banks: micro-relief patterns in semiarid western Australia: *Bulletin of the Geological Society of America*, v. 74, pp. 529-540.

————, 1965, The weathered land surface in Central Australia: *Zeitschrift für Geomorphologie*, Neue Folge, 9, pp. 82-114.

————, 1966, Landforms of the Western MacDonnell Ranges, in: *Essays in Geomorphology*, edited by G. H. Dury, American Elsevier, New York, 404 pp.

————, 1967, Denudation chronology in Central Australia: structure, climate, and landform inheritance in the Alice Spring area, in: *Landform Studies from Australia and New Guinea*, edited by J. Jennings and J. Mabbutt, Cambridge University Press, Cambridge, pp. 144-179.

Macar, P., 1954, L'évolution géomorphologique de l'Ardenne: *Bulletin de la Société Belge de Géographie*, v. 78, pp. 9-33.

————, ed., 1970, New contributions to slope evolution: *Zeitschrift für Geomorphologie*, Supplement-Band 9, 186 pp.

Machatschek, F., 1938-1940, *Das Relief der Erde:* Borntraeger, Berlin, 2 vols., 1159 pp.

Macheus, E., 1966, Zur geotektonischen Entwicklung in West-Afrika: *Zeitschrift der deutschen Geologischen Gesellschaft*, v. 116, pp. 589-597.

Maclaren, M., 1906, On the origin of certain laterites: *Geological Magazine*, v. 3, pp. 536-547.

Mainguet, M., 1968, Le Borkou, aspects d'un modelé éolien: *Annales de Géographie*, v. 77, pp. 295-322.

Markov, K. K., 1948, *General Problems of Geomorphology*: Gosudarstvennoe Isdatel'stvo Geografichnie Literaturi, Moscow, 342 pp.

Markov, K. K., and A. A. Velichko *et al.*, 1969, *Guidebook for Field Excursions Moscow-Upper Volga*: Symposium for Paleography and Pleistocene Periglacial Phenomena, Moscow.

Martonne, E. de, 1910-1911, L'érosion glaciaire et la formation des vallées alpines: *Annales de Géographie*, v. 19, pp. 289-317 and v. 20, pp. 1-29.

————, 1925-1927, *Traité de Géographie Physique:* Armin Colin, Paris, 3 vols., 3070 pp.

Maull, O., 1930, *Vom Itatiaya zum Paraguay, Ergebnisse und Erlebnisse einer Forschungsreise durch Mittelbrasilien:* Hiersemann, Leipzig, 366 pp.

————, 1958, *Handburch der Geomorphologie*: Deuticke, Vienna, 600 pp.

McGee, W. J., 1897, Sheetflood erosion: *Bulletin of the Geological Society of America*, v. 8, pp. 87-112.

Meckelein, W., 1959, *Forschungen in der zentralen Sahara. 1. Klimamorphologie*: Westermann, Braunschweig, 181 pp.

Meinardus, W., 1926, Die hypsographischen Kurven Grönlands und der Antarktis und die Normalform der Inlandeisoberfläche: *Petermanns Geographische Mitteilungen*, v. 72, pp. 97-105.

Menendez, C. A., 1968, Die Fossilen Floren Südamerikas, in: *Biogeography and Ecology in South America*, edited by Fittkau *et al.*, Junk, The Hague, v. 2, pp. 519-561.

Mensching, H., 1950, Schotterfluren und Talauen im Niedersächsischen Bergland: *Göttinger Geographische Abhandlungen*, v. 4, 60 pp.

————, 1953, Morphologische Studien im Hohen Atlas von Marokko: *Würzburger Geographische Arbeiten*, v. 1, 104 pp.

————, 1955, Das Quartär in den Gebirgen Marokkos: *Petermanns Geographische Mitteilungen*, Ergänzungs-Heft, 256, 79 pp.

————, 1957a, Geomorphologie der Hohen Rhön und ihres südlichen Vorlandes: *Würzburger Geographische Arbeiten*, v. 4-5, pp. 47-88.

————, 1957b, *Marokko, die Landschaften im Maghreb*: Geographische Handbücher, Keyser, Heidelberg, 254 pp.

————, 1958a, Glacis-Fussfläche-Pediment: *Zeitschrift für Geomorphologie*, Neue Folge, 2, pp. 165-186.

————, 1958b, Entstehung und Erhaltung von Flächen im seimiariden Klima am Beispiel Nordwest-Afrikas: *Deutscher Geographen-Tag, Würzburg*, v. 31, pp. 173-184.

————, 1960a, Periglazial-Morphologie und quartäre Entwicklungsgeschichte der Hohen Rhön und ihres östlichen Vorlandes: *Würzburger Geographische Arbeiten*, v. 7, 39 pp.

————, 1960b, Bericht und Gedanken zur Tagung der Kommission für Periglazialforschung in der IGU in Marokko vom 19-31 Oktober, 1959: *Zeitschrift für Geomorphologie*, Neue Folge, 4, pp. 159-170.

Mensching, H., 1964a, Die regionale und klimatisch-morphologische Differenzierung von Bergfussflächen auf der Iberischen Halbinsel (Ebrobecken—Nordmeseta—Küstensaum Iberiens): *Würzburger Geographische Arbeiten*, v. 12, pp. 141-158.

———, 1964b, Zur Geomorphologie Südtunesiens: *Zeitschrift für Geomorphologie*, Neue Folge, 8, pp. 424-439.

———, 1968, Bergfussflächen und das System der Flächenbildung in den ariden Subtropen und Tropen: *Geologische Rundschau*, v. 58, pp. 62-82.

———, 1969, Zur Geomorphologie des Hoggar Gebirges (Zentrale Sahara) nach P. Rognon: *Erdkunde*, v. 23, pp. 61-63.

———, 1970a, Geomorphologische Beobachtungen in der Inselberglandschaft südlich des Victoria Sees (Tanzania): *Abhandlungen des 1. Geographischen Instituts der Freien Universität Berlin*, v. 13, pp. 111-124.

———, 1970b, Flächenbildung in der Sudan- und Sahel-Zone (Ober Volta und Niger): *Zeitschrift für Geomorphologie*, Supplement-Band 10, 29 pp.

———, 1973, Pediment und Glacis, ihre Morphogenese und Einordnung in das System der klimatischen Geomorphologie auf Grund von Beobachtungen im Trockengebiet Nordamerikas (USA und Nordmexiko): *Zeitschrift für Geomorphologie*, Supplement-Band 17, pp. 133-155.

Mensching, H., and R. Raynal, 1954, Fussflächen in Ost-Marokko: *Petermanns Geographische Mitteilungen*, v. 98, pp. 171-176.

Mensching, H., K. Giessner, and G. Stuckmann, 1970, Sudan—Sahel—Sahara. Geomorphologische Beobachtungen auf einer Forschungsexpedition nach West- und Nordafrika 1969: *Jahrbuch der Geographischen Gesellschaft Hannover für 1969*, 211 pp.

Messerli, B., 1967, Die Eiszeitliche und die gegenwärtige Vergletscherung im Mittelmeerraum: *Geographica Helvetica*, v. 22, pp. 105-228.

———, 1972, Formen und Formungsprozesse in der Hochgebirgsregion des Tibesti: *Hochgebirgsforschung*, v. 2, pp. 23-86.

Metz, K., 1967, *Lehrbuch der tektonischen Geologie*: Enke, Stuttgart, 338 pp.

Meyer, R., 1967, Studien über Inselberge und Rumpfflächen in Nord-Transvaal: *Münchener Geographische Hefte*, v. 31, 89 pp.

Michel, P., 1967, Les dépôts du quarternaire récent dans la basse vallée du Senegal: *Bulletin de l'Institut Fondamental d'Afrique Noire*, v. 29, no. 2, pp. 853-860.

Miotke, F. D., 1968, Karstmorphologische Studien in der glazial überformten Höhenstufe der "Picos de Europa," Nordspanien: *Jahrbuch der Geographischen Gesellschaft Hannover*, Sonderheft 4, 161 pp.

———, 1973, Die tieferlegung der Oberflächen zwischen Mogoten in Puerto Rico (östlich Arecibo): in: *Neue Ergebnisse der Karstforschung in den*

Tropen und im Mittelmeerraum, edited by A. Semmel, *Erdkundliches Wissen*, v. 23, pp. 34-43.

Molle, H. G., 1971, Gliederung und Aufbau fluviatiler Terrassenakkumulationen im Gebiet des Enneri Zoumri (Tibesti-Gebirge): *Berliner Geographische Abhandlungen*, v. 13, 53 pp.

Monod, T., 1947, *Méharées, explorations au vrai Sahara*: Edition "je sais," Paris, 300 pp.

Morawetz, S., 1930, Beiträge zur Morphologie der Kreuzeck- und Reisseckgruppe: *Veröffentlichung des Geographischen Instituts der Universität Graz*, v. 3.

————, 1967, Zur Frage der asymmetrischen Täler im Grabenland zwischen Raab und Mur: *Mitteilungen des naturwissenschaftlichen Vereins der Steiermark*, v. 97, pp. 32-38.

Mordziol, C., 1927, Rheintalentstehung: *Abhandlung der naturwissenschaftlichen Vereinigung, Koblenz*, pp. 2-24.

Morisawa, M., 1968, *Streams, Their Dynamics and Morphology*: McGraw-Hill, New York, 175 pp.

Mortensen, H., 1927, Der Formenschatz der nordchilenischen Wüste: *Abhandlung der Akademie der Wissenschaften, Göttingen, Mathematisch-Physikalischer Klasse*, v. 12, 191 pp.

————, 1929, Inselberglandschaften in Nord-Chile: *Zeitschrift für Geomorphologie*, v. 4, pp. 123-138.

————, 1930, Einige Oberflächenformen in Chile und auf Spitzbergen im Rahmen einer vergleichenden Morphologie der Klimazonen: *Petermanns Geographische Mitteilungen*, Ergänzungs-Heft 209, pp. 147-156.

————, 1932, Blockmeere und Felsburgen in den deutschen Mittelgebirgen: *Zeitschrift der Gesellschaft für Erdkunde zu Berlin*, pp. 279-287.

————, 1933, Die "Salzsprengung" und ihre Bedeutung für die regional-klimatische Gliederung der Wüsten: *Petermanns Geographische Mitteilungen*, v. 79, pp. 130-135.

————, 1949, Rumpffläche-Stufenlandschaft-Alternierende Abtragung: *Petermanns Geographische Mitteilungen*, v. 93, pp. 1-14.

————, 1953, Neues zum Problem der Schichtstufenlandschaft. Einige Ergebnisse einer Reise durch den SW der USA, 1952: *Nachrichten der Akademie der Wissenschaft in Göttingen, Mathematisch-Physikalischer Klasse*, pp. 3-22.

————, 1956, Über Wandverwitterung und Hangabtragung in semiariden und vollariden Gebieten: *1. Rapport de la Commission de l'Étude de Versants, XIII International Geographical Union*, Amsterdam, pp. 96-104.

Mortensen, H., and J. Hövermann, 1957, Filmaufnahmen der Schotterbewegungen im Wildbach: *Petermanns Geographische Mitteilungen*, Ergänzungs-Heft 262, pp. 43-52.

Müller, F., 1959, Beobachtungen über Pingos: *Meddelelser om Grønland*, v. 153.

Müller, K. H., 1973, Zur Morphologie des zentralen Hintertaunus und des Limburger Beckens. Ein Beitrag zur tertiären Formengenese: *Marburger Geographische Schriften*, v. 58, 112 pp.

Müller, P., and J. Schmithüsen, 1970, Probleme der Genese südamerikanischer Biota, in: *Deutsche Geographische Forschung in der Welt von heute*, edited by H. Wilhelmy, Ferdinand Hirt, Kiel, pp. 109-122.

Murawski, H., 1972, *Geologisches Wörterbuch*: Enke, Stuttgart, 6th ed., 260 pp.

Nansen, F., 1904, *The Norwegian North Polar Expedition 1893-1896: Scientific Results*, Oslo, v. 4.

————, 1921, *Spitzbergen*: Brockhaus, Leipzig, 327 pp.

————, 1922, The strandflat and isostasy: *Videnkapsselskapets Skrifter I, Mathematisch-Naturwissenschaftliche Klasse* 11.

Niedermayer, O. von, 1920, Die Binnenbecken des iranischen Hochlands: *Mitteilungen der Geographischen Gesellschaft München*, v. 14, pp. 9-64.

————, 1940, Krieg in Irans Wüsten, in: *Weltkrieg vor Indiens Toren*, Hanseatische Verlaganstalt, Hamburg, pp. 1-89.

Nye, J. F., 1952, The mechanics of glacier flow: *Journal of Glaciology*, v. 2, pp. 82-93.

Nye, P. H., 1955, Some soil-forming processes in the humid tropics. IV. The action of the soil fauna: *Journal of Soil Science*, v. 6, pp. 73-83.

Oberhausen, R., 1973, Stratigraphisch-paläontologische Hinweise zum Ablauf tektonischer Ereignisse in den Ostalpen während der Kreidezeit: *Geologische Rundschau*, v. 62, pp. 96-106.

Oberlander, T. M., 1974, Landscape inheritance and the pediment problem in the Mohave: *American Journal of Science*, v. 274, pp. 849-874.

Ollier, C. D., 1960, The inselbergs of Uganda: *Zeitschrift für Geomorphologie*, Neue Folge, 4, pp. 43-52.

————, 1965, Some features of granite weathering in Australia: *Zeitschrift für Geomorphologie*, Neue Folge, 9, pp. 285-304.

Pachur, H. J., 1966, Untersuchungen zur morphoskopischen Sandanalyse: *Berliner Geographische Abhandlungen*, v. 4, 35 pp.

————, 1967, Beobachtungen über die Bearbeitung von feinkörnigen Sandakkumulationen im Tibesti-Gebirge: *Berliner Geographische Abhandlungen*, v. 5, pp. 23-24.

————, 1974, Geomorphologische Untersuchungen im Raum der Serir Tibesti (Zentralsahara): *Berliner Geographischer Abhandlungen*, v. 17, 62 pp.

————, 1975, Zur spätpleistozänen und holozänen Formung auf der Nordabdachung des Tibesti-Gebirges: *Erde*, v. 106, pp. 21-46.

Panoš, V., and O. Štelc, 1968, Physiographic and geologic control in the development of Cuban mogotes: *Zeitschrift für Geomorphologie*, Neue Folge, 12, pp. 117-173.

Panzer, W., 1926, Talentwicklung und Eiszeitklima im nordöstlichen Spanien: *Abhandlungen der Senckenbergischen naturforschenden Gesellschaft*, v. 39, pp. 141-182.

————, 1965, *Geomorphologie*: Westermann, Braunschweig, 128 pp.

————, 1967, Einige Grundfragen der Formentwicklung im Rheinischen Schiefergebirge: *Festschrift des 36. Deutschen Geographen-Tags*, pp. 1-15.

Passarge, S., 1895, *Adamaua, Bericht über die Expedition des deutschen Kamerun-Komittees 1893-1894*: Reimer, Berlin, 573 pp.

————, 1903, Bericht über eine Reise im venezolanischen Guyana: *Zeitschrift der Gesellschaft für Erdkunde*, pp. 5-43.

————, 1912, Physiologische Morphologie: *Mitteilungen der Geographischen Gesellschaft Hamburg*, v. 27, pp. 133-337.

————, 1914, Morphologie des Messtischblattes Stadtremda, in: *Morphologischer Atlas*, edited by S. Passarge, Friedrichsen, Hamburg, 221 pp.

————, 1919, Die Vorzeitformen der deutschen Mittelgebirgslandschaften: *Petermanns Geographische Mitteilungen*, v. 65, pp. 41-46.

————, 1923, Die Inselberglandschaft der Massaisteppe: *Petermanns Geographische Mitteilungen*, v. 69, pp. 205-209.

————, 1928-1929, Das Problem der Inselberglandschaft: *Zeitschrift für Geomorphologie*, v. 4, pp. 109-122.

————, 1929, *Morphologie der Erdoberfläche*: Hirt, Breslau, 152 pp.

————, 1933a, Landschaftskundliche Charakteristik der Rhön im Bereich der Messtischblätter Kleinsassen, Gersfeld, Hilders und Sondheim, sowie ihre Bedeutung für die geologische Landesaufnahme: *Mitteilungen der Geographischen Gesellschaft Hamburg*, v. 43, pp. 163-266.

————, 1933b, Wissenschaftliche Ergebnisse einer Reise im Gebiet des Orinoco, Caura und Cuchivero im Jahre 1901-1902: *Abhandlungen aus dem Gebiet der Auslandskunde*, v. 39, 281 pp.

Pécsi, M., 1959, *Development and Morphology of the Danube Valley in Hungary* (in Hungarian): Akademia kiadó, Budapest, 346 pp.

————, 1963, Pediment surface of the Hungarian Middle Mountains (in Hungarian with German summary): *Földrajzi Közlemények*, v. 9, no. 3, pp. 195-212.

————, 1964, *Ten Years of Physicogeographic Research in Hungary*: Studies in Geography, v. 1, Akademia Kiadó, Budapest, 132 pp.

————, 1968, Denudational levels in the Hungarian Middle Mountains with special regard to pediment formation, in: *Geomorphology and Terminology of the Hungarian Mountains and Their Pediments*, Symposium of the Hungarian Academy of Sciences, April 16-20, pp. 24-36.

Pécsi, M., 1970, *Geomorphical Regions of Hungary*: Studies in the Geography of Hungary, v. 6, Akademia Kiadó, Budapest, 45 pp.

Pécsi, M., and J. Szilard, 1968, Über einige Forschungs- und Benennungs-Probleme der Einebnungsflächen, in: *Geomorphology and Terminology of the Hungarian Mountains and Their Pediments*, Symposium of the Hungarian Academy of Sciences, April 16-20, pp. 1-23.

Penck, A., 1894, *Morphologie der Erdoberfläche*: Engelhorn, Stuttgart, 2 vols.

———, 1906, *Beobachtung als Grundlage der Geographie:* Borntraeger, Berlin, 63 pp.

———, 1910, Versuch einer Klimaklassifikation auf physiogeographischer Grundlage: *Sitzungs-Berichte der Akademie der Wissenschaften, Berlin, Physikalische-Mathematischer Klasse*, v. 1, pp. 236-246.

———, 1924, Das unterirdische Karstphänomen, in: *Cvijić Festschrift, Zbornik Radova* (Recueil de Travaux), edited by P. Vujević, Drzhavna Shtamparija, Belgrad, pp. 175-197.

———, 1933, Eustatische Bewegungen des Meeresspiegels während der Eiszeit: *Geographische Zeitschrift*, v. 39, pp. 329-339.

———, 1936, Europa zur letzten Eiszeit, in: *Länderkundliche Forschung, Festschrift für N. Krebs*, edited by H. Louis and W. Panzer, Engelhorn, Stuttgart, pp. 222-237.

Penck, A., and E. Brückner, 1901-1909, *Die Alpen im Eiszeitalter:* Verlag Tauchnitz, Leipzig, 3 vols., 1396 pp.

Penck, W., 1924, Die morphologische Analyse: *Geographische Abhandlungen*, v. 2, 283 pp.

Péwé, T. L., 1970, Altiplanation terraces of early Quaternary age near Fairbanks, Alaska: *Acta Geographica Lodziensia,* v. 24, pp. 357-363; also: Arizona State University Department of Geology Reprint Series, v. 86, pp. 357-363.

Pfeffer, K.-H., 1967, Beiträge zur Geomorphologie der Karstbecken im Bereich des Monte Velino (Zentralapennin): *Frankfurter Geographische Hefte*, v. 42, 86 pp.

———, 1968, Charakter der Verwitterungsresiduen im tropischen Kegelkarst und ihre Beziehung zum Formenschatz: *Geologische Rundschau*, v. 58, pp. 408-426.

———, 1969, Neue Beobachtungen im Kegelkarst von Jamaika: *Deutscher Geographen-Tag, Bad Godesberg*, pp. 345-358.

———, 1969, Kalkkrusten und Kegelkarst: *Erdkunde*, v. 23, pp. 23-236.

Pietsch, K., 1962, *Geologie von Sachsen*: Deutscher Verlag der Wissenschaften, Berlin, 870 pp.

Pippan, T., 1957, Geomorphologische Studien im Stubachtal in den Hohen Tauern: *Mitteilungen der Geographischen Gesellschaft Wien*, v. 99, pp. 204-223.

————, 1965, Diskussionsbemerkungen zur Morphologie der mittleren Tauerntäler. *Mitteilungen der Österreichischen Geographischen Gesellschaft*, v. 107, pp. 218-221.

Pissart, A., 1964, Contribution expérimentale à la connaissance de la genèse des sols polygonaux: *Annales de la Société Géologique de Belgique*, v. 87, pp. 213-222.

————, 1966, Expériences et observations à propos de la genèse des sols polygonaux triés: *Revue Belge de Géographie*, v. 90-91, pp. 55-73.

————, 1974, Discussion of "Slopes," by A. Young: *Zeitschrift für Geomorphologie*, Neue Folge, 18, pp. 119-120.

————, 1979, Geomorphological observations in the Canadian Arctic: unpublished lecture, July 3.

Pissart, A., and P. Macar, 1968, Symposium réunissant la commission pour l'étude de l'évolution des versants et la commission de géomorphologie périglaciaire: *Zeitschrift für Geomorphologie*, Neue Folge, 12, pp. 505-508.

Popov, A. I., 1969, *Problems of Cryolithology*, v. 1, Moscow University Press, Moscow.

Poser, H., and J. Hövermann, 1952, Beiträge zur morphometrischen und morphologischen Schotteranalyse: *Abhandlungen der Braunschweigen Wissenschaftlichen Gesellschaft*, v. 4, pp. 12-36.

Prescott, J. A., and R. L. Pendleton, 1952, Laterite and lateritic soils: *Commonwealth Buried Soils Science, Technical Communication*, v. 47, 51 pp.

Pressel, K., 1928, *Experimentelle Methode der Vorausbestimmung der Gesteinstemperatur im Inneren eines Gebirgsmassivs:* Oldenbourg, Munich, 58 pp.

Pugh, J. C., 1956, Fringing pediments and marginal depressions in the inselberg landscape of Nigeria: *Transactions of the Institute of British Geographers*, v. 22, pp. 15-31.

Quenstedt, A., 1842, Das schwäbische Stufenland, in: *Schwaben wie es war und ist*, edited by L. Bauer, Karlsruhe, pp. 270-374.

Quitzow, H.-W., 1969, Die Hochflächenlandschaft beiderseits der Mosel zwischen Schweich und Cochem: *Geologisches Jahrbuch*, Beiheft, v. 82, 79 pp.

Rahn, P. H., 1967, Sheetfloods, streamfloods and the formation of pediments: *Annals of the Association of American Geographers*, v. 57, pp. 593-604.

Rapp, A., 1960a, Talus slopes and mountain walls at Tempelfjorden, Spitzbergen: *Norsk Polarinstitutts Skrifter*, v. 119, 96 pp.

————, 1960b, Recent development of mountain slopes in Kärgevagge and surroundings, northern Scandinavia: *Geografiska Annaler*, v. 42, pp. 65-200.

————, 1965, Some methods of measuring the rate of periglacial denudation on steep slopes, in: *Vorträge des Fridtjof-Nansen-Gedächtnis Symposions, Ergebnisse der Stauferland-Expedition*, edited by J. Büdel and A. Wirthmann, v. 3, pp. 15-19.

Rathjens, C., 1951, Der Hochkarst im System der klimatischen Morphologie: *Erdkunde*, v. 5, pp. 310-315.

———, 1954, Zur Frage der Karstrandebene im Dinarischen Karst. Karsterscheinungen in der klimatisch-morphologischen Vertikalgliederung des Gebirges: *Erdkunde*, v. 8, pp. 114-115, 120.

———, 1960, Beobachtungen an hochgelegene Poljen im südlichen Dinarischen Karst: *Zeitschrift für Geomorphologie*, Neue Folge, 4, pp. 141-151.

———, 1968a, Ein Rundgesprach über Flächenbildung in Saarbrücken: *Zeitschrift für Geomorphologie*, Neue Folge, 12, pp. 470-489.

———, 1968b, Schichtflächen und Schnittflächen im Trockenklima: *Regio Brasiliensis*, v. 9, pp. 162-169.

———, 1970, Gedanken und Beobachtungen zur Flächenbildung im tropischen Indien: *Tübinger Geographische Studien*, v. 34, pp. 155-161.

———, 1971, Klimatische Geomorphologie: *Wege der Forschung*, v. 218, 485 pp.

———, 1973, Subterrane Abtragung (Piping): *Zeitschrift für Geomorphologie*, Supplement-Band 17, pp. 168-176.

Raynal, R., 1961, Plaines et piedmonts du bassin de la Moulouya (Maroc Oriental): Dissertation, Rabat, 608 pp.

Rehbock, T., 1927, Abfluss, Bettbildung und Energiehaushalt der Wasserläufe: *Petermanns Geographische Mitteilungen*, v. 73, p. 299.

———, 1929, Abfluss, Bettbildung und Geschiebebewegung bei Wasserläufen: *Zeitschrift der Deutschen Geologischen Gesellschaft*, v. 81, pp. 497-534.

Reichelt, G., 1961, Über Schotterformen und Rundungsgradanalyse als Feldmethode: *Petermanns Geographische Mitteilungen*, v. 105, pp. 15-24.

Reinwarth, O., and G. Stäblein, 1972, Die Kryosphäre. Das Eis der Erde und seine Untersuchung: *Würzburger Geographische Arbeiten*, v. 36, 71 pp.

Rich, J. L., 1935, Origin and evolution of rock fans and pediments: *Bulletin of the Geological Society of America*, v. 46, pp. 999-1024.

Richmond, G. M., 1965, Glaciation of the Rocky Mountains, in: *The Quaternary of the U.S.*, edited by H. E. Wright and D. G. Frey, Princeton University Press, Princeton, pp. 217-242.

Richter, H., 1956, Die Vergrusung des Gneises (Erzgebirge): *Wissenschaftliche Zeitschrift der Karl-Marx-Universität Leipzig, Mathematisch-Physikalische Reihe*, v. 5, pp. 562-564.

———, 1963, Das Vorland des Erzgebirges. Die Landformung während des

Tertiärs: *Wissenschaftliche Veröffentlichungen des Deutschen Instituts für Länderkunde*, Neue Folge, v. 19-20, pp. 5-231.

Richter, H., G. Haase, and H. Barthel, 1963, Die Bildung von Gebirgs-fussflächen im Gobi-Altai: *Lehre und Forschung*, pp. 187-213.

Richter, K., 1929, Studien über fossile Gletscherstruktur: *Zeitschrift für Gletscherkunde*, v. 17, pp. 33-46.

———, 1951, Die stratigraphische Bewertung periglazialer Umlagerungen im nördlichen Niedersachsen: *Eiszeitalter und Gegenwart*, v. 1, pp. 130-142.

———, 1958, Fluorteste quartärer Knochen in ihrer Bedeutung für die ab-solute Chronologie des Pleistozäns: *Eiszeitalter und Gegenwart*, v. 9, pp. 18-27.

Richthofen, F. von, 1886, *Führer für Forschungreisende, Anleitung zu Beobachtungen über Gegenstände der physischen Geographie und Geologie*: R. Oppenheim, Berlin, 745 pp.

Ridanović, J., 1967, Neue Beobachtungen über die Eiszeitwirkungen im Orjen-Gebirge (Jugoslavien): *Würzburger Geographische Arbeiten*, v. 20, 23 pp.

Rittmann, A., 1960, *Vulkane und ihre Tätikgkeit*: Enke, Stuttgart, 2nd ed., 336 pp.

Robin, Q., and J. Weertman, 1973, Cyclic surging of glaciers, *Journal of Glaciology*, v. 12, p. 64.

Roche, A., 1956, Sur la date de la dernière inversion du champ magnétique terrestre: *Academie Sciences, Paris, Comptes Rendus*, pp. 243 and 812.

Roglić, J., 1940, Geomorphologische Studie über das Duvansko Polje in Bosnien: *Mitteilungen der Geographischen Gesellschaft in Wien*, v. 83, pp. 152-177.

———, 1960, Das Verhältnis der Flusserosion zum Karstprozess: *Zeitschrift für Geomorphologie*, Neue Folge, 4, pp. 116-128.

Rognon, P., 1967a, *Le massif de l'Atakor et ses bordures (Sahara Central)*: Centre National de la recherche scientifique, Paris, 559 pp.

———, 1967b, Climatic influences on the African Hoggar during the Quaternary, based on geomorphologic observations: *Annals of the Association of American Geographers*, v. 57, pp. 115-127.

Rohdenburg, H., 1965, Die Muschelkalk-Schichtstufe am Ostrand des Sollings und Bramwaldes: *Göttinger Geographische Abhandlungen*, v. 33, 91 pp.

———, 1968, Zur Deutung der quartären Taleintiefung in Mitteleuropa: *Erde*, v. 99, pp. 297-304.

———, 1969, Hangpedimentation und Klimawechsel als wichtigste Faktoren der Flächen- und Stufenbildung in den wechselfeuchten Tropen an Beispielen aus Westafrika, besonders aus dem Schichtstufenland Südost-Nigerias: *Göttinger bodenkundliche Berichte*, v. 10, pp. 57-152.

———, 1970, Morphodynamische Aktivitäts- und Stabilitätszeiten statt Pluvial- und Interpluvialzeiten: *Eiszeitalter und Gegenwart*, v. 21, pp. 81-96.

Rudberg, S., 1954, Västerbottens Berggrundsmorphologie: *Geographica*, v. 25, 457 pp.

———, 1970, Recent quantitative work on slope processes in Scandinavia: *Zeitschrift für Geomorphologie*, Supplement-Band 9, pp. 44-56.

———, 1973, Glacial erosion forms of medium size—a discussion based on four Swedish case studies: *Zeitschrift für Geomorphologie*, Supplement-Band 17, pp. 33-48.

Rudovitz, J. L., 1947, Quantity of glaciations, of boreal transgressions, etc.: *Bulletin de la Commission pour l'Étude du Quaternaire*, v. 9.

Rühl, A., 1928, *Vom Wirtschaftsgeist in Spanien*: Quele und Meyer, Leipzig, 90 pp.

Ruellan, A., 1969, Quelques réflexions sur le rôle des sols dans l'interprétation des variations bioclimatiques du Pleistocène marocain: *Revue Géographique du Maroc*, v. 15, pp. 129-140.

Rütimeyer, 1874, *Über Thal- und Seebildung in den Alpen, Beiträge zum Verständnis der Oberfläche der Schweiz*: Schweighauser, Basel, 144 pp.

Rust, U., 1970, Beiträge zum Problem der Inselberglandschaften aus dem mittleren Südwestafrika: *Hamburger Geographische Studien*, v. 23, 280 pp.

Rutte, E., 1953, Der Albstein in der miozänen Molasse Südwestdeutschlands: *Zeitschrift der Deutschen Geologischen Gesellschaft*, v. 105, pp. 360-383.

———, 1957, *Einführung in die Geologie von Unterfranken*: Würzburger Naturwissenschaftliches Verein, Würzburg, 158 pp.

———, 1968, Kalkkrusten in Spanien: *Neues Jahrbuch für Geologie, Paleontologie und Mineralogie, Abhandlungen*, v. 106, pp. 52-138.

Ruxton, B. P., 1958, Weathering and subsurface erosion in granite at the piedmont angle (Balos, Sudan): *Geological Magazine*, v. 95, pp. 353-377.

———, 1967, Slopewash under mature primary rainforest in Northern Papua, in: *Landform Studies from Australia*, edited by J. Jennings and J. Mabbutt, Cambridge University Press, Cambridge, pp. 85-94.

Ruxton, B. P., and L. Berry, 1957, Weathering of granite and associated erosional features in Hong Kong: *Bulletin of the Geological Society of America*, v. 68, pp. 1263-1292.

Sabelberg, U., 1972, Comment in discussion on "Spezielle Untersuchungen an der Mittelterrasse im Enneri Tabi, Tibesti-Gebirge," by D. Jäkel and

E. Schultz: *Zeitschrift für Geomorphologie*, Supplement-Band 15, p. 141.

Salomon, W., 1916, Die Bedeutung der Solifluktion für die Erklärung deutscher Landschafts- und Bodenformen: *Geologische Rundschau*, v. 7, pp. 30-41.

Salomon-Calvi, W., 1918, Tote Landschaften und der Gang der Erdgeschichte: *Sitzungs-Bericht der Heidelberger Akademie der Wissenschaften, Mathematisch-Naturwissenschaftliche Klasse, Abteilung A*, v. 1, pp. 3-10.

Sapper, K, 1927, *Vulkankunde, Petrographische Einleitung*: Engelhorn, Stuttgart, 424 pp.

———, 1935, *Geomorphologie der feuchten Tropen*: Teubner, Leipzig, 154 pp.

Schaefer, I., 1950, Die diluviale Erosion und Akkumulation: *Forschung zur deutschen Landeskunde*, v. 49, 154 pp.

———, 1951, Über methodische Fragen der Eiszeitforschung im Alpenvorland: *Zeitschrift der Deutschen Geologischen Gesellschaft*, v. 102, pp. 287-310.

Scheffer, F., and P. Schachtschabel, 1973, *Lehrbuch der Bodenkunde*: Enke, Stuttgart, 8th ed., 448 pp.

Scheidegger, A. E., 1961, Mathematical models of slope development: *Bulletin of the Geological Society of America*, v. 72, pp. 37-50.

Scheidl, L. *et al.*, 1969, *Luftbild-Atlas Österreich*: Freitag-Berndt and Artaria, Vienna, 198 pp.

Schenk, E., 1955a, Die Mechanik der periglazialen Strukturböden: *Abhandlungen des Hessischen Landes-Amtes für Bodenforschung*, v. 13, pp. 1-92.

———, 1955b, Die periglazialen Strukturbodenbildungen als Folgen der Hydratationsvorgänge im Boden: *Eiszeitalter und Gegenwart*, v. 6, pp. 170-184.

Schindewolf, O. H., 1954, Über die möglichen Ursachen der grossen erdgeschichtlichen Faunenschnitte: *Neues Jahrbuch für Mineralogie, Geologie und Paläontologie*, pp. 457-465.

———, 1960, Darwins Abstammungslehre in der Sicht eines Paläontologen: *Universitas*, v. 15, pp. 649-661.

———, 1963, Neokatastrophismus?: *Zeitschrift der Deutschen Geologischen Gesellschaft*, v. 114, pp. 429-444.

Schipull, K., 1974, Geomorphologische Studien im zentralen Südnorwegen (mit Beiträgen über Regelungs- und Steuerungssysteme in der Geomorphologie): *Hamburger Geographische Studien*, v. 31, 91 pp.

Schmidt, W. F., 1959, Der morphogenetische Werdegang der Insel Cypern: *Erdkunde*, v. 13, pp. 179-201.

Schmidt-Thomé, P., 1943, Karrenbildung in kristallinen Gesteinen: *Zeitschrift der Deutschen Geologischen Gesellschaft*, v. 95, pp. 53-56.

Schmitthenner, H., 1926, Die Entstehung der Dellen und ihre morphologische Bedeutung: *Zeitschrift für Geomorphologie*, v. 1, pp. 3-28.

―――. 1956, Probleme der Schichtstufenlandschaft: *Marburger Geographische Schriften*, v. 3, 87 pp.

Schneider, S., 1967, Probleme der Luftbildinterpretation in der geographisch-landeskundlichen Arbeit in Deutschland, in: *25 Jahre amtliche Landeskunde*, Bundesanstalt für Landeskunde und Raumforschung, pp. 179-194.

Schnitzer, W. A., 1956, Die Landschaftsentwicklung der südlichen Frankenalb im Gebiet Denkendorf-Kösching nördlich von Ingolstadt: *Geologica Bavarica*, v. 28, 47 pp.

Schönhals, E., 1959, Der Basalt-Tuff von Kärlich als Leithorizont des Würm-Hochglazials: *Fortschritte in der Geologie von Rheinland und Westfalen*, v. 4, pp. 313-322.

Schönhals, E., H. Rohdenburg, and A. Semmel, 1964, Ergebnisse neuerer Untersuchungen zur Würmlöss-Gliederung in Hessen: *Eiszeitalter und Gegenwart*, v. 15, pp. 199-206.

Schuman, W., 1969, Geochronologische Studien in Oberbayern auf der Grundlage von Bändertonen: *Abhandlungen der Bayerischen Akademie für Wissenschaften, Mathematisch-Naturwissenschaftliche Klasse*, v. 134, 98 pp.

Schunke, E., 1968, Die Schichtstufenhänge im Leine-Weser-Bergland in Abhängigkeit vom geologischen Bau und Klima: *Göttinger Geographische Abhandlungen*, v. 43, 219 pp.

Schwarzbach, M., 1961, *Das Klima der Vorzeit; Eine Einführung in die Paläoklimatologie*: Enke, Stuttgart, 275 pp.

―――, 1964, Kreislauf des Wassers und Grundwasser, in: *Lehrbuch der allgemeinen Geologie*: edited by R. Brinkmann, Vol. 1, Enke, Stuttgart, pp. 101-128.

―――, 1967, Isländische Wasserfälle und eine genetische Systematik der Wasserfälle überhaupt: *Zeitschrift für Geomorphologie*, Neue Folge, 11, 377-417.

―――, 1968, Das Klima des rheinischen Tertiärs: *Zeitschrift der Deutschen Geologischen Gesellschaft*, v. 118, pp. 33-68.

―――, 1970, *Berühmte Stätten geologischer Forschung*: Wissenschaftliche Verlagsgesellschaft, Stuttgart, 322 pp.

―――, 1972, Die primäre Ursache der Eiszeiten: *Naturwissenschaftliche Rundschau*, v. 25, pp. 306-307.

Schwarzbach, M., and H. Noll, 1971, *Geologischer Routenführer durch Island*: Sonderveröffentlichung des Geologischen Instituts der Universität Köln, v. 20, 105 pp.

Schwenzner, J., 1936, Zur Morphologie des zentralspanischen Hochlandes: *Geographische Abhandlungen*, Reihe 3, v. 10, 128 pp.

Sdzuy, K., 1960, Zur Wende Präkambrium/Kambrium: *Paläontologische Zeitschrift*, v. 34, pp. 154-160.

Seedorf, H. H., 1955, Reliefbildung durch Gips und Salz im niedersächsischen Bergland: *Schriften der Wirtschaftswissenschaftlichen Gesellschaft zur Studium Niedersachsens*, v. 56, 113 pp.

Seefeldner, E., 1964, Zur Morphogenese der mittleren Tauerntäler: *Mitteilungen der Österreichischen Geographischen Gesellschaft*, v. 106, pp. 45-53.

――――, 1973, Zur Frage der Korrelation der kalkalpinen Hochfluren mit den Altformenresten der Zentralalpen: *Mitteilungen der Österreichischen Geographischen Gesellschaft*, v. 115, pp. 106-123.

Seibold, E., 1964, Das Meer, die Meeresregionen, in: *Lehrbuch der allgemeinen Geologie*, edited by R. Brinkmann, Vol. 1, Enke, Stuttgart, pp. 280-500.

――――, 1973, Vom Rand der Kontinente: *Abhandlungen der Akademie der Wissenschaften und der Literatur Mainz*, v. 2, pp. 1-23.

――――, 1975, Tiefseesedimente als Klimazeugen: unpublished lecture delivered to the Würzburger Geographische Kolloquium, January 16.

Sekyra, J., 1960, Pusobeni mrazu na pudu. Kryopedologie se zvlástnim zretelem k CSR: *Geotechnica*, Czechoslovakian Academy of Science, v. 27, pp. 1-164.

Semmel, A., 1961, Die pleistozäne Entwicklung des Weschnitztales im Odenwald: *Frankfurter Geographische Hefte*, v. 37, pp. 425-492.

――――, 1963, Intramontane Ebenen im Hochland von Godjam (Äthiopien): *Erdkunde*, v. 17, pp. 173-189.

――――, 1969, Verwitterungs- und Abtragungserscheinungen in rezenten Periglazialgebieten (Lappland und Spitzbergen): *Würzburger Geographische Arbeiten*, v. 26, 95 pp.

――――, 1972, *Geomorphologie der Bundesrepublik Deutschland*: Franz Steiner Verlag, Wiesbaden, 149 pp.

Semmel, A., and G. Stäblein, 1971, Zur Entwicklung quartärer Hohlformen in Franken: *Eiszeitalter und Gegenwart*, v. 22, pp. 23-34.

Seuffert, O., 1968, Klimatische und nichtklimatische Faktoren der Fussflächenentwicklung im Bereich der Gebirgsvorländer und Grabenregionen Sardiniens: *Geologische Rundschau*, v. 58, pp. 98-110.

――――, 1970, Die Reliefentwicklung der Grabenregion Sardiniens (Ein Beitrag zur Frage der Entstehung von Fussflächen und Fussflächensystemen): *Würzburger Geographische Arbeiten*, v. 24, 129 pp.

――――, 1973, Die Laterite am Westsaum Südindiens als Klimazeugen: *Zeitschrift für Geomorphologie*, Supplement-Band 17, pp. 242-259.

Sharp, R. P., 1954, Glacier flow, a review: *Bulletin of the Geological Society of America*, v. 65, pp. 821-838.

――――, 1957, Geomorphology of Cima Dome, Mohave Desert, California: *Bulletin of the Geological Society of America*, v. 68, pp. 273-290.

――――, 1960, *Glaciers*: Oregon State System of Higher Education, Eugene, pp. 1-78.

Shepard, F. P., 1948, *Submarine Geology*, Harper and Brothers, New York, 338 pp.

Sidorenko, A. W., and B. M. Michailov, 1973, *Map of the Bauxite Deposits of the USSR* (in Russian): Ministry of Geology, Moscow.

Simon, W., and H. J. Lippold, 1968, Isotopischen Zeitmessung und Erdgeschichte, in: *Vom Erdkern bis zur Magnetosphäre*, Umschauverlag, Frankfurt, pp. 39-66.

Sioli, H., 1950, Das Wasser im Amazonas-Gebiet: *Forschung und Fortschritt*, v. 26, pp. 274-280.

――――, 1956, Über Natur und Mensch im brasilianischen Amazonas-Gebiet: *Erdkunde*, v. 10, pp. 89-109.

――――, 1968, Zur ökologie des Amazonas-Gebietes, in: *Biogeography and Ecology in South America*: Vol. 1, edited by E. J. Fittkau, J. Illies, and H. Klinge, Junk, The Hague, pp. 137-170.

Slupetzky, H., 1968, Glaziologische und glazialmorphologische Untersuchungen im obersten Stubachtal (Hohe Tauern): Dissertation, Vienna.

Soloviev, P. A., 1973, Thermokarst phenomena and landforms due to frost heaving in central Yakutia: *Biuletyn Peryglacjalny*, v. 23, pp. 135-155.

Späth, H., 1969, Die Grossformen im Glocknergebiet: Neue Forschungen im Umkreis der Glocknergruppe: *Wissenschaftliche Alpenvereins-Heft*, v. 21, pp. 117-141.

――――, 1973, Morphologie und morphologische Probleme in den Hassbergen und im Coburger Land: *Würzburger Geographische Arbeiten*, v. 39, 321 pp.

Spiridonov, A. I., 1956, Über den Gegenstand und die wichtigsten Methoden der Geomorphologie, in: *Geomorphologische Probleme*, edited and transl. by H. Taubert, Hermann Haack, Geographische-Kartographische Anstalt, Gotha, pp. 9-26.

Spöcker, R. G., 1952, Zur Landschaftsentwicklung im Karst des oberen und mittleren Pegnitzgebietes: *Forschungen zur Deutschen Landeskunde*, v. 58, pp. 1-53.

Spönemann, J., 1966, Geomorphologische Untersuchungen an Schichtkämmen des Niedersächsischen Berglandes: *Göttinger Geographische Abhandlungen*, v. 36, 167 pp.

Spreitzer, H., 1951, Die Grossformung im oberen steirischen Murgebiet: *Geographische Studien (Sölch Festschrift)*, Geographische Gesellschaft, Vienna, pp. 132-144.

――――, 1957, Zur Geographie des kilikischen Ala Dag im Taurus, in:

Festschrift zur Hundertjahrfeier der Geographischen Gesellschaft in Wien, 1856-1956, edited by K. Wiche, Geographische Gesellschaft, Vienna, pp. 414-459.

———, 1960, Hangformung und Asymmetrie der Bergrücken in den Alpen und im Taurus: *Zeitschrift für Geomorphologie*, Supplement-Band 1, pp. 211-236.

Stäblein, G., 1968, Reliefgenerationen der Vorderpfalz: *Würzburger Geographische Arbeiten*, v. 23, 189 pp.

———, 1970, Grobsediment-Analyse als Arbeitsmethode der genetischen Geomorphologie: *Würzburger Geographische Arbeiten*, v. 27, 203 pp.

———, 1973, Rezente und fossile Spuren der Morphodynamik in Gebirgsrandzonen des Kastilischen Scheidegebirges: *Zeitschrift für Geomorphologie*, Supplement-Band 71, pp. 177-194.

Steinert, H., 1968, Das Grosse Rätsel der Alpen gelöst: Bohrung Urmannsau, Bestätigung der Decken-Theorie des Alpenaufbaus: *Frankfurter Allgemeine Zeitung*, April 3, no. 80, p. 25.

Stickel, R., 1927, Zur Morphologie der Hochflächen des linksrheinischen Schiefergebirges und angrenzender Gebiete: *Beiträge zur Landeskunde des Rheinlands*, v. 5, 104 pp.

Stille, H., 1924, *Grundfragen der vergleichenden Tektonik*: Borntraeger, Berlin, 443 pp.

———, 1929, Der Stammbaum der Gebirge und Vorländer: *Comptes Rendus du XIVe. Congrès Géologique Internationale*, Madrid, pp. 1-24.

———, 1945, Betrachtungen zum Werden des europäischen Kontinents: *Zeitschrift der Deutschen geologischen Gesellschaft*, v. 97, pp. 7-29.

———, 1950, Nochmals die Frage der Episodität und Gleichzeitigkeit der orogenen Vorgänge: *Geologischer Rundschau*, v. 38, pp. 108-111.

Stingl, H., 1969, Ein periglazialmorphologisches Nord-Süd-Profil durch die Ostalpen: *Göttinger Geographische Abhandlungen*, v. 49, 134 pp.

Stoddart, D. R., 1969, World erosion and sedimentation, in: *Water, Earth and Man: A Synthesis of Hydrology, Geomorphology, and Socio-Economic Geography*, edited by R. J. Chorley, Methuen, London, pp. 43-64.

Sunartadirdja, M. A., and H. Lehmann, 1960, Der tropische Karst von Maros und Nord-Bone in SW Celebes (Sulawesi): *Zeitschrift für Geomorphologie*, Supplement-Band 2, pp. 49-65.

Sundborg, A., 1956, The river Klarälven. A study of fluvial processes: *Geografiska Annaler*, v. 38, pp. 127-316.

Sweeting, M. M., 1972, *Karst Landforms*: Macmillan, London, 362 pp.

Szabo, P. Z., 1964, Neue Daten und Beobachtungen über Paläokarsterscheinungen in Ungarn: *Erdkunde*, v. 18, pp. 135-142.

Székely, A., 1969, Possible origin of surface of planation in Tertiary volcanic mountains by taking as example the intra-carpathian volcanic belt: *Studies in Geomorphology*, v. 3, pp. 17-30.

———, 1970, Landforms of the Matra Mountains and their evolution with special regard to surfaces of planation: *Studies in Hungarian Geography*, v. 8, pp. 41-54.

———, 1975, Alt- und Fussflächen des Matra-Gebirges: lecture delivered at the Gordon Conference on relief generations at the Bavarian Academy of Sciences, Feb. 22.

Tator, B. A., 1952-1953, Pediment characteristics and terminology: *Annals of the Association of American Geographers*, v. 42, pp. 295-317, v. 43, pp. 47-53.

Thauer, W., 1954, Morphologische Studien im Frankenwald und Franken-waldvorland: *Mitteilung der Fränkischen Geographischen Gesellschaft*, v. 1, 232 pp.

Thiedig, F., and U. Lehmann, 1973, Die Entstehung von Muren als säkulares Ereignis auf Spitzbergen (Svalbard) und ihre Bedeutung für die Denu-dation in der Frostschuttzone: *Mitteilungen aus dem Geologisch-Pa-leontologischen Institut der Universität Hamburg*, v. 42, pp. 71-80.

Thomas, M. F., 1965, Some aspects of the geomorphology of domes and tors in Nigeria: *Zeitschrift für Geomorphologie*, Neue Folge, 9, pp. 63-81.

———, 1966, Some geomorphical implications of deep weathering patterns in crystalline rocks in Nigeria: *Transactions of the Institute of British Geographers*, v. 40, pp. 173-193.

———, 1967, A bornhardt dome in the plains near Oyo, Western Nigeria: *Zeitschrift für Geomorphologie*, Neue Folge, 11, pp. 239-261.

———, 1974, *Tropical Geomorphology*: Macmillan, London, 132 pp.

Thorarinsson, S., 1944, Tefrokronologiska stùdier på Island: *Geografiska Annaler*, v. 26, 217 pp.

———, 1954, The Tephra-Fall from Hekla on March 29th, 1947, in: *The Eruption of Hekla 1947-1948*, Vol. 3, Visindafélag Islendinga, pp. 1-68.

———, 1960, Der Jökulsá-Canyon und Asbyrgi: *Petermanns Geographische Mitteilungen*, v. 104, pp. 154-162.

———, 1964, Additional notes on patterned ground in Iceland with a par-ticular reference to ice-wedge polygons: *Biuletyn Peryglacjalny*, v. 14, pp. 237-336.

Thorarinsson, S., T. Einarsson, and G. Kjartansson, 1960, On the geology and geomorphology of Iceland: *International Geographical Congress, Norden Excursion E.*, I., pp. 135-169.

Thorbecke, F., 1921, Die Inselberglandschaft von Nord-Tikar, in: *Zwölf Länderkundliche Studien*, edited by F. Thorbecke, Hirt, Breslau, pp. 215-242.

———, ed., 1927, Morphologie der Klimazonen: *Düsseldorfer Geogra-phische Vorträge*, v. 3, 100 pp.

Thornwaite, C. W., and J. R. Mather, 1957, Instructions and tables for computing potential evapotranspirations and the water balance: *Publications in Climatology*, v. 10.

Thorp, M. B., 1967, Closed basins in younger granite massifs, Northern Nigeria: *Zeitschrift für Geomorphologie*, Neue Folge, 11, pp. 459-480.

Tienhaus, R., 1964a, Itabiritische Eisenerzlagerstätten der Erde—ein Überblick: *Schriften der Gesellschaft Deutscher Metalhütten und Bergleute*, v. 14, pp. 1-9.

———, 1964b, Verwitterungsprofile über Itabiriten von Afrika und Indien: *Schriften der Gesellschaft Deutscher Metalhütten und Bergleute*, v. 14, pp. 89-101.

Tillmann, H., 1956, Zur Geologie des Oberpfälzer Tertiärs und seiner Lagerstätten, in: *50 Jahre Bayerische Braunkohlen Industrie*, edited by W. Kristl, Bayerische Braunkohlen-Industrie, Munich, pp. 109-127.

Tollmann, A., 1974, Geologie im Umbruch: *Mitteilungen der Österreichischen Geographischen Gesellschaft*, v. 116, pp. 53-78.

Tongiorgi, E., 1936, Vegetation und Klima der letzten Eiszeit und des Postglazials in Mittelitalien: *Verhandlungen der III Internationalen Quartär-Konferenz, Wien*, pp. 280-282.

Trendall, A. F., 1962, The formation of "apparent peneplains" by a process of combined lateritisation and surface wash: *Zeitschrift für Geomorphologie*, Neue Folge, 6, pp. 183-197.

Tricart, J., 1949-1952, *La partie orientale du Bassin de Paris*: Société d'édition d'enseignement supérieure, Paris, 2 vols., 476 pp.

———, 1956, Types de fleuves et systèmes morphogénétiques en Afrique occidentale: *Bulletin de la Section de Géographie du Comité des Travaux historiques et scientifiques*, Paris, pp. 303-345.

———, 1961, Las características fundamentales del sistema morfogenético de los países tropicales húmedos: *Revista Geográfica*, v. 3, pp. 5-52.

———, 1963, *Géomorphologie des Régions Froides*: Presses Universitaires de France, Paris, 282 pp.

———, 1965a, Le modelé des régions chaudes, forêts et savanes, in: *Traité de Géomorphologie*, edited by J. Tricart and A. Cailleux, Vol. 5, Société d'édition d'enseignement supérieure, Paris, 322 pp.

———, 1965b, *Principes et méthodes de la géomorphologie*, Masson, Paris, 496 pp.

———, 1975, Influences des oscillations climatiques récentes sur le modelé en Amazonie Orientale (Région de Santarém) d'après les images radar latéral: *Zeitschrift für Geomorphologie*, Neue Folge, 19, pp. 140-163.

Tricart, J., and A. Cailleux, 1961-1964, *Le modelé des régions sèches*: Centre de Documentation Universitaire, Paris, 2 vols., 308 pp.

Tricart, J., and A. Cailleux, 1965-1972, *Introduction to Climatic Geomorphology*, Longman Group Limited, London, 295 pp.

Tricart, J., R. Raynal, and F. Besançon, 1972, Cônes rocheux, pédiments, glacis: *Annales de Géographie*, v. 443, pp. 1-24.

Trinkler, E., 1932, *Geographische Forschungen im westlichen Zentralasien und Karakorum: Wissenschaftliche Ergebnisse der Trinklerischen Zentralasien-Expedition I*, Berlin.

Troll, C., 1924, Der diluviale Inn-Chiemsee-Gletscher: *Forschungen zur Deutschen Landes- und Volkskunde*, v. 23, 121 pp.

————, 1926, Die jungglazialen Schotterfluren im Umkreis der deutschen Alpen: *Forschungen zur deutschen Landes- und Volkskunde*, v. 24, pp. 159-256.

————, 1931, Die Landschaftsgürtel der tropischen Anden: *Verhandlungen und wissenschaftliche Abhandlungen des Deutschen Geographen-Tags zu Danzig*, v. 24, pp. 264-270.

————, 1936, Termiten-Savannen: *Landeskundliche Forschungen*, pp. 275-312.

————, 1941, Studien zur vergleichenden Geographie der Hochgebirge der Erde: *Bericht der 23. Hauptversammlung der Gesellschaft von Freunden der Universität Bonn*, pp. 1-95.

————, 1944, Strukturböden, Solifluktion und Frostklimate der Erde: *Klimaheft Geologische Rundschau*, v. 34, pp. 545-694.

————, 1947, Die Formen der Solifluktion und die periglaziale Bodenabtragung: *Erdkunde*, v. 1, pp. 162-175.

————, 1948, Der asymmetrische Aufbau der Vegetationszonen und Vegetationsstufen auf der Nord und Südhalbkugel: *Geobotanisches Forschungs-Institut Rübel, Zurich*, pp. 46-83.

————, 1951, Das Pflanzenkleid der Tropen in seiner Abhängigkeit von Klima, Boden und Mensch: *28. Deutscher Geographen-Tag, Frankfurt*, pp. 35-66.

————, 1952, Die Lokalwinde der Tropengebirge und ihr Einfluss auf Niederschlag und Vegetation: *Bonner Geographische Abhandlungen*, v. 9, pp. 124-182.

————, 1959, Die tropischen Gebirge. Ihre dreidimensionale klimatische und pflanzengeographische Zonierung: *Bonner Geographische Abhandlungen*, v. 25, 93 pp.

————, 1962, Geographische Luftbildinterpretation: *Archives Internationeaux de Photogrammétrie*, v. 14, pp. 266-275.

————, 1963, Quanat-Bewässerung in der Alten und Neuen Welt: *Mitteilungen der Österreichischen Geographischen Gesellschaft*, v. 105, pp. 313-330.

————, 1969a, Inhalt, Probleme und Methoden geomorphologischer Forschung (mit besonderer Berücksichtigung der klimatischen Fragestellung): *Beiheft Geologisches Jahrbuch*, v. 80, pp. 225-257.

————, 1969b, Selenographie und Geographie: *Erdkunde*, v. 23, pp. 326-328.

————, 1973a, Glazial- fluvioglazial- periglazial-Morphologie: unpublished lecture delivered at the Würzburger Geographischer Festkolloquium, October 30.

————, 1973b, Rasenabschälung (Turf exfoliation) als periglaziales Phänomen der subpolaren Zonen und der Hochgebirge: *Zeitschrift für Geomorphologie*, Supplement-Band 17, 32 pp.

Troll, C., and R. Schottenloher, 1939, Ergebnisse wissenschaftlicher Reisen in Äthiopien: *Petermanns Geographische Mitteilungen*, v. 85, pp. 217-237, 265-277.

Troll C., and U. Schweinfurth, 1968, Die Karte des Khumbu-Himalaya (Ostnepal) ¹:50,000 (um den Mt. Everest): *Erdkunde*, v. 22, pp. 29-33.

Tuan, Yi-Fu, 1959, Pediments in southeastern Arizona: *University of California, Publications in Geography*, v. 13, pp. 1-140.

Twidale, C. R., 1962, Steepened margins of inselbergs from northwestern Eyre Peninsula, South Australia: *Zeitschrift für Geomorphologie*, Neue Folge, 6, pp. 51-69.

————, 1967, Origin of the piedmont angle as evidenced in South Australia: *Journal of Geology*, v. 75, pp. 239-311.

Tyutiunov, I. A., 1964, *An Introduction to the Theory of the Formation of Frozen Rocks*: Pergamon Press, Oxford, 94 pp.

Valentin, H., 1952, Die Küsten der Erde. Beiträge zur allgemeinen und regionalen Küstenmorphologie: *Petermanns Geographische Mitteilungen*, Ergänzungs-Heft 246, 118 pp.

————, 1955, Die Grenze der letzten Vereisung im Nordseeraum: *30. Deutscher Geographen-Tag, Hamburg*, pp. 359-366.

————, 1957, Glazial morphologische Untersuchungen in Ostengland: *Abhandlungen des Geographischen Instituts der Freien Universität Berlin*, v. 4, 86 pp.

Van Wambeke, A. R., 1962, Criteria for classifying tropical soils by age: *Journal of Soil Science*, v. 13, pp. 124-132.

Verstappen, H. T., 1974, *A Geomorphological Reconnaissance of Sumatra and Adjacent Islands*: Vol. 1, Royal Netherlands Geographic Society, 224 pp.

Vincent, P., 1963, Les volcans tertiaires et quaternaires du Tibesti occidental et central (Sahara du Tschad): *Mémoirs du Bureau de Recherches Géologiques et Minières*, v. 23, 307 pp.

Vogel, C., 1975, Neue Aspekte zur Evolution des Menschen: *Nova Acta Leopoldina*, N. F. 42, 218, pp. 253-269.

Vogt, J., 1959, Badlands du Nord Dahomey: *Actes du Quatre-vingt-quatrième Congrès Nationale des Sociétés Savantes, Section de Géographie*, pp. 227-239.

Vogt, J., and R. Black, 1963, Remarques sur la géomorphologie de l'Air: *Bulletin du Bureau de Recherches Géologiques et Minières*, v. 1, pp. 1-29.

Vogt, J., and P. Vincent, 1966, Terrains d'altération et de recouvrement en zone intertropical: *Bulletin du Bureau de Recherches Géologiques et Minières*, v. 4, 111 pp.

Wadia, D. N., 1966, *Geology of India*, Macmillan, London, 3rd edition, 536 pp.

Wagner, Georg, 1950, *Einführung in die Erd- und Landschaftsgeschichte mit besonderer Berücksichtigung Süddeutschlands:* Rau, Öhringen, 2nd ed., 864 pp.

Wagner, Günther, 1960, Bodenabtrag und Kleinformenschatz in ihrer historischen Entwicklung in der Landschaft des Taubertals: Dissertation, Würzburg.

———, 1965, Klimatologische Beobachtungen in Südostspitzbergen 1960: *Ergebnisse der Stauferland-Expedition*, v. 10, 69 pp.

Waibel, L., 1928, Die Inselberglandschaft von Arizona und Sonora: *Zeitschrift der Gesellschaft für Erdkunde zu Berlin, Sonder-Band zur Hundertjahrfeier der Gesellschaft*, pp. 68-91.

Walter, H., 1970, *Vegetationszonen und Klima*: Ulmer, Stuttgart, 244 pp.

Walther, J., 1915, Laterit in West-Australien: *Zeitschrift der Deutschen Geologischen Gesellschaft*, v. 67, pp. 113-132.

———, 1916, Das geologische Alter und die Bildung des Laterits: *Petermanns Geographische Mitteilungen*, v. 62, pp. 1-7 and 46-53.

———, 1924, *Das Gesetz der Wüstenbildung*: Quelle und Meyer, Leipzig, 4th ed., 342 pp.

Washburn, A. L., 1956, Classification of patterned ground and review of suggested origins: *Bulletin of the Geological Society of America*, v. 67, pp. 823-865.

———, 1965, Geomorphic and vegetational studies in the Mesters Vig District, Northeast Greenland: *Meddelelser om Grønland*, v. 16/1, 60 pp.

———, 1967, Instrumental observations of masswashing in the Mesters Vig District, Northeast Greenland: *Meddelelser om Grønland*, v. 16/4, 318 pp.

———, 1969, Weathering, frost action and patterned ground in the Mesters Vig District, Northeast Greenland: *Meddelelser om Grønland*, v. 17/6, 303 pp.

———, 1973, *Periglacial Processes and Environments*: Arnold, London, 328 pp.

Wegener, A. L., 1915, *Die Entstehung der Kontinente und Ozeane:* Friedrich Vieweg, Braunschweig, 5th ed., 135 pp.

Weidenbach, F., H. Graul, and H. Kiderlen, 1950, Übersichtskarte des Iller-Riss Gebietes: *Württembergisches stätistische Landesamt*, Stuttgart.

Weise, O., 1967, Reliefgenerationen am Ostrand des Schwarzwaldes: *Würzburger Geographische Arbeiten*, v. 21, 158 pp.

————, 1970, Zur Morphodynamik der Pediplanation mit Beispielen aus Iran: *Zeitschrift für Geomorphologie*, Supplement-Band 10, pp. 64-87.

Weise, O., 1974, Zur Hangentwicklung und Flächenbildung im Trockengebiet des iranischen Hochlandes: *Würzburger Geographische Arbeiten*, v. 42, 328 pp.

Went, F. W., and N. Stark, 1968, Mycorrhizia: *Biological Science*, v. 18, pp. 1035-1039.

Wentworth, C. K., 1928, Principles of stream erosion in Hawaii: *Journal of Geology*, v. 37, pp. 385-410.

————, 1943, Soil avalanches on Oahu, Hawaii: *Bulletin of the Geological Society of America*, v. 54, pp. 53-64.

Wenzens, G., 1972, Morphologische Entwicklung der "Basin Ranges" in der Sierra Madre Oriental (Nordmexiko): *Zeitschrift für Geomorphologie*, Supplement-Band 15, pp. 39-54.

————, 1974, Morphologische Entwicklung ausgewählter Regionen Nordmexikos unter besonderer Berücksichtigung des Kalkkrusten- Pediment- und Poljeproblems: *Düsseldorfer Geographische Schriften*, v. 2, 330 pp.

————, 1976, Flächentypen der Iberischen Halbinsel: unpublished lecture delivered at the Geographical Colloquium, Würzburg, Feb. 17.

Werdecker, J., 1955, Beobachtungen in den Hochländern Äthiopiens auf einer Forschungsreise 1953-1954: *Erdkunde*, v. 9, pp. 305-317.

————, 1958, Untersuchungen in Hochsemién: *Mitteilungen der Geographischen Gesellschaft Wien*, v. 100, pp. 58-66.

————, 1968, Das Hochgebirgsland von Semién: *Erdkunde*, v. 22, pp. 33-39.

Werner, D. J., 1972, Beobachtungen an Bergfussflächen in den Trockengebieten NW-Argentiniens: *Zeitschrift für Geomorphologie*, Supplement-Band 15, 20 pp.

Weyl, R., 1956, Neue Erkenntnisse über den geologischen Aufbau der Ozeanböden: *Nachrichten der Giessener Hochschulgesellschaft*, v. 25, pp. 60-73.

Wiche, K., 1953, Pleistozäne Klimazeugen in den Alpen und im Hohen Atlas: *Mitteilungen der Geographischen Gesellschaft Wien*, v. 95, pp. 143-166.

————, 1955, Fussflächen im Hohen Atlas: *Sitzungs-Berichte der Österreichischen Akademie der Wissenschaften, Mathematisch-Physikalische Klasse, Abteilung I*, v. 164, pp. 389-426.

————, 1959, Klimamorphologische Untersuchungen im westlichen Karakorum: *32. Deutscher Geographen-Tag, Wiesbaden*, pp. 190-203.

————, 1963, Fussflächen und ihre Deutung: *Mitteilungen der Geographischen Gesellschaft Wien*, v. 105, pp. 519-532.

Wiche, K., 1970, Die Flächentreppe des mittleren Burgenlands: *Wissenschaftliche Arbeiten, Burgenland*, v. 44, pp. 5-38.

Wiegand, G., 1965, Fossile Pingos in Mitteleuropa: *Würzburger Geographische Arbeiten*, v. 16, 152 pp.

————, 1970, Zur Entstehung der Oberflächenformen in der westlichen und zentralen Turkei, zugleich ein Beitrag zur Hangentwicklung und Pediplanation: *Würzburger Geographische Arbeiten*, v. 30, 97 pp.

————, 1972, Hydrodynamische Formbildungskräfte in Kleinasien: *Zeitschrift für Geomorphologie*, Supplement-Band 15, pp. 55-65.

Wilhelm, F., 1965, Jungere Gletscherschwankungen auf der Barents-Insel in SO-Spitzbergen, in: *Vorträge des Fridtjof-Nansen-Gedächtnis Symposions*, edited by J. Büdel and A. Wirthmann, *Ergebnisse der Stauferland-Expedition,* v. 3, pp. 73-85.

————, 1966, *Hydrologie, Glaziologie: Das Geographische Seminar*: Westermann, Braunschweig, 143 pp.

Wilhelmy, H., 1951, Die eiszeitliche und nacheiszeitliche Verschiebung der Klima- und Vegetationszonen Südamerikas: *28. Deutscher Geographen-Tag, Frankfurt*, pp. 121-132.

————, 1957, Das Grosse Pantanal in Mato Grosso: *31. Deutscher Geographen-Tag,Würzburg*, pp. 45-71.

————, 1958a, *Klimamorphologie der Massengesteine*: Westermann, Braunschweig, 238 pp.

————, 1958b, Umlaufseen und Dammuferseen tropischer Tieflandflüsse: *Zeitschrift für Geomorphologie*, Neue Folge, 2, pp. 28-54.

————, 1966, *Kartographie in Stichworten*: Verlag F. Hirt, Kiel, 333 pp.

————, 1971-1972, *Geomorphologie in Stichworten*: F. Hirt, Kiel, 3 vols., 510 pp.

————, 1974, *Klima-Geomorphologie in Stichworten*: F. Hirt, Kiel, 375 pp.

————, 1975, Verwitterungs-Kleinformen als Anzeichen stabiler Grossformung: lecture delivered at the Gordon Conference on relief generations at the Bavarian Academy of Sciences.

Wilson, A. T., 1964, Origin of the ice ages: an ice sheet theory for the Pleistocene glaciation: *Nature*, v. 201, pp. 147-149.

Winkler-Hermaden, A., 1951, Die juntertiären Ablagerungen an der Ostabdachung der Zentralalpen und das inneralpine Tertiär, in: *Geologie von Österreich*, edited by F. X. Schaffer, Deuticke, Vienna, pp. 414-524.

————, 1957, *Geologisches Kräftespiel und Landformung*: Springer, Vienna, 822 pp.

Wirthmann, A., 1957, Die geographische Grundlagen des Obstbaus im Gebiet des Unterlaufs der fränkischen Saale: *Würzburger Geographische Arbeiten*, v. 4-5, pp. 157-240.

————, 1961, Zur Geomorphologie der nördlichen Oberpfälzer Senke: *Würzburger Geographische Arbeiten*, v. 9, 41 pp.

————, 1962, Zur Morpologie der Edge-Insel in Südostspitzbergen: *33. Deutscher Geographen-Tag, Wiesbaden*, pp. 394-399.

————, 1964, Die Landformen der Edge-Insel in Südost-Spitzbergen: *Ergebnisse der Stauferland-Expedition*, v. 2, 53 pp.

————, 1965, Die Reliefentwicklung von Neu-Kaledonien: *Deutscher Geographen-Tag, Bochum*, v. 35, pp. 323-335.

————, 1968, Über Talbildung und Hangentwicklung auf Hawaii: *Würzburger Geographische Arbeiten*, v. 22, 22 pp.

————, 1970, Zur Geomorphologie der Peridotite auf Neu-Kaledonien: *Tübinger Geographische Studien*, v. 34, pp. 191-201.

————, 1973, Reliefentwicklung auf Basalt unter tropischen Klimaten: *Zeitschrift für Geomorphologie*, Supplement-Band 17, pp. 223-241.

Wissmann, H. von, 1951, Über seitliche Erosion: *Colloquium Geographicum*, v. 1, 71 pp.

Woldstedt, P., 1958, Eine neue Kurve der Würm-Eiszeit: *Eiszeitalter und Gegenwart*, v. 9, pp. 151-154.

————, 1960, Die letzte Eiszeit in Nordamerika und Europa: *Eiszeitalter und Gegenwart*, v. 11, pp. 148-165.

————, 1962, Über die Gliederung des Quartärs und Pleistozäns: *Eiszeitalter und Gegenwart*, v. 13, pp. 115-124.

————, 1954-65, *Das Eiszeitalter. Grundlinien einer Geologie des Quartärs*: Enke, Stuttgart, 3 vols.

Wurm, A., 1961, *Geologie von Bayern:* Borntraeger, Berlin, 555 pp.

Young, A., 1972, *Slopes:* Oliver and Boyd, Edinburgh, 288 pp.

Zonneveld, J.I.S., 1968, Quaternary climatic changes in the Caribbean and N. South America: *Eiszeitalter und Gegenwart*, v. 19, pp. 203-208.

————, 1972a, Sulas and sula complexes: *Göttinger Geographische Abhandlungen*, v. 60, pp. 93-105.

————, 1972b, Comments in the discussion to H. Bremer, "Flussarbeit, Flächen- und Stufenbildung in den feuchten Tropen": *Zeitschrift für Geomorphologie*, Supplement-Band 14, p. 36.

Zötl, J., 1960, Zur Frage der Niveaugebundenheit von Karstquellen und Höhlen: *Zeitschrift für Geomorphologie*, Supplement-Band 2, pp. 100-102.

————, 1964, Fossile Grossformen im ostalpinen Karst: *Erdkunde*, v. 18, pp. 142-145.

————, 1974, *Karsthydrogeologie*: Springer, Vienna, 291 pp.

Author Index

Italic numbers refer to pages with Figures or Photographs.

Geographical Index

Italic numbers refer to pages with Figures or Photographs.

Subject Index

Italic numbers refer to pages with Figures or Photographs.

Library of Congress Cataloging in Publication Data

Büdel, Julius.

Climatic geomorphology.

Translation of: Klima-Geomorphologie.
Bibliography: p.
Includes indexes.
 1. Climatic geomorphology.   I. Title.
GB447.B8313   551.3   81-47909
ISBN 0-691-08294-4   AACR2
ISBN 0-691-08295-2 (pbk.)

Julius Büdel is Professor Emeritus of Geography
at the University of Würzburg and was awarded
the Victoria Medal of the Royal Geographical Society
(London) in 1981.